THE BIOLOGICAL EFFICIENCY OF
PROTEIN PRODUCTION

THE BIOLOGICAL EFFICIENCY OF PROTEIN PRODUCTION

EDITED BY
J. G. W. JONES

CAMBRIDGE
AT THE UNIVERSITY PRESS
1973

Published by the Syndics of the Cambridge University Press
Bentley House, 200 Euston Road, London NW1 2DB
American Branch: 32 East 57th Street, New York, N.Y. 10022

© Cambridge University Press 1973

Library of Congress Catalogue Card Number: 72–93672

ISBN: 0 521 20179 9

Printed in Great Britain
at the University Printing House, Cambridge
(Brooke Crutchley, University Printer)

CONTENTS

Editor's Preface — page ix

Foreword — xi
by C. R. W. SPEDDING

Part I – INTRODUCTION

1. The purpose of protein production — 3
   by K. L. BLAXTER
2. The meaning of biological efficiency — 13
   by C. R. W. SPEDDING
3. Future demand for protein foods — 27
   by A. A. WOODHAM
4. Factors affecting demand for protein products — 37
   by J. C. MCKENZIE
5. Economics of protein production — 45
   by K. E. HUNT

Part II – THE BIOLOGICAL EFFICIENCY OF PROTEIN PRODUCTION BY PLANTS

6. Biochemical aspects of the conversion of inorganic nitrogen into plant protein — 69
   by A. J. KEYS
7. The potential of cereal grain crops for protein production — 83
   by R. N. H. WHITEHOUSE
8. Plants as sources of unconventional protein foods — 101
   by N. W. PIRIE
9. Potential protein production of temperate grasses — 119
   by TH. ALBERDA

Discussion report — 131
by J. P. COOPER AND P. F. WAREING

Part III – THE BIOLOGICAL EFFICIENCY OF PROTEIN PRODUCTION BY ANIMALS

10. Considerations of the efficiency of amino acid and protein metabolism in animals *page* 141
 by P. J. BUTTERY AND E. F. ANNISON

11. Possibilities for changing by genetic means the biological efficiency of protein production by whole animals 173
 by J. C. BOWMAN

12. Factors affecting the efficiency of protein production by populations of animals 183
 by R. V. LARGE

13. The biological efficiency of protein production by animal production enterprises 201
 by P. N. WILSON

Discussion report 211
 by V. R. FOWLER AND C. C. BALCH

Part IV – THE BIOLOGICAL EFFICIENCY OF PROTEIN PRODUCTION BY ECOSYSTEMS

14. The biological efficiency of protein production by grazing and other land-based systems 217
 by J. PHILLIPSON

15. The biological efficiency of protein production by stall-fed ruminants 237
 by T. HOMB AND D. C. JOSHI

16. Ecological factors affecting amounts of protein harvested from aquatic ecosystems 263
 by H. A. REGIER

Discussion report 281
 by G. WILLIAMS AND P. A. JEWELL

Part V – THE BIOLOGICAL EFFICIENCY OF INDUSTRIAL SYSTEMS OF PROTEIN PRODUCTION

17. Conversion of agricultural produce for use as human food 285
 by F. AYLWARD AND B. J. F. HUDSON

18. Aspects of protein production by unicellular organisms 303
 by M. T. HEYDEMAN

CONTENTS

19. Protein production by unicellular organisms from hydrocarbon substrates *page* 323
    by T. WALKER

20. Protein production by micro-organisms from carbohydrate substrates 339
    by J. T. WORGAN

Discussion report 363
   by E. J. ROLFE AND A. SPICER

List of participants 371

Index 377

EDITOR'S PREFACE

This book reports the proceedings of a symposium held at the University of Reading between 19 and 24 September 1971. Parts II to V each represent the result of a day's papers and discussions, but the papers in Part I were interspersed throughout the symposium and were not presented in a single session.

The Committee responsible for organising the symposium, which consisted of Professor J. C. Bowman, Professor E. H. Roberts, Professor C. R. W. Spedding and myself, sets on record its gratitude to all those whose efforts contributed to the success of the symposium. In addition to the authors of papers and the reporters of discussions which are published in this book, it is appropriate here to acknowledge the contributions of the Chairmen of sessions who included, besides members of the Committee, Dr C. C. Balch, Professor A. N. Duckham, CBE, Professor J. A. F. Rook and Professor R. H. Tuck.

Generous financial support was forthcoming from a number of sources and it is a pleasure to record the Committee's gratitude to:

> The Royal Society
> Imperial Chemical Industries Ltd
> The Nickerson Group of Companies
> Pauls & Whites Ltd
> Rothwell Plant Breeders Ltd
> Rank Hovis McDougall (Research) Ltd
> Unilever Ltd

I should like to record my personal thanks to Mr E. D. Harriss, Technical Secretary of the Department of Agriculture, and his staff, but especially to Mrs S. Smith who played a leading role in the organisation of the symposium.

Finally, I have to acknowledge the patience and tolerance of the Cambridge University Press.

July 1972

J. G. W. JONES
Department of Agriculture
University of Reading

FOREWORD

By C. R. W. SPEDDING

The Grassland Research Institute, Hurley, Berkshire, and
Department of Agriculture, University of Reading

The main purpose of the symposium was to consider the relative biological efficiency of alternative methods of protein production, by both animals and plants, at the level of the tissue, the individual and the population, and in a variety of environments ranging from terrestrial to aquatic. Such a large canvas requires contributors from many disciplines, interested either in the picture as a whole or in the role of their own specialism within the whole picture. Any gathering of this kind is likely to find some communication problems and, partly for this reason, introductory papers dealt with the reasons why we are interested in protein production and what meanings should be attached to the term 'biological efficiency'.

In the event, the contacts established between biochemists, nutritionists, agriculturalists, biologists, ecologists and economists proved valuable in themselves and differences in terminology were discussed in a wholly constructive fashion. Although the main papers dealt with protein production in an agricultural framework, the emphasis was placed on an understanding of the underlying biological processes. This represented a sufficiently wide field for one symposium but, in order to provide a realistic backcloth to the discussions, three contributors were invited to deal with economic and social aspects of protein production in contextual papers which are included in Part I.

Since the primary intention was to achieve some synthesis of all the contributions, discussion leaders were invited to guide the discussions and subsequently to report on them. It is their reports that are published here, in deliberate preference to recording individual contributions. In some important respects, the symposium was experimental as to both form and content, and these proceedings reflect both.

PART I

INTRODUCTION

I

THE PURPOSE OF PROTEIN PRODUCTION

By K. L. BLAXTER

The Rowett Research Institute, Bucksburn, Aberdeen

Man uses the protein which he harvests from plants and animals in a variety of ways. Silk fibroin and hair and wool keratins are used for fabrics. Hides and skins are used for a variety of purposes where their unique physical properties make them particularly valuable. Collagen from animal bones and casein from milk are used in adhesive manufacture: feather quills have a limited use as plectra. All these uses of protein are being eroded by the adoption of synthetic materials. Originally introduced as substitutes, many synthetic materials have now been devised which have attributes superior to those of natural materials derived from animal and plant protein. It seems probable that the production of plants and animals for these uses will decline as technology advances, in much the same way that the million acres of England devoted to the production of plants used as a source of dyestuffs dwindled with the advent of the aniline dyestuffs industry (Ernle, 1936).

The most important use which man makes of the protein from plants and animals, however, is as food. Protein is essential in his diet as a source of the amino acids he cannot synthesise from simple sources of nitrogen and energy for growth, reproduction and maintenance. It is possible that synthetic sources of amino acids may at some future time replace those that he at present obtains from plants and animals. Such a development may take place in a way similar to that currently taking place with farm animals, namely through supplementation of natural protein-containing materials with amino acids. From what will emerge later, the scale of synthesis of amino acids to achieve complete independence of the United Kingdom population from plant and animal sources of protein is certainly within the realms of possibility. It is only about 75 000 tons per annum, which is less than our current industrial production of organic phthalates. Nevertheless, for a number of years to come, plant and animal protein will necessarily be the major source of essential amino acids and non-specific nitrogen in man's diet. For this reason it seems desirable to consider the magnitude of man's needs for protein.

One might think that in the 154 years that have elapsed between the publication of Magendie's 'Elementary Compendium' in 1816–17, in which he distinguished the fundamental nutritional importance of protein, observation and experiment would have resulted in a clear definition of the average protein needs of man and an indication of the biological variation associated with such an average estimate. Strange as it might seem this is not so. A working party (Ministry of Health, 1964) when considering man's protein needs calculated from published metabolic data a minimal requirement of protein. They also calculated from dietary survey data the actual quantities consumed by people in the United Kingdom. These two quantities they were honest enough to describe as the lower and upper limits of knowledge and they regarded the physiological requirement as falling somewhere in between. A Panel of the Department of Health and Social Security in 1969, considering experiments which dealt in effect with what the 1964 working party had regarded as the lower limit of knowledge, were equally honest and stated 'the question of whether the minimum requirements are adequate to maintain health for a lifetime can only be answered by long-term studies which have not yet been made'. It may be stated that the difference between the lower limit of knowledge and the upper is two-fold.

This element of uncertainty is not new. It dates from the late nineteenth century and the early years of the present century: Carl Voit (1876), the pupil of Justus von Liebig, studied from his laboratory in Munich labourers and garrisoned soldiers to conclude that a man weighing 70 kg required 118 g protein each day. For a man at work the need would increase to 145 g per day. Voit's pupils continued his work and perpetuated his view. Max Rubner (1902) at the University of Berlin suggested a value of 127 g per day as the daily protein requirement, and Atwater (1903) in the United States advised 125 g per day for a sedentary man and 150 g per day for a working man. These studies were largely based on what healthy people consumed. They were supported by careful studies of nitrogen metabolism, and it was indeed found by nitrogen balance methods that these amounts of protein would not result in a net loss of protein from the body. Eventually the obvious question was asked, 'What is the minimal amount of protein which will preserve such a nitrogen equilibrium?' Two Scandinavians, Siven (1900) and Landergren (1903), demonstrated that nitrogen equilibrium could be preserved on much lower intakes of protein than those suggested by the German school. Siven's experiments were with himself. He weighed 65 kg and found he could maintain nitrogen equilibrium on 25–31 g protein daily.

The American, Chittenden (1904), took these studies further. Suffering from bilious attacks and sick headaches Chittenden attributed these to his current diet. He therefore undertook investigations on himself which consisted of a reduction of his protein and calorie intake. Initially he lost weight, but eventually on a diet providing 36–40 g protein each day kept his weight stable and was convinced that he was in better health. He followed these studies with similar experiments with twenty-four adults, all male students and members of the staff of the University of Yale, and showed that despite a wide range of physical activity the protein requirements of these men were at least 50 per cent lower than the Atwater and Voit standards of need. His subjects co-operated for nine months of each year; presumably the summer vacation was sacrosanct. Chittenden's views were attacked and the nature of the attacks of the early twentieth century reflects the same problem with which we are confronted today. Crichton-Browne (1909) simply stated that Chittenden's nine-month experiments with the twenty-four Yale students and staff were too short to give conclusive results and that human experience supported the view that a liberal protein intake was desirable for health. McCay (1912), following his Indian Army experience, with imperial arrogance ascribed the small size and limited capacity for physical work of the Bengali to low protein intake and concluded: 'Voit stands today absolutely vindicated while the earnest plea put forward by Chittenden...cannot be regarded as longer possible in the light of the ill effects that follow in the train of chronic underfeeding.'

Since those times there has been this dichotomy of view about human protein needs. Hegsted (1964) summarised the situation admirably when criticising the practice (probably dating back to Sherman, 1920) adopted by FAO of arbitrarily increasing estimates of physiological minima reached through critical study by 50 per cent to provide safe practical allowances. He said: 'It is somewhat ludicrous to go through the effort of calculating the requirements in detail and then double or treble the value obtained since the "guesstimate" added is as large as the estimate. The factual data become a kind of scientific window dressing which lends authority to the final figure.'

Admittedly there have been attempts to rationalise these margins of safety. First, it has been recognised that the nitrogen balance technique ignores losses of nitrogen from the skin in desquamated epithelium, in sweat and in the growth of nails and hair. A nitrogen balance of zero estimated from the intake and the faecal and urinary losses is small at about 250 mg nitrogen per day for a 70 kg man (see summary by Irwin & Hegsted, 1971*a*) corresponding to between 2 and 3 g of dietary protein each day for protein of average quality. Similarly, arguing that a dietary allowance should take into account variability in protein need, the margin of safety can be expressed statistically. Garrow's (1969) analysis of the nitrogen balance data of Sherman (1920) and of Hegsted et al. (1946) indicated a standard deviation of ±6–7 per cent of the mean values. A safety margin such that 95 per cent of individuals would receive *not less than* their requirements, corresponding to a probability in the Fisher sense of 0.975, would result in a safety margin 18–21 per cent of the mean. The standard deviation includes analytical and instrumental error as well as biological error, but even so it would suggest that 95 per cent of all people would receive enough protein if the average minimal need was multiplied by 1.2. Even when this is done the resultant minimal allowance plus safety margin is considered not safe to recommend. The most recent British recommended daily intakes which apply to an average mixed diet are 75 g for a moderately active man. The calculated requirement increased by the statistically defined safety margin is only 45 g. Hegsted's (1964) stricture still applies.

No mention has yet been made of quality of protein; the estimates of protein need discussed above apply to mixed diets containing protein from several sources. The Magendie Commission of 1841, often referred to as the Gelatin Commission, had found that gelatin alone would not nourish animals (see McCollum, 1957) but it was not until Willcock & Hopkins (1906) showed the essentiality of tryptophan that it was realised that proteins differ in their nutritive value and that this variation reflects in the main their amino acid composition. Many studies on the amounts of different proteins needed by man have now been made. Protein intakes necessary to achieve nitrogen equilibrium have been established by experiment. They vary from 30 g for an adult 70 kg man given high quality protein, notably the proteins of milk and whole egg, to more than three times this value for certain plant proteins deficient in essential amino acids. The value of a protein is expressed by the product of its biological value (BV) and true digestibility or its Net Protein Utilisation (NPU) value which in effect measures, under conditions in which protein is limiting, the increment in nitrogen retention which takes place in the body per g nitrogen supplied in the diet. A protein with an NPU of 100 can be used without waste and dietary needs of protein then represent the obligatory losses from the body of nitrogen × 6.25 which amount to 25–30 g per day for an adult man. Replacement with a protein of NPU 70 implies that dietary protein needs should increase by 100/70 to 36–43 g per day. This approach implies that the amino acid requirements of man can be assessed from the analysis of proteins which have a biological value of 100, an approach pioneered by Mitchell (1954). Ideal amino acid patterns in foods for man have been devised in this way, the data on analysis of proteins with BVs near 100 being supplemented by information which has accrued since Rose's (1957) classical studies on the amino acid needs of man. Examples are the FAO reference pattern (Joint FAO–WHO Expert Group on Protein Requirements, 1957, 1965), Swaminathan's ideal pattern (1963), and Oser's (1951) essential amino acid index.

Expression of man's protein needs in terms of amino acids thus seemed to be a solution to problems related to definition of the amounts of proteins of different qualities required to establish nitrogen equilibrium. There is, however, some evidence to suggest that the levels of essential amino acids derived in this way are too high. First, Kofrányi & Müller-Wecker (1961) and Kofrányi & Jekat (1964) diluted proteins with non-essential nitrogen. Egg protein could be diluted by up to 60 per cent with non-essential nitrogen without affecting its biological value. Similar results have been found by Scrimshaw et al. (1969). This predicates questions of considerable importance. How little nitrogen may be provided by essential amino acids, and what is man's requirement for non-essential nitrogen? Certainly the data of Rose & Wixom (1955) suggested that balance could be achieved at intakes of 22 g protein per day of which only 40 per cent was as essential amino acids, suggesting a requirement of less than 10 g essential amino acids per day.

The second source of evidence comes from the determinations of the amounts of essential amino acids which are required (in the physiological sense) by man. Rose (1957) in his series of studies took the best estimate of minimal requirement to be not the mean value for his subjects but the highest estimate. Even so, the summation of his estimates of the essential amino acid needs of man amounts to about 6 g amino acids. More recent data admirably reviewed by Irwin & Hegsted (1971 b) suggest the summation to be slightly less than 4 g essential amino acids per day. The amino acids required are tryptophan, lysine (but not arginine and probably not histidine), leucine, isoleucine, threonine, valine, methionine (or methionine plus cystine) and phenylalanine (or phenylalanine plus tyrosine). As Irwin & Hegsted point out, 4 g essential amino acids may be supplied by 8–10 g high quality protein, amounts very considerably less than the amounts of high quality protein required to achieve nitrogen equilibrium (about 30 g). It would thus seem that the best quality proteins supply amino acids in amounts very considerably in excess of man's minimal needs. Conversely, it may be argued that economy in the use of high quality protein could be achieved by its dilution with diammonium citrate (Swendseid, Harris & Tuttle, 1960) or other simple source of non-essential nitrogen.

This short review poses a number of serious problems which have direct relevance to agriculture and food production. Do we have to consider planning for the purpose of meeting the essential amino acid needs of man which can be met from 8 g protein per day, or for the mean minimal protein needs of man given average mixed diets, namely 30 g protein per day, or this value increased by a statistically defined safety margin of 20 per cent, namely 43–45 g per day, or do we use the upper limit of knowledge of what healthy people habitually consume, and accept values which are of the order of the Voit estimates of seventy years ago, that is, approach 100 g per day?

We can perhaps neglect consideration of the absolute minima of essential amino acid needs since very much more investigational work is required before the implications of current work can be fully explained let alone put to practical use. This leaves the old question of optima and minima or of upper and lower limits of knowledge to be resolved. Application of the concept of the mean minimum calculated by factorial methods as an adequate estimate of human protein needs would imply that in large parts of Asia and Africa diets are adequate in protein content, for the ratio of net dietary protein calories to total calories in most Asian diets is in slight excess of dietary net protein needs. It can be and has been argued that since muscular work does not increase the protein metabolism, if Asians did more physical work and consumed more food, low protein staples would be adequate for virtually all classes save perhaps the weaned infant (Payne, 1972). This would suggest that no major protein problem of any global significance occurs save in those societies which consume the 'empty calories' of cassavas and yams. On the other hand, if the safe estimates of need are adopted and it is stated that protein calories should be 10 per cent of total calories for diets with biological values of 70–75 per cent, and adjustments made for variations in biological value, then protein deficiency is widespread in man in developing countries and no amount of physical work will improve matters.

In support of the higher recommendations about protein needs three factors can be adduced. First, there is the argument of Crichton-Browne (1909) already referred to that no one knows the effects on man of long-term subsistence at the minimum. Secondly, the association of protein in foods with other dietary essentials, notably B complex vitamins, phosphorus and trace metals can be considered, and thirdly, it is human experience – certainly Western experience – that diets of low protein content are not very palatable. Indeed, many affluent sectors of Western societies consume amounts of protein far in excess of the upper limit of knowledge. The last two factors have little to do with the defined problem of estimating human protein needs; they may well, however, be of importance in the design of nutritional policies.

The above remarks about the uncertainty associated with the protein needs of the human subject all refer to evidence accumulated for adult man. The presence of protein-calorie malnutrition in infants and in children in many developing countries of the world, however, surely indicates that here there is no uncertainty about an absolute need for protein, and that dietaries adequate for adults in protein content are inadequate for the young. Again, however, there is uncertainty about the actual amount of protein to be supplied. The minimum requirement of protein with a BV of 70 for children six to twelve months of age has been estimated to be 1.25 g per kg body weight per day by the Joint FAO–WHO Expert Group on Protein Requirements (1965). Chan & Waterlow (1966) arrived at a similar figure in experiments on Jamaican infants, though they realised this was a high value. Calculations by the Department of Health & Social Security (1969) of minimal protein needs, and inclusion of a 20 per cent margin of safety, led to values of 1.7 to 1.8 g per kg. The recommended daily intake in the United Kingdom, however, is assessed as 10 per cent of dietary calories as protein leading to allowances of up to 3 g per kg body weight. In Russia, estimates of requirements are even higher at 3.5 to 4.0 g per kg (Roubstein, Birger & Freid, 1936).

Uncertainties thus apply to the protein needs of young children similar to those we have for the adult subject. The minimal estimates of protein needs of young children, however, when expressed as a proportion of the total calories are about 50 per cent greater than those of adults similarly computed. Any planning of agricultural production and food provision must take this into account.

Finally, the efficiency of a productive process involves a numerator of output and a denominator of input. Eventually we must place values on each of these if we are to ascertain optima. The uncertainties that surround man's need for protein immediately create difficulties when these wider issues are considered. I hope that Professor Spedding in the paper that follows will not ignore them.

REFERENCES

ATWATER, W. O. (1903). The demands of the body for nourishment and dietary standards. *15th Annual Report of the Storrs Agricultural Experiment Station, Connecticut*, 123–46.

CHAN, H., & WATERLOW, J. C. (1966). The protein requirements of infants at the age of about one year. *British Journal of Nutrition* **20**, 775–82.

CHITTENDEN, R. H. (1904). *Physiological Economy in Nutrition*. New York: F. A. Stokes Co.

CRICHTON-BROWNE, J. (1909). *Parsimony in Nutrition*. London and New York: Funk & Wagnalls.

DEPARTMENT OF HEALTH & SOCIAL SECURITY (1969). *Recommended Intakes of Nutrients for the United Kingdom.* Report on Public Health and Medical Subjects No. 120. London: HMSO.
ERNLE, LORD (1936). *English Farming Past and Present.* New (5th) edition, edited by Sir A. D. Hall. London: Longmans Green & Co.
GARROW, J. S. (1969). *Individual variation in minimum requirements for nitrogen balance.* Report on Public Health and Medical Subjects No. 120, Appendix 2, 36–7. London: HMSO.
HEGSTED, D. M. (1964). Protein requirements. In: Munro, H. N. & Allison, J. B. (Eds.), *Mammalian Protein Metabolism,* volume 2, 135–71. New York and London: Academic Press.
HEGSTED, D. M., TSONGAS, A. G., ABBOTT, D. B., & STARE, F. J. (1946). Protein requirements of adults. *Journal of Laboratory and Chemical Medicine* **31**, 261–84.
IRWIN, M. I., & HEGSTED, D. M. (1971a). A conspectus of research on protein requirements of man. *Journal of Nutrition* **101**, 385–430.
IRWIN, M. I., & HEGSTED, D. M. (1971b). A conspectus of research on amino acid requirements of man. *Journal of Nutrition* **101**, 539–66.
JOINT FAO–WHO EXPERT GROUP ON PROTEIN REQUIREMENTS (1957). *FAO Nutrition Studies No. 16.* Rome: FAO.
JOINT FAO–WHO EXPERT GROUP ON PROTEIN REQUIREMENTS (1965). *FAO Nutrition Meeting Report Series No. 37.* Rome: FAO.
KOFRÁNYI, E., & JEKAT, F. (1964). Zur Bestimmung der biologischen Wertigkeit von Nahrungsproteinen. IX. Der Ersatz von hochwertigen Eiweiss durch nichtessentiellen Stickstoff. *Hoppe-Seyler's Zeitschrift für physiologische Chemie* **338**, 154–167.
KORFÁNYI, E., & MÜLLER-WECKER, H. (1961). Zur Bestimmung der biologischen Wertigkeit von Nahrungsproteinen. V. Der Einfluss des nichtessentiellen Stickstoffs auf die biologische Wertigkeit von Proteinen und die Wertigkeit von Kartoffelproteinen. *Hoppe-Seyler's Zeitschrift für physiologische Chemie* **325**, 60–4.
LANDERGREN, E. (1903). Untersuchungen über die Eiweissumsetzung des Menschen. *Skandinavisches Archiv für Physiologie* **14**, 112–75.
MCCAY, D. (1912). *The Protein Element in Nutrition.* London and New York: Longmans Green & Co.
MCCOLLUM, E. V. (1957). *A History of Nutrition. The Sequence of Ideas in Nutritional Investigations.* Boston: Houghton Mifflin Co.
MAGENDIE, F. (1816–17). *Précis Elémentaire de Physiologie,* 2 volumes. Paris: Méguinon-Mervais.
MINISTRY OF HEALTH (1964). *Requirements of Man for Protein.* Report on Public Health and Medical Subjects No. 111. London: HMSO.
MITCHELL, H. H. (1954). The dependence of the biological value of food proteins upon their content of amino acids. *Wissenschaftliche Abhandlungen der Deutschen Akademie der Landwirtschaftswissenschaften zu Berlin,* Band V/2, 279–325.
OSER, B. L. (1951). Method for integrating essential amino acid content in nutritional evaluation of protein. *Journal of the American Dietetic Association* **27**, 396–402.
PAYNE, P. R. (1972). The nutritive quality of Asian dietaries. *Proceedings of the 1st Asian Congress of Nutrition, Hyderabad, India* (in Press).
ROSE, W. C. (1957). The amino acid requirements of adult man. *Nutrition Abstracts and Reviews* **27**, 631–47.
ROSE, W. C. & WIXOM, R. L. (1955). The amino acid requirements of man. XVI. The role of the nitrogen intake. *Journal of Biological Chemistry* **217**, 997–1004.

ROUBSTEIN, L., BIRGER, L., & FREID, R. (1936). Influence de différentes rations albuminoïdes sur la croissance et le développement des enfants de 1½ à 2 années et demie. *Bulletin de Biologie et de Médecine Expérimentale de l'URSS* **1**, 235–6.

RUBNER, M. (1902). *Die Gesetze des Energieverbrauchs bei der Ernährung*. Berlin u. Wien: F. Deuticke.

SCRIMSHAW, N. S., YOUNG, V. R., HUANG, P. C., THANANGKUL, O., & CHOLAKOS, B. V. (1969). Partial dietary replacement of milk protein by nonspecific nitrogen in young men. *Journal of Nutrition* **98**, 9–17.

SHERMAN, H. C. (1920). Protein requirements of maintenance in man and the nutritive efficiency of bread protein. *Journal of Biological Chemistry* **41**, 97–109.

SIVEN, V. O. (1900). Über das Stickstoffgleichgewicht beim erwachsenen Menschen. *Skandinavisches Archiv für Physiologie* **10**, 91–148.

SWAMINATHAN, M. (1963). Amino acid and protein requirements of infants, children and adults. *Indian Journal of Pediatrics* **30**, 189–99.

SWENDSEID, M. E., HARRIS, C. L., & TUTTLE, S. G. (1960). The effect of sources of nonessential nitrogen on nitrogen balance in young adults. *Journal of Nutrition* **71**, 105–8.

VOIT, C. (1876). Über die Kost in öffentlichen Anstalten. *Zeitschrift für Biologie* **12**, 1–59.

WILLCOCK, E. G., & HOPKINS, F. G. (1906). The importance of individual amino acids in metabolism. Observations on the effect of adding tryptophane to a dietary in which zein is the sole nitrogenous constituent. *Journal of Physiology* **35**, 88–102.

2

THE MEANING OF BIOLOGICAL EFFICIENCY

By C. R. W. SPEDDING

The Grassland Research Institute, Hurley, Berks, and
the Department of Agriculture, University of Reading

The object of this paper is to discuss the meanings that can most usefully be attached to the term 'biological efficiency', in the context of this symposium.

One major characteristic of this context is that the efficiency of biological processes is being considered over a wide range of levels of organisation of both plants and animals, from tissues to whole populations, and thus from several points of view, including those of biologists and agriculturalists. It is therefore unusually easy to see that whilst the *meaning* of efficiency may remain constant, its *expression* must vary widely.

The implications are that the most useful definition of 'efficiency' will be a wide one but that any particular expression of it should be very precise.

Since it is obviously undesirable to increase the number of different definitions of any expression, it should be emphasised that this paper is oriented to this particular symposium. Blaxter (1968), for example, described biological efficiency as a measure of the ability of a species to reproduce, survive, and maintain its numbers in a given environment. This is a perfectly valid use of the term, although probably not the only one. In general, biological efficiency could sensibly relate to any measure of the efficiency with which a biological process is carried out for a *biological purpose*. In case the word 'purpose' may be unacceptable, however, it should be noted that it is used here to mean 'ends' and the previous sentence could conclude 'for biological ends'.

I am therefore accepting that a distinction can usefully be made between the 'efficiency of a biological process' and the much narrower term 'biological efficiency': in this sense the symposium title is using the latter as a shorthand version of the former.

It is proposed to define efficiency, therefore, simply as a ratio of units, usually functionally related, generally equivalent to a rate. The ratio Output/Input exemplifies this relationship and appears appropriate for all processes, whether a product is involved or not, provided that time is accepted as an input (see Spedding, 1971).

There are then two main ways of defining biological efficiency: clearly a biological process must be involved but its efficiency may be expressed in either biological or non-biological terms. It is probably a more accurate use of words to restrict the expression to biological terms or, at least, to terms that have a biological significance. It seems more generally useful, however, to impose no restriction at all and to accept, initially at any rate, that expression in physical or even financial terms may sometimes be useful in the consideration of the efficiency of biological processes.

If man is included in the picture, then such expressions are of direct relevance: for example, money offers a more useful summary of all aspects of the value of food to man than any single term available for use with other animals. (The fact that money frequently reflects many other factors as well demonstrates its flexibility rather than any inadequacy.)

There are probably some risks in accepting these broad interpretations but none that outweigh the advantages.

The discipline that has to be equally accepted, however, is that the particular expression used must be sensibly related to the purpose for which the calculation is made. This purpose needs to be clearly stated, therefore, and the terms used to express efficiency have to be justified as the most useful for that purpose.

It is necessary then to be clear about the purpose of the symposium, which is to enlarge our understanding of protein production as a biological process, whether by plants or animals, individuals or populations. The purpose of protein production has already been discussed, and there are many reasons why we should wish to understand the production processes better. Some of our interests, of course, go well beyond the biology of the processes but distinctions between economic and non-economic interests are much less clear than is generally supposed.

Within any applied interest, such as agriculture, economic assessments are of major importance, not only in their own right, but as a means of determining the relative importance of component biological processes. On the other hand, by themselves they cannot confer understanding of these very biological processes, on which the economic output depends.

It is to be expected, then, that understanding of the biological processes involved in protein production may require a variety of expressions of efficiency, which may be judged neither right nor wrong but only useful or inappropriate.

To illustrate these ideas, the biological efficiency of egg production will be considered briefly. (The subject is chosen because of its biological and agricultural importance and to avoid overlap with subsequent contributions.)

THE BIOLOGICAL EFFICIENCY OF EGG PRODUCTION

Eggs represent an important stage in the life of a wide range of animals (see Table 1 for some examples of this range) and, in the case of some species, an important agricultural product, either for direct consumption or as a starting point for a further growth process.

There is, therefore, considerable similarity between the biological and economic expressions of efficiency for egg production. In both cases, the egg may have a value to a consumer (man or some other animal) and the relevant measures of output will be some combination of number, content and food value. In both cases, also, the egg may be important as a starting point for further development and, again, the output will be expressed as egg number and content (in the sense that this may determine size, rate of development, potential for growth, survival,* etc.). In passing, it is worth noting how much easier it is to summarise *potential* values for economic purposes, i.e. in monetary terms, than it is for biological purposes.

There are much greater differences in the relevant inputs, however. For economic purposes inputs may include money and labour; for biological purposes such components as energy and protein may be more relevant; in both cases, land area, the size of the breeding population and the food they consume are likely to be important inputs on some occasions.

Efficiency of egg production may usefully be assessed, therefore, by considering ratios of these inputs and outputs but, for protein production, the most appropriate are likely to be as follows:

Agriculturally: Edible protein (amount or value of eggs or products derived from them) per unit of nitrogen, money, energy, breeding population, food or land.

Biologically: Egg (number produced or reared, weight, protein content) per unit of population, food or land.

In the second category, the biological significance of the reproductive rate is such that egg number per breeding female, for example, seems of obvious interest: the relevance of the amount of protein produced may be less obvious. Quite apart from the possible influence of the protein content of the egg on survival and subsequent growth, however, there is the possibility that the quantity of protein might be amongst the more useful expressions of population size, for a wide range of populations. Clearly the number of individuals may not be very useful and total weight (wet or dry) also suffers from several disadvantages.

All of these ratios must relate to some period of time, obviously the same

* The properties of the shell may be important in this context.

Table 1. *Nitrogen outputs of eggs for various species (Table compiled by Miss Angela Hoxey, Grassland Research Institute)*

Figures in parentheses refer to the relevant references in List A or B.

Species	Whole egg weight (g)	N content of egg (g)	N content of parental unit[a] (g)	N output per egg, as % of N content of parental unit	Annual egg output (no.)	Annual N output of eggs, as % of N content of parental unit
Hen (*Gallus domesticus*)	58	Edible 1.08 (6) Whole 1.11 (12)	42.2 (8)	2.50 2.60	240 (8)	609 626
Duck (*Anas domesticus*)	74	Edible 1.45 (6) Whole 1.49[b]	—	—	300 (2)	—
Goose (*Anser domesticus*)	165	Edible 3.22 (3, 6) Whole 3.43[b]	115 (1, 8)	2.80 3.00	50 (4, 5)	140 149
Goose (*Anser domesticus*)	197	Edible 3.93 (6) Whole 4.10[b]	230 (1, 8)	1.70 1.80	30 (4, 5)	51 53
Slug (*Agriolimax reticulatus*)	0.005	0.04 mg (4)	0.033 (4)	0.12	500 (10, 11)	61
Locust (*Schistocerca gregaria*)	0.01	0.45 mg (3, 4)	0.18[c] (4)	0.25	504 (1, 7) (Total of 3 generations)	43 (Calculated for 3 generations per year)

N = nitrogen.

[a] The parental unit consists of one female and a fraction of a male, varying from 0 for the slug and $\tfrac{1}{12}$ for the hen to 1 for the locust.

[b] Calculated as for the hen, i.e. shell contains 3% of the protein in the whole egg.

[c] % N in locusts 3.56 (Grassland Research Institute, mean of 4 locusts). Mean weight of locusts 3.11 g (females), 1.89 g (males).

for both numerator and denominator. For comparisons of different ratios, either the same or comparable time periods can be used, e.g. a year, one complete reproductive cycle or one lifetime.

It is proposed to illustrate this discussion with two examples: the first is concerned with egg production by different birds and the second compares the egg production of two pests, a slug (*Agriolimax reticulatus*) and a locust (*Schistocerca gregaria*). Amongst the reasons for choosing these examples are availability of data, the probability that such contrasting animals offer a worthwhile challenge to definitions of efficiency, and the belief that understanding of a process is often increased by looking at it over a wide range of biological material.

Egg production in birds

Table 2 gives the basic data available for nitrogen production and consumption by domestic hens and geese and Table 3 shows some of the efficiency ratios that can be derived from them.

Clearly, all these ratios and others, and their comparisons, have some interest and it would be absurd to argue that one expression is right and another wrong. The table is given to show the variety of expressions that can be used and the variety of ways in which different species can then be compared. Nitrogen has been used, rather than protein, because it assumes nothing about the form in which it is present.

One of the interesting results of these comparisons is that egg production in the hen appears to be similar in efficiency (0.22) to that of growing those same eggs on to meat (0.27), in terms of food nitrogen use by the whole population. Since this seems extremely unlikely for very low egg production and, indeed, appears not to be so at all for geese, it suggests that the efficiency of both egg and meat production for protein increases with number of eggs per bird per annum, but that the relationships are quite different for these two products.

So far as our calculations have been taken, this conclusion seems well-founded (see Figure 1): there seems to be no theoretical reason why the efficiency of egg protein production should not overtake that for meat in both hens and geese, though at quite different numbers of eggs per bird per year. Clearly, these values would be quite different on an energy basis, and it is possible that, in any event, a more useful relationship could be established with egg weight (per unit weight of bird) than with egg number.

If eggs are thought of solely as part of the mechanism for population increase in 'natural' conditions, then the number of progeny that can be reared from them is a more relevant measure. This is illustrated in Table 4,

Table 2. *Egg production, meat production and food consumption of hens and geese*
(Table compiled by Miss Angela Hoxey, Grassland Research Institute)

	Hen			Goose		
	Egg production	Broiler production	References in List A from which information used in calculations	Light (egg production)	Heavy (meat production)	References in List A from which information used in calculations
1. Weight of N per egg and [ ] per edible portion (g)	1.11 [1.08]	—	(3, 6, 12)	3.43 [3.22]	—	(4)
2. No. of eggs per year	240	180	(8)	50	30	(4, 5)
3. Weight of N produced per year as eggs (g)	266	—	—	171.5	—	—
4. Mean weight of female bird (kg)	2.0	3.0	(8)	4.5	9.0	(3)
5. Mean weight of parental unit (kg)[a]	2.2	3.4	(8, 13)	6.0	12.0	(4)
6. N content of parental unit (g)	42.2	65.3	(8)	115	230	(1, 8)
7. Weight of food (DM)[b] consumed per parental unit[c] per year (kg)	41.4	49.5	(8)	109.1	192[d]	(9, 10)
8. Weight of food N per parental unit per year (kg)	1.2	1.4	(8)	3.0	5.3	(7, 10, 11)
9. Food energy per parental unit per year (Mcal)	167	199	(8)	367	647	(7, 10, 11)
10. Land area required per parental unit per year[e] (ha)	0.02	—	[e]	0.06	—	[e]
11. No. of progeny at slaughter	—	108	(8)	—	18	(4, 5, 10)
12. Weight of N in one progeny at slaughter (g)	—	35	(8)	—	148	(1, 8)
13. Weight of N in edible carcass (g)	—	26	(1, 8)	—	100	(1)

14. Weight of food (DM) eaten by one progeny from hatching to slaughter (kg)	—	3.6	(8)	34.5	(9, 10)
15. Weight of N in DM eaten by one progeny from hatching to slaughter (g)	—	115.2	(8)	998	(9, 11)
16. Energy in DM eaten by one progeny from hatching to slaughter (Mcal)	—	15.3	(8)	124	(7, 10)

N = nitrogen.   DM = dry matter.

[a] The parental unit consists of one female and a fraction of a male. Ratio of males to females: hens, egg production 1:12, broiler production 1:10; geese, light 1:4, heavy 1:3.
[b] DM taken as 90% of food as fed, i.e. air-dry.
[c] For the purposes of these calculations it has been assumed that the male bird has a similar intake to that of a productive female.
[d] Estimated.
[e] Calculated from a ration of 72% maize, 18% soyabean meal (44% crude protein), 10% minerals and vitamins, using European yields quoted in FAO (1969) *Production Yearbook*, **23**.

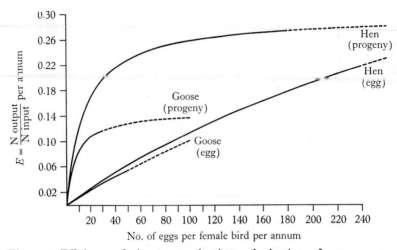

Figure 1. Efficiency of nitrogen production as body tissue from progeny and as eggs, by the hen and the goose. Curves for progeny are those for ratio 5, and for egg production those for ratio 1, in Table 3. Solid lines refer to levels of egg production that are readily achieved on a flock basis.

Table 3. *Efficiency ratios for egg and meat production by hens and geese (Table compiled from data in Table 2 by Miss Angela Hoxey, Grassland Research Institute)*

	Items in Table 2 to which ratio relates	Hen	Goose
Egg production efficiency			
1. $\dfrac{\text{Egg N}}{\text{Food N}}$	3/8	0.22	0.06
2. $\dfrac{\text{Egg N}}{\text{Food energy}}$	3/9	1.59	0.47
3. $\dfrac{\text{Egg N}}{\text{N in parental unit}}$	3/6	6.30	1.49
4. $\dfrac{\text{Egg N}}{\text{Area of land required for parental unit}}$	3/10	13 300	2858
Meat production efficiency			
5. $\dfrac{\text{N in progeny slaughtered}}{\text{Food N for parental unit} + \text{progeny}}$	$\dfrac{12 \times 11}{8 + (15 \times 11)}$	0.27	0.11
6. $\dfrac{\text{N in progeny slaughtered}}{\text{Food energy for parental unit} + \text{progeny}}$	$\dfrac{12 \times 11}{9 + (16 \times 11)}$	2.04	0.92
7. $\dfrac{\text{N in progeny slaughtered}}{\text{N in parental unit}}$	$\dfrac{12 \times 11}{6}$	58	11.58
8. $\dfrac{\text{N in edible carcass of progeny}}{\text{Food N for parental unit} + \text{progeny}}$	$\dfrac{13 \times 11}{8 + (15 \times 11)}$	0.20	0.08
9. $\dfrac{\text{N in edible carcass of progeny}}{\text{Food energy for parental unit} + \text{progeny}}$	$\dfrac{13 \times 11}{9 + (16 + 11)}$	1.50	0.63

N = nitrogen.

both for numbers and for the amount of nitrogen added to that of the original population. Actual population increase would need to take into account mortality rates, but the figures given have some bearing on the potential rate of increase in the size (biomass) of the population, measured as nitrogen. Clearly, the result of the calculation is much influenced by the number of eggs that one bird can rear.

Table 4. *Reproductive rate of three species of birds under conditions of natura rearing (Table compiled by Miss Angela Hoxey, Grassland Research Institute)*

	Hen[a]	Goose[b]	Owl[c]
Mean clutch size	13	9.5	2.5
Weight of eggs per kg of bird per year (g)	312	263	200[d]
Probable no. of eggs that one bird can rear per year	10	6	1.3
Weight of N in progeny after one year (g), a	576	806	—
Weight of N in one parental unit (g), b	52	172	—
Ratio of a/b	11.1	4.7	—

N = nitrogen.

[a] Medium size hen, 2.5 kg liveweight, egg weight 60 g.
[b] Medium size goose, 6.5 kg liveweight, egg weight 180 g.
[c] Southern, H. N. (1970). The natural control of a population of Tawny Owls (*Strix aluco*). *Journal of Zoology: Proceedings of the Zoological Society of London*, **162**, 197–285.
[d] Cott, H. B. (1954). Palatability of the eggs of birds. *Proceedings of the Zoological Society of London*, **124**, 335–463.

Egg production in invertebrate pests

The egg production of such animals as slugs and locusts is of considerable importance and interest to man, since they are pests and their numbers matter, and to other animals that feed on them (as eggs or at other stages), as well as, presumably, to the slugs and locusts themselves.

Since, in both these animals, hatching and rearing are independent of the female, the number of eggs available for natural rearing is not different from that produced in total.

Table 5 summarises the relevant data relating to egg production and food consumption and Table 6 shows the results of similar calculations to those made for birds.

The comparison of nitrogen production as body tissue and as eggs is not so relevant in the case of slugs and locusts, except where both eggs and adults are the food of other animals: nevertheless, the comparison is shown in Figure 2, in relation to the number of eggs per unit of the breeding population per year.

Table 5. *Egg production and food consumption of a locust and a slug*
(*Table compiled by Miss Angela Hoxey, Grassland Research Institute*)

Figures in parentheses indicate references in List B from which information was obtained

		Locust (*Schistocerca gregaria*)	Slug (*Agriolimax reticulatus*)
(a)	Weight of N per egg (mg)	0.45 (3, 6)	0.04 (6)
(b)	No. of eggs per year	504 (1, 7)	500 (10, 11)
(c)	Weight of N as eggs per year (mg)	226.8	20.0
(d)	Mean weight of female (g)	3.1 (4, 12)	0.33 (2)
(e)	Mean weight of parental unit (g)[a]	5.0	0.33
(f)	N content of parental unit (mg)	178[b] (6)	33 (6)
(g)	Weight of DM consumed per parental unit per year (g)	121.8[c] (1, 4, 7)	3.3 (9)
(h)	Weight of N in food consumed by parental unit (g)	4.26[d]	0.102 (5)
(i)	Energy in food consumed by parental unit (Mcal)	0.536[e]	0.014 (8)[e]
(j)	No. of progeny maturing per year	209	234
(k)	Weight of N in one progeny (mg)	89	33
(l)	Weight of N in progeny per year (g)	18.601	7.7
(m)	Food N consumed by progeny per year (g)	148.4	11.9
(n)	Energy in food consumed by progeny per year (Mcal)	18.6	1.6

N = nitrogen.   DM = dry matter.

[a] Account has been taken of the weight of the female plus necessary males, i.e. locusts: 1 female + 1 male; slugs: 1 hermaphrodite.
[b] Assuming male locust to have the same % N as female.
[c] This figure is simply double that for three females: it is probably high because the smaller male would eat less than the female.
[d] Assuming 20 % DM and 3.5 % N.
[e] Assuming 4.4 kcal per g DM.

Discussion

Since the data presented are given primarily to illustrate the varied uses of various expressions of biological efficiency, it is not intended to discuss the results in detail. The following tentative conclusions appear worth stating, however.

1. Efficiency of nitrogen production in the hen and the goose, whether as eggs or as body tissue, increases with increasing number of eggs per bird per year.

2. Beyond a certain number of eggs per bird per year, efficiency of nitrogen production may be greater for eggs than for body tissue.

3. The slug (*Agriolimax reticulatus*) appears to be very efficient in the

THE MEANING OF BIOLOGICAL EFFICIENCY

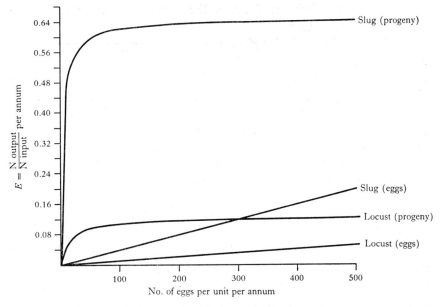

Figure 2. Efficiency of nitrogen production as body tissue from progeny and as eggs, by the locust and the slug. Curves for progeny are those for ratio C, and for egg production those for ratio A, in Table 6.

Table 6. *Efficiency ratios for egg and progeny production by locusts and slugs* (*Table compiled from data in Table 5 by Miss Angela Hoxey, Grassland Research Institute*)

Efficiency ratio (calculated over 1 year)	Items in Table 5 to which ratio relates	Equivalent ratio in Table 3	Locust (*Schistocerca gregaria*)	Slug (*Agriolimax reticulatus*)
(A) $\dfrac{\text{Egg N}}{\text{Food N}}$	$\dfrac{c}{h}$	1	0.05	0.20
(B) $\dfrac{\text{Egg N}}{\text{Food energy}}$	$\dfrac{c}{i}$	2	0.42	1.43
(C) $\dfrac{\text{N in progeny maturing}}{\text{Food N of parental unit} + \text{progeny}}$	$\dfrac{l}{h+m}$	5	0.12	0.64
(D) $\dfrac{\text{N in progeny maturing}}{\text{Food energy of parental unit} + \text{progeny}}$	$\dfrac{l}{i+n}$	6	0.97	4.8
(E) $\dfrac{\text{N in progeny maturing}}{\text{N in parental unit}}$	$\dfrac{l}{f}$	7	34.8	233

N = nitrogen.

use of nitrogen, relative to the hen, the goose and the locust, for the production of body tissue.

4. The slug shows similar efficiency for egg nitrogen per unit of food nitrogen (ratios 1 (Table 3) and A (Table 6)) to that of the hen, whilst the locust is much poorer: but the slug is much more efficient at adding nitrogen to the population (ratios 7 and E) than either hen or goose.

These tentative conclusions suggest a multitude of interesting questions and hypotheses, many of which derive from having compared egg-producing species that differ fundamentally in other respects. It might be extremely illuminating to go further and contrast the efficiency of protein use for reproduction in plants and animals; this would certainly illustrate the need to select the most relevant efficiency ratios. It is precisely these kinds of comparison that this symposium provides an opportunity to make.

The ratios employed appear to be the most useful for the purposes of this particular discussion, but one of the difficulties that emerges is the need to decide what is meant by production. Agriculturally, a product is implied, and a value can always be attached to a product, but in non-agricultural situations productivity does not necessarily result in one or even several well-defined products (see Macfadyen, 1948, 1957). In these circumstances, the same argument can be applied to productivity as was used in relation to efficiency: the appropriate definition will vary with the particular purpose for which the calculation is made, but when used each definition should be stated precisely.

Dr Blaxter has already pointed out that there are many difficulties in deciding precisely what the purpose of protein production may be, in terms of the balance of amino acids required for human nutrition.

The argument put forward here simply insists that measures of efficiency can only reflect the terms employed in their calculation. Since it is unlikely that one expression of protein production will be equally satisfactory for all purposes, it is clear that the number of efficiency ratios that are relevant may be quite large.

Each of these ratios may be used to define terms such as 'energetic efficiency' and 'feed conversion', but it will generally be better to state precisely which ratio is being used, rather than to rely wholly on these numerous subsidiary definitions.

It is also clear that no single value for efficiency can be attributed to any particular species, since the value will vary with a number of characteristics, including levels of performance (such as reproductive rate). Efficiencies are therefore better compared in terms of response curves relating them to changes in the value of the most important factors influencing them.

REFERENCES

BLAXTER, K. L. (1968). Relative efficiencies of farm animals in using crops and byproducts in production of foods. *Proceedings of the 2nd World Conference on Animal Production, Maryland*, 31–40.

MACFADYEN, A. (1948). The meaning of productivity in biological systems. *Journal of Animal Ecology* **17**, 75–80.

MACFADYEN, A. (1957). *Animal Ecology*. London: Pitman.

SPEDDING, C. R. W. (1971). *Grassland Ecology*. Oxford University Press.

Sources of information used in hen and goose calculations (List A)

1. AGRICULTURAL RESEARCH SERVICE. *Composition of Foods*. Agriculture Handbook No. 8. United States Department of Agriculture.
2. BOLTON, W. (1970). The future of animals as sources of human food. The role of poultry. *Proceedings of the Nutrition Society* **29**, 253–62.
3. BRODY, S. (1945). *Bioenergetics and Growth*. New York: Reinhold Publishing Corporation.
4. MCARDLE, A. A. (1966). *Poultry Management and Production*. 2nd edition. Sydney: Angus & Robertson.
5. MINISTRY OF AGRICULTURE, FISHERIES AND FOOD (1962). *Goose Production*. MAFF Advisory Leaflet 112. London: HMSO.
6. MINISTRY OF AGRICULTURE, FISHERIES AND FOOD (1967). *Poultry Nutrition*. MAFF Bulletin 174. London: HMSO.
7. MONACHON, G. (1968). Institut national de la Recherche agronomique, Section de Recherches Avicoles, Domaine Expérimental d'Artiguères, Benquet (Landes). Rapport D. 594.
8. MORRIS, T. R. (1971). Personal communication.
9. ONTARIO DEPARTMENT OF AGRICULTURE AND FOOD. *Duck and Goose Raising*. Publication 532.
10. Personal communications with goose farmers.
11. ROBERSON, R. H., & FRANCIS, D. W. (1963). The effect of energy and protein levels of the ration on the performance of White Chinese Geese. *Poultry Science* **42**, 867–71.
12. STURKIE, P. D. (1954). *Avian Physiology*. New York: Comstock Publishing Associates.
13. WATSON, J. A. S., & MOORE J. A. (1949). *Agriculture. The Science and Practice of British Farming*. 9th edition. Edinburgh: Oliver & Boyd.

Sources of information used in slug and locust calculations (List B)

1. BALLARD, E., MISTIKAWI, A. M., & ZOHEIRY, M. S. EL (1932). *The Desert Locust, Schistocerca gregaria Forsk., in Egypt*. Cairo: Government Press.
2. BARNES, H. F., & WEIL, J. W. (1945). Slugs in gardens: their numbers, activities and distribution. Part 2. *Journal of Animal Ecology* **14**, 71–105.
3. CHEU, S. P. (1952). Quoted by Uvarov, B. (1966). *Grasshoppers and Locusts*, volume 1, 250. Cambridge University Press.
4. DAVEY, P. M. (1954). Quantities of food eaten by the Desert Locust *Schistocerca gregaria* (Forsk.) in relation to growth. *Bulletin of Entomological Research* **45**, 539–51.
5. DAVIES, W. (1960). *The Grass Crop*. London: Spon.
6. GRASSLAND RESEARCH INSTITUTE (1971). Unpublished data.
7. HAMILTON, A. G. (1955). Quoted by Uvarov, B. (1966). *Grasshoppers and Locusts*, volume 1, 320. Cambridge University Press.

8. HOXEY, A. (1971). Unpublished data. Grassland Research Institute.
9. PALLANT, D. (1970). A quantitative study of feeding in woodland by the grey field slug (*Agriolimax reticulatus* Muller). *Proceedings of the Malacological Society of London* **39**, 83–7.
10. QUICK, H. E. (1960). British slugs. *Bulletin of the British Museum (Natural History)* **6**, 105–226.
11. STEPHENSON, J. W. (1971). Personal communication.
12. WEIS-FOGH, T. (1952). Fat combustion and metabolic rate of flying locusts (*Schistocerca gregaria* Forskal). *Philosophical Transactions of the Royal Society, London,* B **237**, 1–36.

3
FUTURE DEMAND FOR PROTEIN FOODS

By A. A. WOODHAM

The Rowett Research Institute, Bucksburn, Aberdeen

The importance attached to the problem of assessing the future demand for protein on a world-wide basis is indicated by the increasing number of conferences and symposia now being held with a view to studying ways in which supplies of protein foods may be augmented. No one possessing any knowledge of world economic conditions can be unaware of the possibility of catastrophe being brought about by the combination of the 'population explosion' and a totally inadequate rate of expansion of agricultural production. So much has been written on the subject that it becomes increasingly difficult to find something new to say. In the United States, the President's Science Advisory Committee published a three volume report on the world food problem (Panel on the World Food Supply, 1967) and in this country the Faculty Board of Agriculture of the University of Cambridge organised a course of lectures by experts on various aspects of population and food supply in 1966, since published in book form (Hutchinson, 1969). The present author has reviewed the situation recently (Woodham, 1971) and in preparing this present contribution all of these sources and others have been drawn upon extensively so that in a sense one might describe this as a 'review of reviews'.

It will become clear that so many imponderables are involved that a more suitable title might have been 'Difficulties inherent in predicting future demand for protein foods'. A realisation of the difficulties is, however, a necessary preliminary without which it is impossible to forecast even the broad paths along which we must progress. To refrain from action merely because of the size of the problem is one sure way of hastening eventual catastrophe. To avert it, not only must the facts be appreciated by all but they must be repeatedly brought to the notice of those with the power to mobilise resources and influence the mass of people. The problem is not merely to overcome apathy, though the predominant mood within the developed countries is indeed apathetic, but to combat the views of those optimists who maintain that everything will be all right in the end. Man's technological know-how, it is claimed, will enable him continuously to increase protein production and keep abreast of the situation. Such thinking is dangerous and ignores the facts.

Man is already behind in the race. A large proportion of the world population at the present time is either starving or at least suffering from severe undernourishment and malnutrition.

POPULATION INCREASE AND PROTEIN DEMAND

World hunger is no new thing. It is 173 years since Malthus stated that population increases at a greater rate than do food resources. Famines occur from time to time but the effects are generally local and the ripples of interest which are aroused soon subside. The explosive growth of world population is constantly before us. Improved communications have made it possible for the facts to be disseminated widely, and through the medium of television the effects of protein shortage in the underdeveloped countries have been presented dramatically to the more fortunate in their own homes. Through the ages population has increased progressively despite the effects of famine, war and disease. Advances in medical knowledge during the present century have led to a decline in infant mortality and increased longevity. The rate of growth of population has thereby been accelerated, and this acceleration is a direct result of progress. Control of population growth is unlikely to be the answer in itself, for the countries which would benefit most from such control are in fact those where ignorance and religious prejudice pose insuperable barriers. It is interesting to note that some have postulated that a decline in infant mortality, far from increasing the rate of population growth, may actually reduce it. The fear of child deaths, often caused by malnutrition, in underdeveloped countries leads to an increase in family size beyond that actually desired. If people were assured that their children had an increased chance of survival, family size would be diminished voluntarily and the ultimate effect would be to reduce rather than to raise the rate of population growth. Thus reduction of malnutrition could itself be a key factor in solving the population problem (Panel on the World Food Supply, 1967).

It would seem appropriate at this point to consider by how much the present world supply of protein falls short of that required for adequate nutrition of the total world population, and then to see if it is possible to arrive at a reasonable estimate of the world requirement of protein in, say, thirty years' time. Such calculations are of course fraught with difficulties. Considering population alone, it was predicted in 1937 that the population of the United States would reach 150 millions and then decline, but by 1965 it was 190 millions and still rising. Calculations based on studies made by the United States Bureau of the Census suggest that the population may be more than 300 millions by the year 2000 and it may indeed exceed 400 millions. Population doubles in size in 18 to 27 years in the less developed countries but only in 55 to 88 years in the developed countries (Panel on the World Food Supply, 1967). Williams has estimated that 85 per cent of world population growth will occur in Africa, Asia and Latin America and these areas will account for 80 per cent of the world population in the year 2000. Thus, he concludes, while the population in the developed areas will increase by about 50 per cent by the end of this century, in areas where food, and especially protein, is scarce, the population will increase by more than 100 per cent (Williams, 1966).

PROBLEMS OF ASSESSING CURRENT PROTEIN PRODUCTION

The difficulties involved in estimating current world supplies of food and protein have been fully described by Farmer (1969) and only his main points will be summarised here. Crop yields may be underdeclared in order, for example, to evade tax or to conceal produce destined for black market disposal. Some categories of produce may be officially ignored. Allotment and kitchen produce in the United Kingdom represent good examples of this. In less developed countries it may be difficult for a government to assess production from shifting plots in jungle areas. There may be a tendency to exaggerate yields if the estimation is carried out on immature crops, and such overestimation has been a characteristic of postwar agricultural statistics in the USSR. Losses due to storms, lodging, pests and disease may supervene to make nonsense of an early estimate. It has been claimed that as much as one-third to one-half of the grain actually harvested in India is lost for one reason or another, although reports from Delhi suggest an overall loss of only 5–10 per cent. In fact no one knows the true rate of loss and even here it seems possible that there may be vested interests at work which stand to gain by overestimating. Governments may be misled by their own statistics. Clark & Haswell (1967) have pointed out that the government of India mistakenly thought that a grain shortage in the early 1950s was sufficiently dangerous to warrant the use of scarce foreign exchange to alleviate it, basing their belief on erroneous production figures attributable to underdeclaration in a period of black market activity.

Although estimates of food and protein production throughout the world are only approximately correct, the overall differences between the developed and the less developed countries are so large that quite substantial fluctuations due to the operation of the factors discussed above are unlikely to affect the general conclusions. FAO statistics reveal striking differences, for example, between the production of milk and meat in the developed and the less developed countries. In India vast numbers of cattle produce no meat for human consumption and little milk, while in the USA the growth of agricultural production has been deliberately restricted. New Zealand too is faced with the prospect of vanishing markets for plentiful dairy produce as a result of changing world economic conditions. With such artificial restraints in operation it is extremely difficult to arrive at a useful index of potential protein production.

THE ASSESSMENT OF PROTEIN NEEDS

Let us now consider whether some estimate can be made of protein needs in terms of known requirements. Though ideally protein requirement should be calculated in relation to body weight, it is usual to consider adults only. In terms of world needs varying family size would make it difficult to arrive at a precise estimate, in any case, and some simplification is necessary. In view of the widely varying figures quoted for the requirements of adult man ranging from 20 to 100 g protein per day, the additional refinement of taking into account the requirements of an unknown number of children seems hardly important.

In the first paper presented at this symposium, Blaxter (1972) has drawn attention to this extraordinary disparity in the suggested ranges of values for protein requirements. At the upper end of the range are the luxus provisions of those who feel that recommendations should not be minimal, but rather that they should err on the generous side, on the grounds that levelling should be up rather than down, and also that there are perhaps intangible benefits from a protein excess in terms of general wellbeing and disease resistance (Pirie, 1969). At the lower end of the scale are those who indulge in calculations of minimal needs. Blaxter has made such calculations but has stated that such low levels of protein intake, though possibly adequate, are not necessarily desirable. One must bear in mind the possibility, however, that they may one day be mandatory. That health and the ability to do physical work is indeed possible on protein intakes as low as 12–15 g per day is demonstrated by the inhabitants of New Guinea, who have been shown to obtain only that level of protein from their diet of sweet potato, sago and taro (Oomen, 1961; Oomen & Corden, 1970).

As well as these two extremes there are those who prefer the middle course, and committees traditionally and predictably fall into this category, settling for a mean figure somewhere around the middle of the range.

The Joint FAO–WHO Expert Group on Protein Requirements (1965), proposing that protein needs should be based on the amount of food

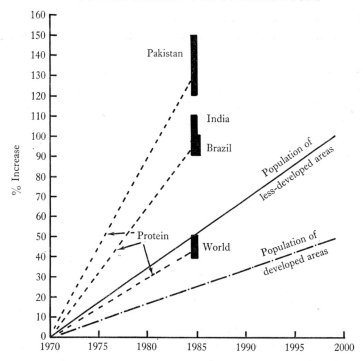

Figure 1. Comparative increases in population and protein requirement for the developed and some developing countries between 1970 and 2000 A.D.

required from a given national food supply which is calculated to be equivalent to the necessary amount of FAO reference protein, arrived at a figure of 35 g per day reference protein. Assuming a practical net protein utilisation (NPU) of 65 per cent, the requirement becomes 54 g per day. This level is achieved by supplementing the normal cereal or cassava diet with fish, milk or eggs. Such a practical diet would in fact provide approximately the same levels of the necessary essential amino acids as Blaxter's (1972) 8 g ideal protein.

Assuming that the FAO–WHO estimate is a reasonably correct one, the extrapolation of the calculations leads to the conclusion that between 1970 and 1985 the world requirement for protein would increase by 40–50 per cent, the requirement for India would increase by 90–110 per cent, for Pakistan by 120–150 per cent and for Brazil by 90–100 per cent (Figure 1). The position is further complicated by the fact that, apart from the FAO–WHO figures, all of the requirements so far considered take no account of protein quality or amino acid availability. In addition to receiving only two-thirds of the total quantity of protein consumed in the developed countries it must be remembered that only 11 per cent of that consumed in the less developed countries is high quality animal protein, compared with over half of the total consumption in the developed countries.

The Joint FAO–WHO Expert Group did not have the temerity to lay down daily requirements for amino acids. The complexity of the problem with human beings of different genetic types, living under different climatic conditions and consuming widely different diets, would daunt even the most assiduous of those manipulators of figures who produce tables of nutrient requirements. There does not exist an agreed set of requirement figures even for the ubiquitous fowl whose diet and environment are within our complete control. How much more difficult is it to produce a set for man, the most uncooperative of experimental animals!

Blaxter (1972) has referred to the experiments of Kofrányi and his colleagues at the Max Planck Institute, Dortmund, in which, for example, potatoes were found to be capable of sparing egg protein in human diets. This phenomenon is well known but little understood. Similar results have been obtained at the Rowett Research Institute when chickens were fed mixtures of proteins. Diets containing proteins of good quality may be further enhanced in value by including protein sources of poor quality, and while this complementation effect can often be explained in terms of better overall amino acid balance, or in terms of improved 'chemical score', this is not invariably so.

One may conclude from such experiments that calculations using reference protein may be unrealistic and that more practical diets containing protein of generally lower quality than FAO reference protein may result in equivalent, or even better performance.

One must also guard against judging proteins solely on the basis of their ability to give rapid growth, maintain nitrogen balance or even provide a feeling of wellbeing in the consumer. A different amino acid make-up, judged poorer in terms of growth-promoting ability, may well be superior when judged on its ability to increase longevity for example, and this might eventually prove to be a more important criterion than those currently used.

Much more research is needed before precise conclusions can be reached and it is appropriate here to quote the problems requiring urgent research listed by the Joint FAO–WHO Expert Group on Protein Requirements (1965) to which reference has already been made.

The determination of human protein requirements on sound physiological bases is not possible until important gaps in knowledge of protein metabolism have been filled. Among the problems on which research is urgently needed are relations between energy intake and protein intake, adjustment of the body to different levels of protein intake, effect of physical activity and climate on protein requirements, mechanism of the body's reaction in terms of protein metabolism to stresses of different kinds, suitability of essential amino acid patterns to meet the needs of different physiological states, and individual variation in protein requirements. To these might be added the question of amino acid availability.

It is clear that a big difference exists between the quantity of protein necessary for mere existence and the quantity which is considered desirable for sustenance in countries where no shortage exists. It seems that there is some leeway should the situation become so serious that world-wide rationing of protein is contemplated. In the meantime, the universal desire for increasing standards of living demands that nutritionists should work upon the basis that standards throughout the world are levelled up rather than down. It should not be the intention here to seek minimum requirements, but rather to discover whether in the future it will be possible to guarantee every living person upon this planet an amount of protein which is considered at the present time to constitute a normal intake within the developed countries. What then can reasonably be done, not only to satisfy the demands which already exist, but also to keep pace as the requirement increases in the future?

PROPOSALS FOR AUGMENTING WORLD PROTEIN SUPPLIES

If the magnitude of current protein supplies as well as accurate estimates of protein requirement are in doubt, it is clearly impossible to say with any certainty what proportion of the present world's population is suffering from protein malnutrition. If, in addition, the overall rate of population growth is uncertain, forecasts of protein needs in the years 1985 or 2000 can be little more than guesswork, and could indeed be dangerous if they erred on the low side. The correct approach would seem to involve considering what avenues might be usefully explored and to pursue vigorously each and every promising line. Included amongst these should be the introduction of improvements in agricultural practice and in land utilisation, the extension of productive areas by irrigation and fertilisers, and the use of modern knowledge in order to farm the desert. Better use should be made of sea and fresh water by the encouragement of scientific fish farming. Waste must be eliminated, the distribution of available protein foods must be accomplished in such a way that local surpluses are avoided, the best possible use must be made of existing conventional protein sources, and finally attempts must be made to introduce new protein sources as and when they become available and circumstances become propitious for absorbing them into existing dietary regimens. It is tempting to look for a dramatic answer to the problem, matching the dramatic nature of the population explosion itself.

Plant life is the primary source for all dietary protein consumed by man. Some plant protein may be converted rather inefficiently into animal protein which is often, but not always, superior to the original plant source in its nutritional characteristics. The cost of this improvement is the wastage of a large amount of protein which could be utilised more efficiently by direct feeding. It must be admitted that in some cases the animal utilises vegetable material which is unsuitable for human feeding directly. For instance, the cost of harvesting and processing would preclude the use of the protein of sparse hill pasture at the present time, and sheep and deer are well suited to make the best of this unpromising material.

Land suitable for growing cereals is also suitable for growing good quality grass for milk production, and Blaxter (1968) has made the point that, because a given area of land will produce approximately the same quantity of cereal or milk, in terms of land use the superior amino acid composition of milk over cereal protein should tip the balance in favour of the former, so far as efficiency of protein production is concerned.

This is, of course, absolutely true if one considers only the traditional forms of land utilisation. If, however, the aim is to get the maximum yield of high quality protein from a given area, one would not consider cereal seed protein which has a notoriously poor amino acid spectrum. Cereals have, after all, been looked upon as a source of energy from time immemorial and their protein content has been largely discounted by nutritionists. The comparison must be made between milk and leaf protein, the latter being of high quality, produced in good yield and not subject to the climatic vagaries which hazard a ripe cereal crop. Multiple cropping could make possible the production of protein in quantity. For example, the growing of two crops of green cereal followed by two of mustard has been pronounced feasible in Great Britain (Byers & Sturrock, 1965). That leaf protein is of excellent quality for supplementing cereal seed diets for both humans and non-ruminants has now been amply demonstrated. Duckworth & Woodham (1961) showed that such material prepared from green cereal leaves was equivalent in quality to soyabean meal and was only slightly inferior to dried skimmed milk.

Some years have passed since it was customary to distinguish between animal and vegetable protein, referring to the former as first class and to the latter as second class. It is now realised that both animal and vegetable proteins may differ very considerably in amino acid composition and consequently in their nutritive value, some vegetable materials being comparable in quality to good quality animal proteins and some of the latter in turn being inferior to many vegetable protein foods. It is, of course, true that animal protein foods are generally considered to be more palatable and more desirable than vegetable proteins and consequently consumers with the requisite purchasing power will continue to include the former in their diet. Food habits do not change overnight, and even in the present situation leaf protein and other novel protein sources must be considered merely as adjuncts to the existing diet and will be of most value in those parts of the world where animal protein is beyond the reach of the ordinary person. Nevertheless, when exigency demands it – and it appears inevitable that it must eventually – we may rest assured that it will be possible to live and to live well on protein foods which are not at present considered to be acceptable dietary components.

It seems certain that in the short term man will continue to rely on conventional plant protein sources such as oilseeds, pulses and legumes. Though generally deficient in sulphur amino acids, they are capable of supplementing the poorer quality cereal seeds satisfactorily. Supplementation with amino acids can often improve the mixture, but even without such additions a good standard of nutrition can be maintained. The optimum utilisation of these high-protein materials can be improved in two ways – by increasing the quantity and by improving the quality. Plant geneticists have been concerned for many years with problems associated with increasing the yield of legumes and oilseeds. It seems likely that in the future the major impacts upon the problem of world protein shortage will come from increased production. This will be achieved not only by selecting improved strains of oilseeds and legumes which yield larger quantities of protein, but also it is to be expected that strains will be produced which will be capable of growing successfully in areas where the production of such high-protein plant sources cannot be considered at present.

CONCLUSIONS

It has to be admitted that no tangible figures for future protein demand have been presented. Rather has it been the object of this paper to indicate the areas of ignorance and to suggest that the shortfall in protein supplies is so great that, as every effort will be needed to make up the present deficit and provide for the future, precise target figures are unnecessary. Long before the world production of protein approaches the point where demand is satisfied, more accurate information on protein requirements, and on the relative quality of alternative protein sources, should be available. By that time dare we hope that the over-riding problem of forecasting and controlling population growth will also be solved? Is it too much to ask that the gravity of the situation may become so apparent to all nations that the establishment of really effective means of balancing population growth and planetary resources may become a reality? International cooperation has been achieved recently in certain fields – the International Geophysical Year and the International Biological Programme. How much more pressing than either is the need for an effective International Food Programme?

REFERENCES

BLAXTER, K. L. (1968). The animal harvest. *Science* **4**, 53–9.

BLAXTER, K. L. (1972). The purpose of protein production. In: Jones, J. G. W. (Ed.), *The Biological Efficiency of Protein Production*. Cambridge University Press, 3–11.

BYERS, M., & STURROCK, J. W. (1965). The yields of leaf protein extracted by large-scale processing of various crops. *Journal of the Science of Food and Agriculture* **16**, 341–55.

CLARK, C., & HASWELL, M. R. (1967). *The Economics of Subsistence Agriculture*. 3rd edition. London: Macmillan.

DUCKWORTH, J., & WOODHAM, A. A. (1961). Leaf protein concentrates. The effect of source of raw material and method of drying on protein value for chicks and rats. *Journal of the Science of Food and Agriculture* **12**, 5–15.

FARMER, B. H. (1969). Available food supplies. In: Hutchinson, J. B. (Ed.). *Population and Food Supply*. Cambridge University Press, 75–95.

HUTCHINSON, J. B. (Ed.) (1969). *Population and Food Supply*. Cambridge University Press.

JOINT FAO–WHO EXPERT GROUP ON PROTEIN REQUIREMENTS (1965). *FAO Nutrition Meeting Report Series No. 37*. Rome: FAO, pp. 51 and 61.

OOMEN, H. A. P. C. (1961). The nutrition situation in Western New Guinea. *Tropical and Geographical Medicine* **13**, 321–35.

OOMEN, H. A. P. C., & CORDEN, M. W. (1970). *Metabolic Studies in New Guineans. Nitrogen metabolism in sweet potato eaters*. South Pacific Commission Technical Paper No. 163.

PANEL ON THE WORLD FOOD SUPPLY (1967). *The World Food Problem. Report of the President's Science Advisory Committee*. Washington, DC: The White House.

PIRIE, N. W. (1969). After 1984 – What? Protein foods of the future. *Getting the Most out of Food* **5**, 107–17. Van den Berghs Ltd, Kildare House, Dorset Rise, London, E.C.4.

WILLIAMS, R. M. (1966). Food and population in the scales. In: Farber, S. M., Wilson, Nancy L., & Wilson, R. H. L. (Eds.), *Food and Civilisation*. Springfield, Ill.: Thomas, 257–73.

WOODHAM, A. A. (1971). The world protein shortage: prevention and cure. *World Review of Nutrition and Dietetics* **13**, 1–42.

4

FACTORS AFFECTING DEMAND FOR PROTEIN PRODUCTS

By J. C. McKENZIE

Food and Drink Research Limited, Centre House,
114/116 Charing Cross Road, London WC2H 0JR

'People can only consume products which are available.'

Such a statement may seem a particularly naive way in which to introduce a paper on factors affecting demand for protein products, but in essence perhaps the only clear and explicit assertions which can be made concerning demand by the consumer concern those aspects of what we tend to call availability. Availability of food means that it is physically there and we have the means to buy it. As such we can move from daydreaming desire to *effective* demand – demand that can be activated. Whatever we may want to eat, we can only consume products which are physically available, are fresh or have been preserved in such a way that they have not completely deteriorated, and which we can afford to buy.

Unfortunately, whilst it is on such limited bases that most scientists have developed new food products and supplements, in reality to the individual consumer they only reflect the starting point from which selection of foods is made. They indicate foods that *can* be chosen, not necessarily which *will* be chosen. Thus, by analogy, availability may be portrayed as a table full of all the possible foods from which one may choose, but the mechanics by which individuals select particular items, if any, are a good deal more complex.

It is these less tangible demand criteria with which I wish to concern myself today. Others at this symposium have projected estimates concerning protein requirements in both the developed and the developing world and it is for the scientist and politician to make sure that these requirements are matched by availability.

But what are the criteria which lead us to choose particular foods and reject others? What are the ways by which we can encourage individuals to go for particular items?

Let us get one thing straight right from the start. Choice does not adequately reflect nutritional needs. Humans do not normally go for the foods that will do them good and which will satisfy their deficiencies, either by intuition or as a result of learning. People do not usually crave for suitable foods or go for the right ones because they have been told to do so.

The basic criteria on which demand is fashioned are threefold – taste, fashion, habit. People have inherent preferences for particular tastes, others they learn to like over time, others corrupt their palate with experience. But all have some influence on food selection, if only in an arbitrary way distinguishing sweet or sour, bland or spiced.

The sociological phenomena involved with fashion are considerably more complex. In selecting a particular product, for example, one may be:

'Keeping up with the Joneses';

Satisfying some cultivated need or instruction put forward by advertising;

Projecting the image of oneself one would like others to see; or

Conforming to social patterns or rebelling against them.

One knows that nutritional requirements have only a very small role to play in selection of food, but one cannot help conjecturing somewhat cynically, as one explores this area, as to whether or not food has anything to do with the issue at all!

Lastly, selection is most strongly hidebound by habit. What we select today is almost totally predetermined by what we selected yesterday. If a decision-making process ever took place in our minds it has usually been forgotten long since; at 11 o'clock we now simply ask for or weakly accept a cup of coffee; we no longer contemplate why we act in this way, we just do.

The points I have made here directly relate to experience in Western Europe, but they are equally relevant for developing countries. The same sort of symbolism was very aptly demonstrated by Miss Whiteman (1966) in a table some years ago (Table 1). Thus the position is clear in terms of the problems affecting demand for protein in the coming years. We can easily make foods available, but we cannot ensure that the products are selected.

Thus to my mind a major problem in both developing and developed society is not only to make protein foods available but also to actually ensure that they are consumed, that is, that they taste right, are fashionable and become part of the food habits of the community. How can we ensure that a demand exists for these products?

First, we could best achieve the situation by the ideal of providing what people want, namely, the highly palatable animal protein foods they go for whenever possible. Some years ago I conducted a major survey amongst members of the public who were asked, amongst other things, to identify the foods they would like to have more of and the food they find it difficult

Table 1. *The comparison of social functions of food in the United States and the Trobriand Islands*

Function	American society	Trobriand society
For prestige status and political power	The giving of cocktail parties, banquets, dinner parties, etc., on a scale suited to one's social position; agricultural programmes, etc., related to national and international politics	The public display of yams grown or given away; receiving of urigubu* by chief and its redistribution by him; display of best yams in open store houses and in prominent places in the village by men of importance
Economic purposes and the use of food like a currency	The giving of Christmas presents, business dinners, etc., in order to establish favourable business relationships between people	The barter of vegetables for fish and vice versa; gifts of food in return for services, articles of value, articles requiring skill in their making or repayment for gifts received; all these act as an incentive for the economic production of food
Fulfilling social and kinship obligations; establishing friendly relations	Day to day family meals which are a social event in most households; meals eaten with friends at home or out; wedding receptions, etc.	Urigubu payments of food by brother to his sister's husband; in the case of a chief a brother's kin group help him with his urigubu payment; urigubu demonstrates the ties a married woman has with her natural group.
Recreation	Pleasure of eating with friends	Pleasure at seeing the display of food
Aesthetic and creative satisfaction	Banquets, private dinner parties, etc.; the cooking of special dishes	Sight of public display of food; satisfaction of careful cultivation of gardens which also demonstrate the skill of the gardener if they produce well; bragging about one's skill as a gardener
Religious significance and satisfaction	Jewish, Mormon, Seventh Day Adventist and Roman Catholic observances regarding food	Public displays of food at harvest time to please the visiting ancestral spirits

Table 1 (cont.)

Function	American society	Trobriand society
Symbolism, mythology and superstition	Communion bread and wine; some fattening and slimming foods, when without scientific justification; fish being a brain food; oysters causing sexual potency	Funeral rites; mythology of origin of garden magic and land-ownership
Magical significance	Throwing salt over the shoulder to ward off bad luck	Special kinds of fish required for the performance of some magical rites
Ceremonial	Christmas and Thanksgiving dinners	Funeral feasts
Legal		Urigubu initial payment sanctions marriage and is a form of endowment; the boy and girl eating together in public is an indication that they are married
Medicinal	Senna pod tea; slimming foods	For diarrhoea, bananas in their skins are washed and boiled and the cooking water drunk

* Urigubu is the name given by the Trobriand Islanders for the system of gifts of goods and services that a man gives to his sister throughout her married life. Usually about half the produce of his garden is given away ceremonially in this fashion to the husband of his sister. In return the sister and her husband produce and bring up children who are members of the brother's kin group and who will eventually inherit from their uncle.

Table 2. *Foods people would buy more of if they had more money*

		Social class			Children		Area	
					With children 16 and under (%)	Without children under 16 (%)		
Item	Total (%)	AB (%)	C1 (%)	C2DE (%)			South (%)	North (%)
Meat	51	46	48	53	61	42	49	54
Chicken	40	27	38	43	43	39	34	52
Apples	28	17	26	30	34	23	28	29
Oranges	28	20	24	30	34	22	27	29
Butter	21	12	15	24	20	21	16	30
Fish	19	19	16	20	18	20	18	21
Milk	14	6	12	16	14	14	12	17
Eggs	14	7	13	15	13	14	12	17

to do without for any length of time (McKenzie, 1970). The results are shown in Tables 2 and 3. This in fact takes one step forward projections based quite simply on the poundage (or I suppose, more scientifically, in 1971 on the grams) of various food types we eat each year and on the ways in which the proportions consumed are changing.

Table 3. *Foods people hate to do without for a month or two*

Item	Total (%)	Social class			With children 16 and under (%)	Without children under 16 (%)
		AB (%)	C1 (%)	C2DE (%)		
Butter	58	64	62	56	57	59
Milk	58	59	61	56	57	58
Tea	54	45	52	55	49	57
Meat	50	58	55	47	56	44
Eggs	47	53	53	44	51	43
Sugar	29	27	25	30	33	26
Bread	28	23	28	29	29	28
Cheese	27	30	29	26	26	27
Potatoes	17	14	16	17	20	13
Apples	16	21	22	14	16	16
Fish	16	22	18	14	14	17
Coffee	15	26	19	12	18	13
Oranges	14	19	17	12	14	14

Comparable studies for the developing world might easily be made, and, indeed, in terms of volume turnover per product category have already become commonplace (McKenzie, 1966).

If we cannot give people simply what they want, the next easiest step is to provide protein or other nutrient supplements which may be added to existing foods. This has a number of advantages:

(*a*) It enables people to continue with their established, well tried food patterns.

(*b*) It enables one to overcome inherent deficiencies in the diet without forcing them to change from the foods they like – protein-supplemented Coca Cola rather than no Coca Cola!

(*c*) It is easier to make people wear crash helmets than it is to stop them riding motor cycles; in other words one is making positive recommendations for additions to a diet rather than negative recommendations to stop eating things. And most people react better to the first line of approach.

(*d*) 'What the eye does not see the heart does not grieve for' – it may be possible to add supplements 'secretly' so that nobody knows. This may not necessarily imply deceit but simply lack of general publicity; how many people are really aware that we add supplements to flour?

Once past this stage one is really in the nitty gritty of persuading people either to eat new foods or products already around but which they have not regarded as food for human consumption. Both are considerably more difficult tasks to succeed with than were the earlier suggestions.

There are many reasons for this. The advantages of the *status quo* are usually self-evident. Moreover,

> it must be recognized that if change involves anything more than a switch of two identical brands of a particular type of food, then it sets off a series of complex interrelated movements. In essence all foods are competitive with each other. At a physiological level once in terms of bulk we have enough to eat, increased consumption of one food tends to lead to the reduction in consumption of another.
>
> Similar interrelationships exist both in economic and socio-psychological terms. Most people have a relatively fixed amount of money which they are prepared to spend on food, at least in the short run. A change in the amount spent on one set of food products automatically leads to a change in the amount of money available for other food products. Similarly, a change in one food may change the whole meal pattern because only certain foods are regarded as acceptable in combination with others.
>
> The marketing implications of the two issues outlined are clear. To sell a product it must be shown to satisfy needs at least as effectively as foods already consumed. In addition, because one item in the diet cannot be changed without a whole series of repercussions on overall choice, some support must be provided to justify these resultant total responses. This is why commercial involvement is so important – these sorts of problems frequently face the businessman. He is always working in competitive situations. Most cigarettes marketed at a given price and using a particular type of tobacco have a similar appearance and taste. Most tins of canned fruit will have the same basic constituents as each other. Thus once price has been settled and the taste found to be acceptable, the job of selling has only just begun. Complex advertising themes will be required to encourage the consumer to believe he needs these products and that this packet of cigarettes or that tin of canned fruit is the best brand to buy. When the problem involves a totally new type of product the task becomes of even greater importance (McKenzie, 1969).

How can we successfully achieve the requisite change in these

Table 4. *Key information required before initiating sale of new product*

Area	Type of information required
1 Product usage	When is the product likely to be used, and with what foods will it be combined? Will it blend well with these foods?
2 Product competitiveness	With what other foods will the product effectively be competing? Can it successfully demonstrate some acceptable superiority? If the product is a supplement to be added to existing foods during preparation, then can methods of justifying this addition be substantiated?*
3 Impact on food habits	Does its acceptance require any major or minor modification in eating habits? If so, can these changes in consumer behaviour be successfully achieved?
4 Equipment required	Is any new equipment or cooking procedure necessary? If so, can the housewife be easily persuaded to make this change?
5 Storage	Does the product require any special type of storage arrangements? If so, will these special facilities be available and acceptable?
6 Advertising copy	How can advertising copy be designed to indicate effectively the fulfilment of a need, overcome competitive products, and justify changes in behaviour?†
7 Name	What, from the consumer's viewpoint, would be a suitable name for the product?
8 Packaging	What sort of packaging would be most acceptable to the consumer?

* The superiority or justification will not be in health or nutritional terms but in terms of appeal based on taste, better fulfilment of psychological needs, etc.
† This will imply general knowledge of the consumers' motivations.

circumstances? It will of course help a good deal once we recognise the problem and all too often people seem to have difficulty in doing this. Ever since the Cuernevaca Conference in 1960 we have had a text book identifying the problems but this has been frequently ignored (Burgess & Dean, 1962). Over and above this we must get the facts about the food habits of a particular community before we can hope to effect any change. Such essential data must cover the areas identified in Table 4. Then we shall be aware of the real problems to be dealt with and indeed whether any change of plan on our part can ease the situation.

Once the task is tackled scientifically in this way many of the apparently insoluble problems disappear, and the commercial world is being increasingly adroit in dealing with the rest; thus in an earlier paper (McKenzie, 1965) I have suggested ways in which conflicting needs may be accommodated. However, it must still be acknowledged that, on occasions, the task is too great and then we would do better to accept defeat and try a different line of approach rather than labour on trying to persuade a community to modify its food habits in a way they find unacceptable.

One last word however; often the food industry has been conspicuously unsuccessful in selling desirable nutritional products, especially in developing countries. The reasons for this may be many. But in many instances

I believe that if we carry out a post-mortem on attempts to sell nutrient supplements or change food habits, it is evident that even the businessman has often not studied the social and psychological problems involved when planning his marketing strategy. Perhaps this has been partly due to the false assumption that these problems are less significant and these patterns are less complex in developing society. Perhaps it is that some salesmen have for the first time become really convinced that the products they are trying to sell are really good and that in these circumstances they imagine they will be successful without persuasive encouragement. Whatever the reason, it seems that in this context too often the marketing men have at worst left at home or at best skipped a couple of pages in their text book on how to sell products (McKenzie, 1969).

The implications of what I have said, if you find it acceptable, are inevitable. Hard as we labour to provide acceptable protein products and other nutrients over the coming years, the crucial decision in attempts to provide a better-nourished community will not rest with the scientist or politician but with the individual consumers. This is both refreshing and sad, refreshing in that in our increasingly mechanistic world the individual can still have a significant role to play, sad in that at least a proportion will not decide in the best interests of their health and welfare.

REFERENCES

BURGESS, A., & DEAN, R. F. A. (Eds.) (1962). *Malnutrition and Food Habits*. London: Tavistock Publications.
MCKENZIE, J. C. (1965). Recent developments in social science related to nutrition and dietetics. *Proceedings of the Ninth International Congress of Dietetics, Stockholm*.
MCKENZIE, J. C. (1966). Dietary trends as an aid to prediction. In: Barker, T. C., McKenzie, J. C., & Yudkin, J. (Eds.), *Our Changing Fare*, 135-49. London: MacGibbon & Kee.
MCKENZIE, J. C. (1969). Overcoming resistance to new food products. *Proceedings of the Nutrition Society* **28**, 103-9.
MCKENZIE, J. C. (1970). Poverty and nutrition in disease. In: Townsend, P. (Ed.), *The Concept of Poverty*, 64-84. London: Heinemann.
WHITEMAN, J. (1966). The function of food in society. *Journal of Nutrition* **20**, 4-8.

5
ECONOMICS OF PROTEIN PRODUCTION

By K. E. HUNT

Institute of Agricultural Economics, Oxford

BACKGROUND

It might be possible, with some guessing and by making various assumptions, to set out a comparison in terms of pence per lb of the costs of producing various protein-supplying foodstuffs. Some examples of this are in fact given later: however, the protein problem cannot be narrowly isolated. Such figures are of little guidance unless they are placed within a rather wide setting in concept, place and time. In preparing a contextual paper of this kind one is faced, therefore, from the start with a common problem for economists, namely, the optimum use of scarce resources. In this case the scarce resource is time and the choice concerns the proportion of the available time for this paper which should be spent on providing a setting for the protein supply situation compared with that spent on giving specific cost comparisons. It seems likely that a mix would be more rewarding than concentrating wholly on either 'setting' or 'detail'. But what is the shape of the curve relating 'contribution to the discussion' to the percentage of the time spent on 'setting'? We can probably rely on the curve rising to a single peak somewhere between the 100 per cent 'setting'/0 per cent 'detail' and 0 per cent 'setting'/100 per cent 'detail', but where does this maximum lie? A review of a substantial number of papers on the protein problem suggests that the hump comes rather near the 100 per cent 'setting' end of the graph.

The production of protein is connected with agriculture and with manufacturing industry, with mammals, birds, fishes, insects, termites, snails, with flowering plants, algae and fungi, with land, fresh water and salt water, with developed countries and developing countries. About the only structural connection linking them all is man's economic behaviour, i.e. his behaviour as a decision maker in the use of the scarce inputs. These scarce inputs include the natural resources, the human resources of workforce and special skills, and the capital which he commands. Even protein itself provides a less than perfect common thread since the nitrogen compounds concerned in the various fields may differ in characteristics which are highly important, so that 100 g protein in maize is not equivalent to 100 g protein in milk or snails.

Even within the field of interest of this symposium there are few instances where production of *protein* as such is the dominant, let alone the exclusive, interest in the production field. In most cases we have to consider the production of bundles of useful characteristics. Neither for animal nor for human food can we consider protein apart from the energy, mineral and vitamin constituents of the product except, perhaps, for certain cases of amino acid supplementation. For almost all human foods we must also bring in matters of taste, texture and flavour, and less tangible factors of custom and association as adjuncts of the protein and other nutritional characteristics.

At first sight the production of protein for feeds for animals might be considered as largely uncomplicated, except in extreme cases, by characteristics of products other than their nutritional values. Even if this may be in large measure true, it still remains that growing and manufacturing processes, which yield oilseed and fish byproducts, highly important feeds for livestock, also yield edible oils for human use. The production process cannot be appraised without taking into account the characteristics of the human food, tangible and intangible, which the process yields as a joint product.

Amongst the characteristics appraised by the user of a food and, hence, which must be produced by the production process if the user is to be satisfied, is its presence where it is needed, when it is needed and in the possession of those who need it. These are characteristics which are as real and relevant as are physical ones but seem often to get downgraded in importance, or treated as if they should, in a reasonably ordered world, be virtually costless to produce; a supply of high-grade fish protein located 50 miles away, or due to arrive in two months' time, makes no more contribution to a family's current protein nutrition than a supply of manioc on their plates today. Equipment must be used, work done and thought taken to create these time, place and possession utilities just as they are required to produce physical products at the farm gate or the factory door.

The production processes with which we have to be concerned here may, perhaps, be illustrated by turning briefly to the demand side. A review of production economics for protein must recognise that at least the following groups must be satisfied by production processes in the developing countries. At the bottom end of the scale are those in the traditional rural sector whose food supply is largely determined by what they can grow or collect in their surroundings and by their customary practices. A large proportion of the population of Africa falls into this category.

Urban population might be split into three sectors, low, medium and high income. The situation of the first is dominated by low income resulting from lack of skill and unemployment, and many prefer the foods they used to know in the rural sector. The middle level probably earns enough to make sure that all members of the family are well fed but may not know how to use their income or the foods available to the best advantage. The upper income groups are unlikely to be lacking protein. In the developed areas typical traditional types with diets determined by local environment will be rarer, modern type farmers may eat much as do well-to-do urban people. In urban areas where families are badly fed, it is likely to be due to ignorance rather than destitution.

Against that background, any review of economics of production will need to cover at least three situations, viz:

(a) Self-supplied food or food traded only locally – staple foods, especially grains, can be expected to predominate in these diets.

(b) At the other extreme a wide range of foods, including expensive ones, supplied to well-to-do urban dwellers who value flavour, appearance, etc. For this sector the protein content is largely incidental to the enjoyment.

(c) For those between, foods lending themselves to ready distribution as precisely as possible to those members of the sector who are most vulnerable to protein shortage.

It seems likely that farming will have to supply rather directly group (a) and can well supply group (b) but we might look with hope to the non-traditional methods for a larger contribution to group (c).

Probably the best lead to the essence of production economics for the present purpose is to ask who has the decisions to make and what is their nature? For the present purpose it may be enough to take examples. (It is not to be understood that those concerned actually formulate their problem in the terms used here but these concepts usually apply.)

(a) Subsistence producer. How much labour do I devote to farming compared with leisure? How do I divide it between (i) looking after more land, and (ii) less land more carefully tended, between using this system and that system? (For an extended indication of issues in this field see, for example, Clark & Haswell, 1970; Mosher, 1966.)

(b) Developed crop producer (e.g. British or North American farmer). What mix of crops will most evenly use the labour I have available? Would my return as a result of increasing slightly one crop and decreasing an alternative crop, using the same land and available labour, increase or decrease my profit? At what point as I increase fertiliser application will the resulting increase in output be worth (net) no more than the cost of the fertiliser? What are the chances that yield variation or variation in price of the sale product will make nonsense of this calculation?

Developed livestock producers have similar issues to face plus such matters as which feed mix at given prices of feed ingredient would give the lowest cost and how far they can intensify stocking per unit of capital in buildings before disease builds up.

Food analogue producers will calculate return on investment in the proposed products against the return on alternative investment. In doing so they will try to appraise exactly which market their product will compete in. Will it compete only with a similar product from another firm or will consumers regard it as freely substitutable for, for example, meat? They will also try to appraise the extent to which, if their product is successful, the demand for their raw materials will force the price up, forcing them either to cut their profit margin or to raise retail prices. And they must take a view over a period during which consumers become acquainted with the product.

It is decisions of this kind that are of the essence of production economics. Cost comparisons at any one time are a transient byproduct of the outcome of such assessments. For many of these decisions the data of the biological scientist and of the economist are alike essential and complementary. The decisions will be made badly if biologists and economists cannot get together and integrate their approaches.

Though the criteria which government economic planners use may be drawn from mixed disciplines, they may have rather similar decisions to make. Often, particularly in developing countries, protein nutrition does not figure explicitly in overall plans. If it does, the decisions of planners may be of the form, 'Is it better to encourage private firms to market a protein food of general usefulness and to hope that it reaches vulnerable groups fairly soon, or to try to reach as many as possible directly through clinics? Is it better to try to change the operating environment of subsistence farmers with protein-deficient babies by investment in plant-breeding research, by buying and distributing improved seeds, by investing in education or advice, by edging subsistence farmers into the market economy by improved roads, or by some other means?'

THE WORLD ECONOMIC SETTING FOR PROTEIN PRODUCTION

Since the bulk of the world's food comes from agriculture, the production conditions for protein must be very largely dependent on the economic conditions of the agricultural industries and of the trade in their products. In fact there is little of the world's agriculture, outside pure subsistence types, which is not an artifact of administrators and politicians either in the country concerned or in those with which it trades. Variable and low prices and/or variable and low incomes have made farmers restive. As a group within their country they are often politically powerful either because they are numerous or because national self-sufficiency is valued for security or, latterly, prestige or balance-of-payment reasons. The situation is now complex. Among the reasons why farmers tend to be relatively poor in the wealthier countries is the combination of slow increase in effective demand with substantial technical advances which allows demand for food to be met by progressively fewer people engaged in farming. Farmers are nevertheless reluctant to migrate fast enough to other occupations to keep their income per head in step with incomes outside farming. In many instances, too, farmers are poor because they are technically bad at their job compared with people in other industries, or because each has so few resources in the business that returns are unavoidably low, however big they are per unit of input, and perhaps because their bargaining power in the market is weak. Poor technique, few resources and poor bargaining power are particularly important in developing areas.

Incomes are variable, sometimes because of weather and disease but often the cause is variation in prices received for their produce (Helleiner, 1965). There are many possible reasons for variable prices; often supplies elsewhere vary and affect market prices of products far from the origin of the disturbance. Demand variations resulting from fluctuations in general business activity may be important too.

The outcome is that there are now innumerable examples of national and international intervention. Among them might be cited price support devices of various kinds in the USA, the Common Agricultural Policy in the EEC, the UK price support system, marketing boards for West African produce which skim substantial sums from the export prices before passing the remainder back to the producers of cocoa, oil seeds, etc., and (though not directly for the same reasons) the agricultural planning of the socialist countries. In addition, there are numerous examples of import duties on farm products and various international agreements or working arrangements influencing production and trade in farm products. Furthermore, there is an extensive system of duties on manufactured goods, e.g. textiles. This affects the agricultural sector, for example, by hampering countries which would otherwise sell more manufactures and buy more food abroad.

A byproduct of some of these support and protection methods for national agricultural industries is the growth of surpluses, because policy makers have not usually been successful in providing high enough prices to satisfy farmers without encouraging them to produce more than they can find outlets for within the country. The goods so produced have been used to supply various international food aid schemes. It has been estimated (Allen, 1963) that this procedure is almost as cheap for the donor country as satisfying farmers by paying them not to produce, but the overall benefits to the recipient are not, in the view of many observers, costless.

Some of these measures directly affect protein foods but even those that do not cannot be regarded as irrelevant to the present symposium. Though the links may be distorted by these interventions, most agricultural products are connected by economic considerations. A measure which raises the price of wheat discourages its use for livestock feed and reduces the production of livestock products. A high price for livestock products means a high price for fishmeal and oilcake and less incentive to use these products directly as raw materials for human food supplements. Intervention in the sugar market may mean higher molasses prices and higher costs of raw materials for single cell proteins.

How will the future world food pattern develop?

It is against this picture in both its technical and its political features that we must consider the possible future development of world food supplies. In the light of present conditions, reasonable projections of the future production and demand for food products are given in Table 1.

In other words, partly because of policies of encouragement to farming, and partly because of the outcome of investment by government and industry in fundamental and developmental research, there is active promise of expanding food production in developed countries. Of the supplies surplus to their own needs some are likely to be dried skim milk, some soya beans or its products, eggs, broilers, some less favoured cuts of meat, but much will be cereals. In contrast, the less developed countries, where much of the protein deficiency exists, will have expanding populations and food production rising more slowly than population. Obviously there are many assumptions underlying such calculations. They are of the same nature as Scrooge's experiences with the Spirits of Christmas. Those concerned hope that events will prove them wrong.

The critical question posed by such projections is how, if at all, such commodity transfers are to be effected. If they balanced on an ordinary commercial basis then they would require huge increases in non-food exports by less developed countries to pay for their food imports. What effect does this have on proposals that oilseed exporting countries should keep the oilseed cake and use it to prepare human food concentrates? This would reduce their earnings of foreign currency. Presumably, too,

Table 1. *Projections of food production and demand.*
(US $ × 10⁹) (from Kristensen, 1970)

	Developed countries			Less developed countries		
	1960	1980	2000	1960	1980	2000
(1) Demand	80	113	151	47	89	170
(2) Production	78	125	186	48	77	135
(3) (2)−(1)	−2	+12	+35	+1	−12	−35

many would require foreign exchange to import the equipment to do so. More fundamentally, would the developed countries waive or relax import duties on manufactured goods to give scope for less developed countries earning foreign exchange? (See discussion following Kristensen, 1970.) If the drift of these events leads to continuance of some degree of assisted commodity transfer, it might be worth putting a lot of thought, administrative skill and the results of experience in food fortification into making sure that the surpluses, when they leave the exporting country, provide cheaper protein, and one hopes more of it for a given outlay by donors, than might occur under the impact of forces internal to the exporter alone.

The specifically protein aspects of the prospective food situation was considered by a special UN study group which proposed action across a wide front (United Nations, 1968).

National planning in developing countries

Most development plans for developing countries place promotion of economic growth as a central, if not exclusive, goal. (For a discussion of national planning and the protein problem, see Eicher & Zalla, 1971.) Comparison between countries or families with different income levels shows that the wealthier have diets which are richer in protein and more varied and hence are more likely to have an effectively high biological value. One way ahead towards improved protein diets might, therefore, be to make sure that the economy grows and that demand is reflected back to producers. Their production decisions should then adjust towards higher output of the protein-rich foods.

Whether one considers the likely rate of growth, or the doubtful future suggested in the previous section, or the inevitable inequalities in the food distribution between sections of the population and within households, this *laissez faire* approach seems inadequate. Indeed, it is probable that, in the ordinary course of implementation of national plans, populations will migrate in the hope of employment and a better level of living, and will find themselves adding to the many who replace traditional rural poverty by the poverty of urban slums. Many governments actively appreciate this but the addition of other objectives to the central growth objective complicates the working of the plan.

The way ahead may not be straightforward. If we leave aside supplementation of the diets of vulnerable groups as belonging to a discussion of demand, on the production side the solution may vary from locality to locality; for example, Eicher & Zalla (1971), in stating that in Nigeria as a whole limiting amino acids are methionine and cystine, pose the question of the usefulness of the high-lysine maize often proposed as a major potential weapon in the attack on protein shortage. Elsewhere, they stress the scope for improving the protein status of those dependent on local foods by breeding strains of the staple crops with better amino acid content. It seems necessary to stress here the problem of achieving general replacement of traditional strains by ones which differ only in sophisticated features.

Possibly research into agricultural administration or organisation may be as productively rewarding as research in biology or economics. For example, lack of employment is likely to be at least as grave a problem for many developing countries as lack of protein – politically, if not as a contributor to human suffering. Or again, suppose there is scope for increasing milk production on smallholdings, how can this best be organised to minimise the competition between humans and animals for grain, to give as many farmers as possible a part in the project, while maintaining health controls and reasonable economy in collection, pasteurising and distribution of the milk? How can milking stock be replaced as required with stock at least as good and healthy, bearing in mind the question whether land and labour on the milk holding should also be used for replacement stock and the capital requirements of the smallholder? Another example of an organisation problem, but with a different slant, is posed by egg production in developing countries and, to some extent, poultry meat production.

There is an underlying question as to the extent that rich consumers should be allowed, through the derived demand for grain used to supply them with livestock products, to compete with very poor consumers who can hardly offer the feed price. If we leave that aside, however, there is the equally delicate question as to whether poultry should be kept in smallholder units or whether big producers should be allowed to bid away the trade from the small men as they have tended to do in most countries unless restrained by legislation. The advantage of the small unit is that it will occupy, at least partially, more people and spread involvement physically more widely.

Apart from economic and social arguments in its favour, smallholding production is likely to result in the farm families concerned becoming consumers of the product. Even if they did not do so originally, they are likely to use up cracked eggs, odd supplies of milk, damaged vegetables, etc., when these products are grown for market, and hence to move towards consumption by specific intent, particularly if their business prospers.

The network of influences

It may not be appropriate to pursue the subject exhaustively but equally it would be confusing to discuss more specific possibilities of protein expansion without illustrating the connections, by way of economic links, between foodstuffs. Generally speaking, any step which results in increased supplies of a foodstuff becoming available will lead to lower prices to producers, other things being equal. Thus a new high-yielding strain, or the construction of extensive irrigation facilities, will, or at least should, mean more output and hence lower product prices to producers than they would have had if the old techniques had continued. If the introduction was so rapid that it resulted in supply increases which outpaced demand increases, then producer prices would actually drop. For those who take up the new techniques, the gain in supplies for sale may more than offset increased costs of seed, water, fertilisers, pesticides and harvesting, but those who, from lack of capital or enterprise, cannot take up the new method will be trying to make a living on the old yields with the new prices. This may be another case of the rich getting richer and the poor poorer; indeed, it could mean that a group of farmers who had previously lived tolerably well become protein-deficient while the gain will appear by way of grain cheap enough to feed to livestock to supply already adequately fed sectors of the population.

It might be unwise to generalise too widely about the consequences of economical means of improving the protein content of cereals by supplementation, e.g. with specific amino acids. Since the demand for chemical protein as such by humans, i.e. as distinct from the demand for foods for their taste, texture, etc., is uncertain, it is possible such supplementation may have little effect. In the animal feed sector the position may be different, particularly if the feed industry is highly developed. If a method of effectively 'stretching' the protein in cereals is available the consequences may be varied; there may be a fall in feed prices compared with livestock and an encouragement to increased livestock production with a consequent fall in livestock product prices and hence increased consumption by urban consumers. Whether this proceeds smoothly may depend on the market communication within the industry; thus, in cattle production using concentrated feed, increased financial attractiveness of the final feeding process may merely result in a rapid increase in demand for calves which breeders may be slow to appreciate and react to. Of more note to developing areas is the likelihood that a higher effective protein level in feed grains in developed countries or areas would reduce the need for oilcake, fish and meat meal. This would mean lower prices for oilseeds and fishmeal which could be represented as advantageous to the development of direct human foods from these commodities, if the development of such products had previously been hampered by the total costs of raw materials and processing. However, in fact new foods from oilseed are at least as likely to be limited by the cost of the market research and promotion needed to launch the final food products in each locality. In such cases a falling demand for oilseeds may depress the producers. If they are operating under marketing schemes which divert part of the proceeds away from the producer the actual outcome is unforecastable.

The possible consequences of a technique of converting oilseeds to a human foodstuff under conditions which will yield an enhanced price to oilseed growers are more complicated. If the product is a meat 'extender' it can be expected to lower somewhat the demand for meat. At the other end it will draw off oilseeds from the feed market, pushing up prices and discouraging livestock production.

The human and animal competition may be direct for the monogastric categories. For ruminants, however, there is the possibility that urea will provide an increasing alternative to oilseed. Some believe that important quantities of soya beans will be released for human food concentrate production by this means. However, the forces operating in the world soya bean market are so complex as perhaps to make this but a detail within it.

PRODUCTION ECONOMIC ASPECTS OF IMPROVING PROTEIN SUPPLY

Against this background of economic conditions it is possible to consider the economic aspects of producing more protein by the variety of methods open. Among these are:

(*a*) Increasing output of all foods using present technology with more inputs of fertiliser, labour, capital, etc.

(*b*) Reduction in the losses of production through animal diseases, pests and diseases of growing crops, losses in store and in distribution.

(c) Distribution systems which produce place, time and possession utility with lower inputs of resources.

(d) The breeding and introduction of (i) generally higher-yielding ranges of crops; (ii) crops yielding more protein or better balanced protein.

(e) Production of direct food products from oilseeds.

(f) Production of 'new foodstuffs' – leaf protein, single cell proteins, fungi, etc. – either from inedible raw materials or by using more efficiently materials which would otherwise contribute to livestock feed.

In practice these possibilities are often combined. Thus in the Green Revolution new varieties, more fertilisers, more water, pesticides and, ultimately, better marketing facilities have been introduced together (Brown, 1970).

Relative costs

A general view of relative costs of protein in different forms as estimated by various authors is given in Tables 2 and 3. These provide only an indication of general tendencies. In any specific locality there will be differences

Table 2. *Relative cost of protein in major traditional foods (from Abbott, 1969)*

	Price of product (¢/lb)	Protein content (%)	Price of protein[a] (¢/lb)
Chick-peas[b]	5	20	6
Wheat flour[c]	5	11	11
Beans[d]	9	22	24
Skim milk powder[e]	15	36	31
Fish, dried[f]	35	37	89
Cheese[g]	32	25	111
Chicken[h]	26	19	123
Beef[i]	27	15	164
Pork[j]	24	10	197
Eggs[k]	24	11	204
Lamb[l]	30	12	228

[a] Net of calorie contribution valued *pro rata* with sugar at 4¢ per lb.
[b] Banda, Uttar Pradesh, India, 1966.
[c] Canada, f.o.b., 1965–6.
[d] Dry, Mexico City, 1966.
[e] Commercial exports, average price, f.o.b., 1966.
[f] Export price Norway and Iceland, 1965–6.
[g] New Zealand white, London Provision Exchange, 1966.
[h] US ready-to-cook broilers, 1966.
[i] Australia average first and second export quality, 1966.
[j] Average all weights, Chicago, 1966.
[k] Average monthly price, Danish Export Co-operative, 1966.
[l] New Zealand frozen carcasses, Smithfield, London, 1966.

Table 3. *Relative cost of protein in new processed protein foods and potential protein ingredients (from Abbott, 1969)*

	Price of product (¢/lb)	Protein content (%)	Cost of protein[a] (¢/lb)
Multi-purpose food (India)[b]	18	42	36
Arlac[b]	21	42	40
Incaparina (Guatemala)[b]	19	28	47
ProNutro[b]	15	22	47
Cottonseed flour[c]	7	55	5
Toasted soy protein[b]	8	50	11
Lysine[d]			7
Fish protein concentrate[e]	27	85	25
Protein isolate, Promine D[b]	35	97	32
Lypro[b]	35	65	48
Leaf protein[f]	47–37	50	83–63

[a] Net of calorie contribution valued *pro rata* with sugar at 4¢/lb.
[b] Prices as reported in Orr & Adair (1967).
[c] Ranchers' Cotton Oil Co., Fresno, 1967.
[d] Cost of additional utilisable protein following application of lysine to wheat flour at 0.4 per cent rate on the basis that 1 lb of lysine costs $1.00.
[e] SONAFAP, Morocco, 1966.
[f] Investigation into the production of a high protein concentrate from leaves for inclusion in the diet of infants and children. Scientific Research Council, Jamaica Tech. Report I/C 5, Hope, Kingston (May 1963). (N. W. Pirie, in discussion on the paper, quoted examples of much lower cost figures for 1971, but see pp. 57–8 below in relation to cost comparisons.)

Note. (1) Cost figures given by Parpia (1968), p. 79, for various mixed baby and other foods show the cost of protein in soyabean flour fortified with dl-methionine as of the same order of magnitude as that for Indian Multipurpose Food when calculated on the basis of protein percentage. However, his 'Protein Value Cost Index' (PVCI) calculated as 'protein content in 100 g of food × protein efficiency ratio ÷ cost of 100 g food in US cents' gives a value of 44 for the soya-methionine mixture compared with 30 for the Multipurpose Food and 13 for skim milk powder.

(2) Realistic costs for algae seem unlikely to be available at this stage in the development of the technology of production. Tannenbaum & Mateles (1968), p. 91, state: 'The Japanese are finding that these algae cannot be produced for less than 84 cents per kg and, although they are used in Japan as a flavouring (not as a protein source), they are unpalatable to a Western taste.'

both as to the range of foodstuffs actually or potentially available and as to the detailed costs of individual foods. These will depend on natural conditions, technology in use, entrepreneurial effort and the extent to which local production is protected from competition by producers elsewhere by transport costs, protective customs duties or other restraints on external trade.

Table 4. *Variation in costs of production of milk in England and Wales, 1961–2*

Net farm costs in pence per gallon	Number of herds within cost group
Under 18	9
18.00–19.99	22
20.00–21.99	41
22.00–23.99	67
24.00–25.99	74
26.00–27.99	77
28.00–29.99	61
30.00–31.99	47
32.00–33.99	35
34.00–35.99	28
36.00–37.99	21
38.00 and over	27
Total	509

Source: *Costs of Milk Production in England & Wales, April 1961 to March 1962.* Milk Marketing Board in association with the Ministry of Agriculture, Fisheries and Food on behalf of the Conference of Provincial Agricultural Economists.

Even in a rather narrowly defined situation with good economic communications, the idea of average costs of production is often surprisingly unhelpful because of the range of variation. The example of milk costs in Britain given in Table 4 shows a fairly typical picture for a farm product. Such ranges naturally pose the question how the producers with costs at the top end of the scale survive and why those at the bottom end do not expand and take over. Rigorous explanations are not usually forthcoming; in some commodities the low-cost producers may be supplying a low-price market whilst some may be prepared to accept a lower return from their business activities voluntarily because they have few 'felt' wants or of necessity because they have no other work open to them. Often, though the same spread may be found from year to year, the same herd may appear in different cost groups in different years. All cost comparisons must, therefore, recognise the existence of such cost spreads and the uncertainty which they introduce into the conclusions which might be drawn from the comparison.

It is also highly important to remember that the potential for *change* in costs may be more important than their current comparison. An important benefit to a low-cost producer may be a profit margin big enough to allow him to invest in new technology – its development or its application, or both – leading to an even bigger cost advantage in the future. Or a firm may be in a position to subsidise a product for a while, e.g. if it has other profitable products or if it is credit-worthy. If it firmly believes that technical developments are in the pipeline which will reduce considerably the costs of a protein food in a few years, it may be quite content with a fairly high cost and low margin at present in order to be in a position to exploit new developments. Current comparisons would therefore give little guidance. It may be, too, that the producer of a product which is not too highly differentiated and has prospects of expansion may be content to allow a market to be developed by a rival at cut prices whilst waiting to be able to benefit from the promotion; this strategy is less applicable if the product is highly differentiated by branding. Farm products tend to be rather undifferentiated and it is possible that the launching and promotion of a named special protein food or drink might nonetheless open a market from which other firms could benefit without contributing to the cost in much the same way as if the product had been undifferentiated.

High protein yield cereals

Increased protein may come from higher grain yield without change in protein content, higher protein percentage without gain in yield, or higher content of limiting amino acids without increase in yield or protein percentage, or combinations of these. The subject has been extensively discussed (e.g. see Brown, 1970). However, any appraisal should include three considerations. One is the actual protein and amino acid contents of the grain currently being grown since they vary notably. A second is the overall pattern of food supply for the diet as a whole for the year as a whole – it may be that in the months when grain protein is the principal source the protein intake is higher than at another period of the year.

Requiring greater emphasis, however, is the effect of a proposed change of genetic material on the farmers' decision situation. Specifically, will the improved protein content, quantity or value be reflected in a higher value *as seen by the producer*? High protein and high biological value grains are usually discussed as contributing to improvement in the traditional rural sector (Eicher & Zalla, 1971). However, unless they also have an increase in gross yield the decision situation for comparison with that for the usual grain variety is not likely to encourage change. On the credit side, there is no observed increase in return; on the debit side, possibly increased cash costs for seed and, at the least, the costs represented by the effort of adjusting to the slightly different behaviour of a new variety in respect of seedbed requirements, timing, pest attack, and so forth.

From an economic standpoint improvement in protein alone would not catalyse improvement in the protein supply situation unless it was accompanied by gain in yield or other evidently valuable features. The situation would, however, be different if the farmer was in a market economy and if the new product attracted a higher price than the old. Brown (1970) expresses his view as follows, 'To realise the nutritional promise of high-lysine corn, plant breeders must incorporate it into commercial varieties without sacrificing too much yield potential. They must also eliminate some of its undesirable functional characteristics, such as a bitter taste, which reduces consumer acceptance.'

Amino acid supplementation

It seems that lysine and methionine are available cheaply enough to be used for fortification of staples. The additional quantities required are such that the process must be under close control, either through precision equipment at the main processing plant or by the use of a premix; in other words it lends itself mainly to products processed at large central mills. As Table 3 indicates, given that the goal is known, the economics appear promising. However, the goal is not always clear and among the questions requiring answers are:

(*a*) Is it the aim to channel fortified staples only to vulnerable groups or are these to be included only through a general distribution?

(*b*) As a special case of (*a*), are both urban and rural groups to be covered? Total protein consumption in rural areas may be higher than in urban but animal protein intake may be less (Abbott, 1969). On this basis the main need may be in rural areas.

(*c*) Is the same amino acid limiting at all times of the year even though diets vary with season?

These may be judged matters of demand rather than production, but (*b*) at least underlines the need to include the production of time, place and possession 'utility' in realistic production costs. The costs of fortification cannot be taken at the urban supermarket; they must include distribution costs to the villages, costs which will include the overheads of supplying and retailing. A UN report on protein foods (United Nations, 1968) notes that the raw material cost of the food component in a processed and packaged food item is often only 25 per cent of the retail cost at consumer level in developing countries, and that a 10–12 per cent profit on ex-factory costs may be the minimum that will encourage a manufacturer to incur the bother and the risks.

If this seems a large distribution margin it might be wise to itemise the steps through which a food must pass if packets in bulk containers are to be loaded on to lorries, moved to suitable distribution points, moved into store, moved out in mixed consignments suited to retail demand, moved to retailer, distributed to shelves, moved from shelves across the counter to consumer, money received, together with all the ancillary equipment and accounting. If the consumer's shopping basket is full the retail stage may not be so onerous but often in developing countries retailing is on a microscale (e.g. see Galbraith *et al.* 1955).

Livestock

It has been noted above that protein in livestock products is dearer than that in most traditional foods but that there is more than mere nutrients in meat and in milk too. The problem of organisation and its ambivalent bearing on national policy towards livestock has also been noted.

Situations where livestock have a specifically *protein* supplying role and where diets are critically short of protein are difficult to deal with quantitatively because they concern livestock as scavengers rather than as business machines.

Commercial livestock production may be based on range land, intensive grassland, concentrated feed or combinations of these. However, the more they depend on concentrates the easier, generally speaking, is the production management; feed supply and environmental conditions are much easier to keep steady through the season. This is reflected in the observation by Altschul (1969) that, as animal protein consumption increases from about 10 to 45 lb per head per annum the proportion of grain supplies fed to animals increases from zero to 75 per cent. Linear programming techniques have made useful contributions to economy in concentrated feed usage.

The relationship of feed supply to the nature of the labour required is also important. It is usually a good deal easier to simplify the demands on the bulk of workers in a system based on concentrated feed than in one subject to the fluctuations of fodder growth and climatic exposure. Of course heavy responsibility rests on the supervisors. This may be a further force making for concentration and industrialisation of production in a developing country where farmers have no traditional skill with livestock, but where a demand for livestock can draw grain into use for feed.

In many developing areas where rural diets are based on cereal or root staples there is, in fact, no margin for feeding grain to livestock or diverting land to fodder production (Clark & Haswell, 1970). In these conditions any chickens or mammalian farm stock which can get their living from scraps about the homestead or wayside weeds, grass or bushes, make a contribution to human protein supply which is of almost infinite practical efficiency.

In stressing the need not to lose sight of scavenging livestock and arguing that they are likely to contribute good protein in situations where it is scarce, it must be admitted that milk is less likely than meat and eggs to result from scavenging livestock. And the significance of this is indicated by a statement by Hollingsworth & Greaves (1970) that, 'In practice, no country without an adequate milk supply has yet wholly solved the problem of assuring the consumption of adequate amounts of good quality protein by young children after they are weaned from the breast.'

Fish

The hope that there is as good fish in the sea as ever came out of it buoys up the outlook on this source of protein. The technical situation is difficult to determine; overfishing has been a common experience too often in the past to justify dogmatic statements. Freshwater production seems likely often to be in a similar situation to most land based livestock, i.e. supplying a highly prized dish rather than a nutrient.

Two production points might perhaps be noted in passing. One is the cost of producing the time, place and possession utilities since fish of all kinds goes 'off' very quickly. The consequences of traditional techniques of producing these utilities is indicated by a big price rise as one goes inland.

The second is the possible effects on the production of fish flour and other products of (*a*) the demand for fishmeal for livestock feeding and (*b*) the demand for fish oil, perhaps less often given its full weight. Between 1961 and 1967 world fish oil exports increased from about 400 thousand to about 750 thousand tons. In 1961 and 1967 the total use of oils and fats in margarine production in the United Kingdom was similar at 265–270 thousand tons but marine oils other than whale increased from 41 to 143 thousand tons and the use in compound cooking fat increased from 28 to 57 thousand tons. Peruvian fish oil prices dropped from 191 to 127 dollars per metric ton between 1966 and 1967. There are many reports of heavy investments in oilseed production in various parts of the world – the USSR increased her exports of edible oils from 120 thousand tons in 1961 to 448 thousand tons in 1966, apparently from improved genetic strains of sunflower.

It may be well to weigh the effects of uncertainties of this magnitude on entrepreneurs when we may be inclined to criticise apparently slow developments in desirable directions.

Human protein foods from oilseeds

From many points of view (e.g. see costs in Table 3) oilseeds are amongst the most attractive sources of food protein. Most of the special protein foods such as Incaparina are based on them. The use of quantities of oilseeds as fertiliser in developing countries, not as livestock feed let alone as human food, naturally attracts criticism by nutritionists. There are, however, a number of influences in the situation which are important but difficult to quantify; one concerns the location of crushing. In brief, the import duties levied by importing countries on oilseeds, oil and cake have often been directed towards encouraging crushing in the importing country (Stopforth & O'Hagan, 1967). Developing countries which produce oilseeds would prefer to earn the crushing margin themselves, though it is not very large, and governments have encouraged this in various ways. However, in the importing countries the big companies using vegetable oils usually also crush the seeds and their plant is designed to handle many kinds of seeds and nuts so they can use whatever material is economic at the time and can operate the year round. The plant in exporting countries may have to work to a seasonal peak with many slack months. Moreover, one might expect some advantage in business co-ordination in the importing country in having the crushing plant integrated with the using plant. How far the longer-run tendency will be for processing in the exporting country is arguable.

The preceding paragraph relates to seeds and their products traded internationally. As regards home use of oil in the producing country, there may be quite strong economic forces making for local crushing in country districts. This has, in theory at least, the double disadvantage of using inefficient presses, instead of large-scale units with better extraction rates for oil, and of leaving unco-ordinated the use of the cake. It is under these conditions that use for fertiliser is common.

Among the reasons for continuance of these crushing practices is that the oil produced is more to the taste of village consumers than is the product of the large plant. Further, local mills may be able to buy raw materials at lower prices than can port mills. The effect of transport costs from country to port would tend in this direction but, in some countries where marketing organisations control prices and divert part of the proceeds away from producers, port mills may have to pay export prices while farmers would be content to sell to local mills for a small margin over what the organisation pays them.

The dominant position of the soya bean in the oilseed protein picture and, perhaps, the place of US supplies of soya beans should also be noted, bearing in mind that soya beans appear to be particularly suitable for conversion to products for human use. The three principal oilseeds in terms of world seed production are soya (40 million tons), cottonseed (20 million tons) and groundnuts (16 million tons unshelled). Of the 40 million tons of soya beans, the USA provides about 26 million tons and, with an export of some 7 million tons, accounts for nine-tenths of world trade (Commonwealth Secretariat, 1968).

While the requirements of oilcake for the production of food proteins is small in any area it is not likely to impinge on these features of the world oilseeds situation. If demand expanded, however, these influences might become increasingly important and they could become significant much earlier in particular localities.

New proteins

For practical purposes these protein foods can only be produced under what amount to manufacturing industry conditions. In so far as substantial investment is involved the number of firms operating may be few and it is not to be expected that they will provide an appraisal of the production economic picture over a time that will allow useful cost comparisons to be made. There are, in any event, many unknowns which lie outside technology. Consumer response is an obvious question. Perhaps less obvious but more critical are the possible impacts of legislation. How will such foods have to be described in the ingredients list on the food packet? What precautions will be required before petroleum-based products are marketed for direct human use?

Savings from reductions in losses

Much discussion of this subject is highly unrealistic in economic terms, the implication being that such gains are virtually costless. There are two very important elements of cost, apart from raw materials, for pest control which themselves may have to be bought with scarce foreign exchange. One is the cost of the extension work needed to show traditional farmers how to operate the technique correctly; some spray routines may require half a dozen or more applications each year at precisely determined times. If the producers are in the market economy an evidently sound technique may spread fast, but otherwise extension may be slow. Secondly, there is the uncertainty of outcome. There are the crude effects as, for example, when weather conditions do not favour the disease and the non-spraying farmer gets as good a crop as the one who sprays. More subtle are the cases where, for example, comparisons of amounts of insecticide and intervals between applications on different varieties has shown that the best cost benefit return is shown by different varieties, different rates of application and different intervals in the wet and the dry season. If protection is cheap, such complexity, with its attendant uncertainty for the decision maker, may not be a matter for concern, but where it is dear the problem is significant.

CONCLUDING REMARKS

Any selection of key features from the preceding review must be a highly personal matter since the choice will depend on views about the current situation, the course of technological development and the ideas which will influence national and, above all, international policies affecting trade in foodstuffs between continents. We would, in any event, need to specify where, between avoidance of kwashiorkor on the one hand, and providing a steak in every kitchen whose owner loves good living on the other, we are to focus our attention.

To bring the issues to a head perhaps the following would be a defensible view, assuming no marked policy changes. As to helping with gross protein deficiency in urban societies in developing countries, the best prospective materials seem to be based on oilseeds, with uncertainties attaching to fish protein concentrates. For those suffering from rural poverty, plant breeding and technological means to increase *total* production of grain are likely to be more promising than efforts directed to improving protein or amino acid content specifically. Livestock production based on scavenging should not be discouraged but otherwise livestock and their products have little to offer towards helping the grossly deficient sectors if the livestock production operates under the normal economic conditions without agricultural protection. They have a great deal to offer in the way of enjoyment rather than nutrition to those who can afford to view food as a pleasure. It seems likely that the meat substitutes will be more providers of pleasure than providers of amino acids.

There is, however, a much wider international issue. In our dreams peopled by wholly rational, politically neutral, amicable people who work willingly when the human need is shown to them, we can picture the world's food needs for the next decade or two being met by the developed countries putting in their skill, knowledge and resources to the full, moving the grains, oilseeds and other foods to where they are needed, with some shifting to provide everyone as nearly as possible with the kinds they are used to. However, to such rational beings there would be something faintly illogical in having gross and increasing disparity between the technological levels of different parts of the world's agricultural industries. Consequently, steps would be taken by research in plant breeding and the like, and by agricultural advice and training to gear up the agricultural industries of the developing countries. Within many countries welfare schemes have gained acceptance which have similar ends, but in the real wide world, if we started out beyond the first steps on that road, we would meet fears and strains and confusions. The taxpayers of one country would grumble at paying for the food of another. In the recipient countries the young people would grumble at lack of jobs, the farmers at the disruption of their markets, and all would fear domination. No doubt all this need not happen but new administrative skills, adjustments in the economic system and in people's aims and aspirations tend to lag behind such needs. The key question for the next few decades is how fast we can nudge reality towards the dream.

REFERENCES

ABBOTT, J. C. (1969). Economic factors affecting the distribution of world protein food supplies. In: Milner, M. (Ed.), *Protein-Enriched Cereal Foods for World Needs*, 13–29. St Paul, Minnesota: American Association of Cereal Chemists.

ALLEN, G. (1963). Economics, politics and agricultural surpluses. *Journal of Agricultural Economics* 15, 410–30.

ALTSCHUL, A. M. (1969). Low cost foods, fortified cereals and protein beverages. In: Milner, M. (Ed.), *Protein-Enriched Cereal Foods for World Needs*, 82–96. St Paul, Minnesota: American Association of Cereal Chemists.

BROWN, L. R. (1970). *Seeds of Change. The Green Revolution and Development in the 1970s*. New York: Praeger.

CLARK, C., & HASWELL, M. R. (1970). *The Economics of Subsistence Agriculture*, 4th edition. London: Macmillan.

COMMONWEALTH SECRETARIAT (1968). *Vegetable Oils and Oilseeds*. London: Commonwealth Secretariat.

EICHER, C., & ZALLA, T. (1971). Protein, planners and planning techniques in Africa. *Rural Africana, No. 13*. East Lansing, Michigan: African Studies Center, Michigan State University.

GALBRAITH, J. K., HOLTON, R. H. *et al.* (1955). *Marketing Efficiency in Puerto Rico*. Harvard University Press.

HELLEINER, G. K. (1965). Peasant agriculture, development and export instability: the Nigerian case. In: Steward, I. G., & Ord, H. W. (Eds.), *African Primary Products and International Trade*, 44–64. Edinburgh University Press.

HOLLINGSWORTH, D. F., & GREAVES, J. P. (1970). Nutrition policy with regard to protein. In: Lawrie, R. A. (Ed.), *Proteins as Human Food*, 32–45. London: Butterworths.

KRISTENSEN, T. (1970). The approaches and findings of economists. *Proceedings of the 13th International Conference of Agricultural Economists*, 72–88. Oxford University Press.

MOSHER, A. T. (1966). *Getting Agriculture Moving. Essentials for Development and Modernisation*. New York: Praeger.

ORR, E., & ADAIR, D. (1967). *The Production of Protein Foods and Concentrates from Oilseeds*. London: Tropical Products Institute.

PARPIA, H. A. B. (1968). *Conservation and Technological Production in Agricultural Sciences and the World Food Supply*. Miscellaneous Papers 3. Wageningen, Netherlands: Landbouwhogeschool.

STOPFORTH, J., & O'HAGAN, J. P. (1967). Structure of the oilseed crushing industry and factors affecting its location. Parts 1 and 2. *Monthly Bulletin of Agricultural Economics and Statistics* **16** (4), 1–9 and **16** (5), 1–15.

TANNENBAUM, S. R., & MATELES, R. I. (1968). Single cell protein. *Science Journal* **4** (5), 87–92.

UNITED NATIONS (1968). *International Action to Avert the Impending Protein Crisis*. Report to the Economic and Social Council of the Advisory Committee on the Application of Science and Technology to Development. New York: UN.

PART II

THE BIOLOGICAL EFFICIENCY OF PROTEIN PRODUCTION BY PLANTS

PART II

THE BIOLOGICAL EFFICIENCY OF
PROTEIN PRODUCTION BY PLANTS

6

BIOCHEMICAL ASPECTS OF THE CONVERSION OF INORGANIC NITROGEN INTO PLANT PROTEIN

By A. J. KEYS

Rothamsted Experimental Station, Harpenden, Herts

Plants in the field have three sources of inorganic nitrogen: nitrate, nitrogen gas used through the mediation of symbiotic bacteria, and ammonia or ammonium ions. All three sources ultimately provide ammonium ions in the plant tissues as the form of nitrogen to be brought into organic combination, and present evidence suggests that ammonium ions are first incorporated into glutamic acid and glutamine. The synthesis of the various amino acids and amides needed to make proteins depends on the existence of the corresponding carbon skeletons. The protein-synthesising system converts the amino acids into proteins with structures dictated by the messenger RNA (mRNA) molecules present in the cell. The mRNA is determined genetically to correspond with the requirements of the cell, its stage of growth and the function of the tissue in which the cell exists.

NITRATE REDUCTION

The subject of nitrate reduction in higher plants was reviewed by Beevers & Hageman (1969), who concluded that reduction to ammonia in leaf tissue involves two separate enzymes, nitrate reductase and nitrite reductase. Nitrate reductase catalyses the reduction of nitrate to nitrite by either $NADH$ or $FMNH_2$ ($FADH_2$). A few species of plants have a similar, but less important, enzyme catalysing reduction of nitrate by $NADPH$. Molybdenum and FAD are constituents of the partly purified NAD-specific enzyme. Although the rate at which nitrate is reduced by leaves is stimulated by light, current evidence suggests that nitrate reductase is not very active in chloroplasts. Roots and other plant tissues without chlorophyll show nitrate reductase activity, part of which may be associated with cell particles.

At the time of the review by Beevers & Hageman (1969), nitrite reductase, which catalyses the reduction of nitrite to ammonia *in vitro* by reduced ferredoxin or an appropriate redox dye system, was thought to be associated with chloroplasts. However, it does exist in non-green tissue. Miflin (1970) showed that, although roots mainly contain soluble nitrate and nitrite reductases, some are associated with a particle that seems to be very dense from its behaviour in density gradient centrifugation. Grant, Atkins & Canvin (1970) produced evidence showing that chloroplasts have little nitrite reductase, as they have little nitrate reductase, and they concluded that nitrate and nitrite are mainly reduced at a site remote from the chloroplasts. It follows that with no evidence that ferrodoxin occurs in leaves outside the chloroplast, the natural reductant for the conversion of nitrite to ammonia in plants is not yet established.

Metabolic regulation of nitrate and nitrite reductase activity

One important feature of nitrate reduction in plants is that the two enzymes responsible seem to be substrate-inducible in a manner resembling that for β-galactosidase in *Escherichia coli* (for a recent review, see Marcus, 1971). Hence, seedlings grown without nitrate and dependent on organic nitrogen reserves of the seed, or plants grown by water culture with ammonium ions the only source of nitrogen, have very little nitrate reductase and nitrite reductase activity. Activity increases when nitrate is provided. Nitrite reductase in higher plants is thought to be induced independently of nitrate reductase. It usually appears following induction by nitrate but only after some delay and following a period while nitrite accumulates. The induction of nitrate reductase in leaves depends on light, but induction also occurs in tissues that are not green, such as roots (Wallace & Pate, 1965), tobacco pith cells in culture (Filner, 1965) and the aleurone layer of barley grain during germination (Ferrari & Varner, 1970). It has been suggested that light acts by stimulating nitrate uptake, but recent evidence of Travis, Jordan & Huffaker (1970) with barley leaves supports a different explanation. Taken together with evidence that, to be effective in the induction of nitrate reductase, light must be accompanied by the presence of CO_2 (i.e. photosynthetic fixation) (Kannangara & Woolhouse, 1967), and evidence that the induction in corn seedlings occurs sooner in plants given ammonium salts than in nitrogen-starved plants (Schrader, quoted by Beevers & Hageman, 1969), a supply of new organic carbon and nitrogen seems to be required before the enzyme is synthesised. Evidence that the induction of nitrate reductase in leaves involves *de novo* synthesis of the enzyme protein is discussed by Beevers & Hageman. Ingle (1968) showed protein synthesis to be a necessary factor in the induction of nitrate reductase activity, but has also obtained evidence that the protein synthesised is not nitrate reductase itself.

Another aspect of the effect of light and CO_2, discussed at length by

Beevers & Hageman (1969), is that they must be indirectly responsible for providing the reducing power used to reduce nitrate and nitrite. This cannot be through the transport of photoreduced pyridine nucleotides because the chloroplast membranes prevent such transport (Heber & Santarius, 1965; Ogren & Krogmann, 1965). Various products of photosynthesis, such as triose phosphate, move rapidly from the chloroplasts and the necessary reducing power must be obtained during the metabolism of one or more of these products (Stocking & Larson, 1969). Beevers & Hageman (1969) describe experiments showing that added triose phosphate does stimulate nitrate reduction, but not nitrite reduction, in leaf disks in the dark. However, nitrite was rapidly metabolised by such leaf disks in the light. Nitrite reduction by roots is especially stimulated by adequate aeration.

Although nitrate and photosynthesis stimulate activity of nitrate reductase, activity in leaves does not increase indefinitely; there is evidence that, with various light intensities and nitrate concentrations (Hageman & Flesher, 1960), the tissues develop different amounts of nitrate reductase. Also, removing the nitrate source or transfer from light to darkness decreased nitrate reductase activity. This probably results from the continuous breakdown of nitrate reductase or some protein factor necessary for its activity, so that the activity at any given time depends on the rate at which the active enzyme is formed. This, as we have seen, is controlled by supply of nitrate and photosynthetic products. Other controls also operate on the formation of nitrate reductase and suppress the induction process when the demand for reduced nitrogen is satisfied. For example, mature leaves in the light and containing much nitrate may have very little nitrate reductase activity (Carr & Pate, 1967). In mature leaves the amounts of nitrogenous compounds are decreased by their removal for use in other parts of the plant. The mature leaf may fail to synthesise the enzyme protein because it lacks free amino acids (Carr & Pate, 1967) or because, during the general decline of protein synthesis in the ageing leaf, gene transcription is blocked and the appropriate mRNA is no longer synthesised. Endproduct repression has been observed with plant cells in tissue culture (Filner, 1965), in which amino acids added to the culture medium repress induction of nitrate reductase by nitrate. In *Chlorella* cells (Morris & Syrett, 1963) and in *Lemna minor* (Joy, 1969) both ammonium ions and amino acids act as repressors. With excised seedlings of *Zea mays* such effects were not demonstrated and, when the tissues were rich in free amino acids, synthesis of nitrate reductase was increased (Schrader & Hageman, 1967).

The integration of nitrate reduction into the nitrogen metabolism of the whole plant

Pate (1968) stressed that nitrogen metabolism differs in different plant species and, in particular, there are some plants, such as *Pisum arvense*, in which the roots can actively reduce nitrate, and others such as *Xanthium pennsylvanicum* or *Perilla fruticosa* which usually reduce nitrate almost entirely in the leaves. However, even in plants such as *Pisum arvense*, the main site of nitrate reduction is transferred to the leaves when nitrate is abundant. Beevers & Hageman (1969, p. 508) conclude that, in mature crop plants, nitrate is reduced mainly in the leaves. Plants grown with a constant amount of nitrate in the medium (Wallace & Pate, 1965) maintained a constant activity of nitrate reductase in their roots, but there were large diurnal fluctuations in activity in leaves, correlated with light intensity, as discussed previously, and with a large diurnal fluctuation in the amount of nitrate passing from the root to the shoot. Control over the uptake of nitrate into plant cells has been observed (Heimer & Filner, 1970); uptake was decreased when amino acids were supplied as a source of nitrogen. Nitrate and nitrite reductase activities are clearly determined by various factors, so that the supply of organic nitrogen to the plant as a whole is in keeping with both available nitrate and the demand for ammonia and ammonium ions for the synthesis of amino acids. These conclusions are summarised in Figure 1.

NITROGEN GAS AS A SOURCE OF PLANT NITROGEN

As far as the biochemistry is concerned, it will suffice to deal with the reduction of nitrogen in nodules of leguminous plants, a topic reviewed by Bergersen (1971) whose conclusions are embodied here. Cells forming the nodules on the roots of leguminous plants contain bacteria of species of the genus *Rhizobium* that possess an enzyme system called nitrogenase which catalyses reduction of nitrogen from the atmosphere to ammonia and ammonium ions. Reduction of nitrogen in the intact nodulated plant is stimulated by light because the bacteroids of the nodules depend on a supply of carbon compounds from the host plant (Virtanen, Moisio & Burris, 1955; Bach, Magee & Burris, 1958). These are used by the bacteroids as respiratory substrates and to provide α-oxoglutarate from which glutamate and glutamine are formed by reaction with ammonium ions and ammonia. Other amino acids can be derived by transamination either in the bacterial cell or in the cells of the host plant. Nitrogenase is a complex of two units, the larger consisting of protein associated with

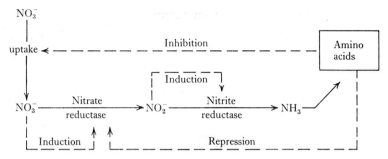

Figure 1. Nitrate reduction.

molybdenum and iron, and the smaller consisting of protein with iron only. For the reduction of nitrogen *in vitro* the combined system shows an absolute requirement for ATP and a reductant. Either sodium dithionite or β-hydroxybutyrate, NAD and benzyl or methyl viologen can serve as the electron donor. It seems that, *in vivo*, electrons for the reduction of nitrogen may be drawn from the respiratory electron transport system of the bacteroids at a point after NAD diaphorase. The natural redox systems transferring electrons to nitrogenase have not been identified.

While fixing nitrogen, nodules require much oxygen, which is thought to be related to the large requirement for ATP shown by the cell-free nitrogenase system. The function of ATP in the reductive process remains unknown. While evolving the symbiotic system, the species of *Rhizobium* concerned seem to have lost the ability to synthesise nitrogenase except in association with the host plant. Apparently the control of fixation by the nodulated root system has passed to the host plant. Nodulation of roots is suppressed when enough nitrogen is supplied in the form of nitrate (Nutman, 1956) and favoured when the carbohydrate supply is enhanced either by photosynthesis or by added sugars. Figure 2 is an outline of the metabolic system of the nodulated root.

AMMONIA AND AMMONIUM IONS AS SOURCES OF NITROGEN FOR PLANTS

Many experiments have used ammonium salts as the source of nitrogen for plants growing in sand or water culture. As ammonia enters the tissue faster than anions, the pH decreases near the root surface and may adversely affect growth. Changes in pH may be avoided on a laboratory scale by frequent changes of culture solutions (Hewitt, 1966), by continuous flow of culture solutions (Weissman & Keys, 1969), or by adding solid calcium carbonate (Hewitt, 1966). When nitrogen is supplied to wheat seedlings

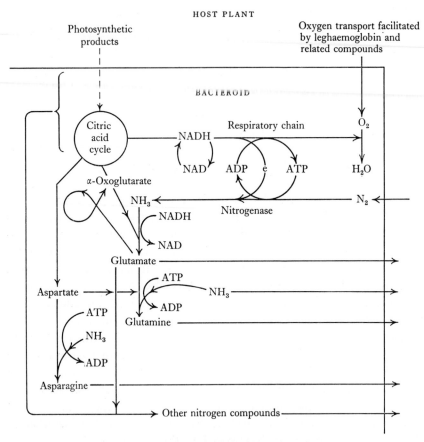

Figure 2. Nodule metabolism and nitrogen fixation.
(Adapted from Bergerson, 1971.)

as ammonium salts, and acidification is prevented, the amide content of the soluble fraction in the shoots increases considerably and the protein content slightly (Weissman, 1959). These may be expected effects of by-passing the controls acting on nitrate reduction and introducing excessive amounts of ammonia into the tissue. *Chlorella* cells, growing with ammonium ions as nitrogen source photosynthesise faster than similar cells growing with nitrate as nitrogen source (Kanazawa, Kanazawa, Kirk & Bassham, 1970). This is presumably because reduction of nitrate to ammonia uses some of the reducing capacity generated in the light-reactions of photosynthesis that otherwise would be used to reduce the carbon fixation products.

SYNTHESIS OF AMINO ACIDS FROM AMMONIA AND AMMONIUM IONS

Ammonia and ammonium ions are thought to enter into organic combination in the plant mainly by reductive amination of α-oxoglutarate to form glutamic acid and the subsequent conversion of glutamic acid to glutamine. That glutamate and glutamine are among the first products of ammonium metabolism in roots (Yemm & Wills, 1956; Cocking & Yemm, 1961) and root nodules (Kennedy, 1966) was established by supplying the tissues with ^{15}N-labelled ammonium salts or nitrogen gas. In leaves the situation is less certain and the role of glutamate synthesis is based on the work of Bassham & Kirk (1964) with *Chlorella*. When this alga photosynthesised in a medium containing ^{15}N-ammonium ions as the nitrogen source, ^{15}N was incorporated into glutamate sooner than into alanine, serine or aspartate. However, the experiments were done with relatively concentrated bicarbonate, which decreases the synthesis of glycine and serine but favours glutamate synthesis (Whittingham, Hiller & Bermingham, 1963). In normal atmospheres containing 0.03 per cent CO_2, glycine, serine, alanine and aspartic acid are the main amino acids into which carbon is rapidly assimilated during photosynthesis by leaves (Hellebust & Bidwell, 1963; Ongun & Stocking, 1965; Hatch & Slack, 1966; Roberts, Keys & Whittingham, 1970); glutamate is a minor product.

I know of no experiments with leaves that have made use of $^{15}NO_3^-$ to determine whether newly assimilated nitrogen enters into any of these amino acids. Interest in glycine and serine arises because the proportion of total carbon fixed during photosynthesis that is converted to these amino acids can be greatly increased by reducing the carbon dioxide, and increasing the oxygen concentrations of the atmosphere.

Lips (1971) suggested that the metabolic pathway leading to the synthesis of glycine and serine provides reduced NAD for the reduction of nitrate. Under these circumstances less photosynthate would be expected to be incorporated into glycine and serine when ammonium ions replaced nitrate as the source of nitrogen. In *Chlorella* the rates of photosynthesis of glycine and serine were slightly increased by replacing nitrate in the culture medium by ammonium ions (Kanazawa *et al.* 1970), and the two amino acids were synthesised at similar rates in wheat leaves whether nitrate or an ammonium salt was the source of nitrogen (Drossopoulos & Keys, unpublished work). The amino acids most rapidly labelled during photosynthesis in detached leaves (glycine, serine and alanine) are also incorporated rapidly into leaf proteins, so that the residues of these amino acids are those most strongly labelled in the protein (Hellebust & Bidwell, 1963;

Ongun & Stocking, 1965; Pate, 1968). Pate (1968) discussed the situation in *Pisum arvense* where in the light glycine, serine, alanine, aspartate and aromatic amino acids seem to be synthesised in the leaves, whereas other amino acid residues in the leaf protein are derived from photosynthate that had first been metabolised in the root. For both nodulated plants and plants grown in the presence of nitrate, 80–90 per cent of the organic nitrogen in root exudate was in glutamine and asparagine. With intact plants, photosynthesising in $^{14}CO_2$, the ^{14}C became more evenly distributed among the various amino acids of protein in the shoot apex when time was allowed for the assimilated carbon to be cycled through the roots. With plenty of nitrate, when leaves are the main site of nitrate reduction, either the leaves can synthesise a greater range of amino acids or a different mechanism makes use of the capacities of the root to synthesise amino acids. Diurnal variations in the composition of the soluble nitrogen fraction of leaves (Noguchi & Tamaki, 1962), in which for example glycine and serine are most concentrated at mid-day, must be studied further in relation to nitrogen metabolism and nutrition of the shoot and root of crop plants. If normal dark respiration is restricted in the light (Jackson & Volk, 1970) there would be a shortage of α-oxoglutarate, the substrate for glutamate synthesis, and other carbon compounds needed for the synthesis of various amino acids. Kanazawa *et al.* (1970) propose that in *Chlorella* the citric acid cycle operates only to supply organic acids for amino acid synthesis. The situation in leaves, with regard to glutamic acid dehydrogenase activity needed to synthesise glutamate from α-oxoglutarate and ammonia, is confused. Two enzymes have been identified, one in mitochondria using NADH as cofactor and the other in the chloroplasts using NADP as the co-factor (Leech & Kirk, 1968). It is not known whether either is important in glutamate synthesis *in vivo* in the light.

Figure 3 shows the metabolites from which the various amino acids are thought to be formed in higher plants and the routes of synthesis. These routes are based largely on conclusions drawn from work with bacteria (Umbarger, 1969). Isotope experiments using plant tissues have tended to confirm them in plants, but with some modifications. The flow of carbon along the pathways is controlled, so that excessive synthesis is prevented and the sites of protein synthesis are presented with a range of amino acids at suitable concentrations. Control mechanisms are probably similar to those in micro-organisms (Umbarger, 1969). The situation in plants has been examined by studying the effects of individual amino acids on the incorporation into amino acids of ^{14}C from either sugar or acetate by embryos (Joy & Folkes, 1965) and excised root tips (Oaks, 1965) in sterile culture, and from the effects of individual amino acids on the growth of

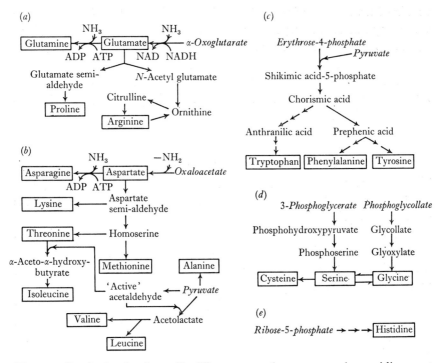

Figure 3. Synthesis of amino acids. The compound or compounds providing most of the carbon for each group of amino acids are *in italics*. The amino acid and amide endproducts important for protein synthesis are shown in boxes.

excised embryos (Miflin, 1969a). The growth of barley embryos in sterile culture with nitrate as the nitrogen source was slowed by the presence of any one of several amino acids, including leucine, valine and isoleucine. Valine and leucine when present together, but without isoleucine, inhibited growth more than would be expected from the effects of each separately. Isoleucine, which on its own inhibited growth less than either leucine or valine, relieved the inhibition produced by leucine and valine when all three were present in the medium. Miflin proposed that the observations resulted from the inhibition, by the endproducts, of an enzyme common to the synthesis of each. The enzyme in this instance is one considered to be involved at the branch point in the pathway for the synthesis of valine and leucine (Figure 3b) and necessary for the synthesis of acetohydroxybutyrate in the separate pathway of isoleucine synthesis from threonine (also Figure 3b). This hypothesis is supported by extraction of the enzyme acetolactate synthetase (also having acetohydroxybutyrate synthetase activity) from barley seedlings and showing that the enzyme *in vitro* (Miflin, 1969b) is also inhibited by leucine, valine and isoleucine in the manner predicted from the *in vivo* results. This type of control, which is common where a synthetic pathway is branched, is known as co-operative inhibition by endproducts. Where synthesis does not involve branching of the pathway simple endproduct inhibition will suffice. Evidence is accumulating for control systems operating by inhibition by endproducts in plants: these must be distinguished from the slower control by induced enzyme formation already described for nitrate reductase.

PROTEIN SYNTHESIS FROM AMINO ACIDS

The biochemical components required for protein synthesis in higher plants are thought to resemble those required in bacteria and animal tissues (Boulter, 1970). The amino acid sequence in each protein is determined by a nucleotide sequence in the DNA of a gene. mRNA is formed by transcription of the DNA code into the corresponding RNA code. The code in the mRNA is used then for translation into the amino acid sequence of the protein molecules. The protein synthesis step requires mRNA, ATP, ribosomes, GTP, transfer RNA (tRNA), amino acids, various enzymes and co-factors. An association of one or more ribosomes with the mRNA forms the main particulate component, a polysome. Amino acids are activated and attached to specific tRNA molecules before they can take part in forming peptide bonds at the site of a ribosome in the polysome complex. An amino acid with its specific tRNA can be used to form peptide bonds only when the ribosome is at a nucleotide triplet sequence on the mRNA strand requiring that amino acid to be added to the growing peptide chain. Following complete translation of the base sequence in the mRNA into an amino acid sequence, the ribosome is discharged from the end of the mRNA strand. The system in higher plants is complicated by the existence of DNA in chloroplasts and mitochondria, where there are also distinctive ribosomes and, possibly, tRNA species involved in the synthesis of some of the proteins in these organelles. Boulter (1970) points out that at the time of his review no *in vitro* system had been described, using a cell-free system from plants, in which indisputable net protein synthesis had occurred. Until such a system can be produced our knowledge of the mechanism of control of protein synthesis in higher plants must remain limited. However, various factors that influence the rate at which protein is synthesised *in vivo* can be indicated and can be connected with biochemical observations.

Protein synthesis is fastest in young cells that are actively growing. This has been recognised in several kinds of study including laboratory experiments with roots (Brown & Broadbent, 1950) and nutritional evaluation of pastures at different seasons of the year. As leaves age their protein content decreases. The relative rates of synthesis of different proteins change during the life of a cell and differ for cells in different tissues. Plant growth hormones may induce or restrict the synthesis of specific proteins when there is a change in the physiological state of cells. Changes in physiological state are accompanied by transient increases in growth hormones (Clark & Heath, 1959).

The synthesis of α-amylase and some other enzymes by the aleurone cells of cereal grain is stimulated by gibberellic acid. Glasziou (1969) lists endo-β-glucanase, proteases, α-amylase, ribonuclease and peroxidase as enzymes produced by the aleurone cells in response to gibberellic acid, and suggests that in this system gibberellic acid acts by removing repressor molecules from sites on both the DNA of the gene and on mRNA, thus controlling synthesis of both the specific mRNA and the protein for which it codes. A model of this sort would explain results obtained when either inhibitors of nucleic acid synthesis, or inhibitors of protein synthesis, are added to aleurone tissue.

Cytokinins also affect the rates at which some proteins are synthesised. Their mode of action may differ from that of gibberellic acid in that cytokinins are constituents of some tRNA molecules and are actually present in these molecules at the sites of binding to the polysome-mRNA complex. It seems (Skoog & Armstrong, 1970) that cytokinins act at the stage of translating mRNA into the amino acid sequence of proteins. Plant growth hormones, substrates such as nitrate that induce enzyme synthesis, and endproducts that repress enzyme synthesis, probably act on the proteinsynthesising system in various ways. In addition to those already mentioned, there are modifications of the stability of mRNA or enzyme protein in the tissue and mechanisms such as those described by the models of Jacob & Monod (1961) (see also Glasziou, 1969; Marcus, 1971).

CONCLUSIONS

Beevers & Hageman (1969) referred to the use of nitrate reductase activity as a criterion in breeding plants for larger protein contents. It seems clear that greater benefits might be obtained if the criteria used were modifications of the controls that limit synthesis of other specific proteins not necessarily involved in the early stages of nitrogen assimilation. This would ensure increased use of amino acids, thereby limiting repressive effects on nitrate reductase activity and endproduct inhibition of amino acid synthesis. It should be recognised that growth hormones and similar substances seem to interact directly with components of the protein-synthesising system; it is at this step that we may wish to change plant metabolism. Further, some species of plants can direct protein synthesis towards forming large quantities of storage protein in seeds. Such a protein, presumably having no enzyme activity that interferes with normal metabolism, and selected for a nutritionally satisfactory amino acid composition, should be the aim in attempts to increase protein production by genetic or other means.

Further experiments are necessary to define the pathway by which ammonia and ammonium ions are incorporated into amino acids in leaves. It has been pointed out that glutamate is not a major product of carbon assimilation in leaves and that the relative amount of carbon accumulated in glycine and serine during photosynthesis is readily modified by changing the atmospheric conditions. There is a need for more work with crop plants to define clearly the role of root and shoot in amino acid synthesis. It would be helpful to know more about diurnal changes in the concentrations of ammonium ions, organic acids, nitrate, amides and amino acids in the leaves and roots of crop plants under field conditions. The object of these researches should be to define more clearly the factors that limit protein synthesis in leaves.

REFERENCES

BACH, M. K., MAGEE, W. E., & BURRIS, R. H. (1958). Translocation of photosynthetic products to soyabean nodules and their role in nitrogen fixation. *Plant Physiology* **33**, 118–24.

BASSHAM, J. A., & KIRK, M. (1964). Photosynthesis of amino acids. *Biochimica et Biophysica Acta* **90**, 553–62.

BEEVERS, L., & HAGEMAN, R. H. (1969). Nitrate reduction in higher plants. *Annual Review of Plant Physiology* **20**, 495–522.

BERGERSEN, F. J. (1971). Biochemistry of symbiotic nitrogen fixation in legumes. *Annual Review of Plant Physiology* **22**, 121–40.

BOULTER, D. (1970). Protein synthesis in plants. *Annual Review of Plant Physiology* **21**, 91–114.

BROWN, R., & BROADBENT, D. (1950). The development of cells in the growing zones of the root. *Journal of Experimental Botany* **1**, 249–63.

CARR, D. J., & PATE, J. S. (1967). Ageing in the whole plant. *Symposia of the Society for Experimental Biology* **21**, 559–600.

CLARK, J. E., & HEATH, O. V. S. (1959). Auxins and the bulbing of onions. *Nature* **184**, 345–7.

COCKING, E. C., & YEMM, E. W. (1961). Synthesis of amino acids and proteins in barley seedlings. *New Phytologist* **60**, 103–16.

FERRARI, T. E., & VARNER, J. E. (1970). Control of nitrate reductase activity in barley aleurone layers. *Proceedings of the National Academy of Sciences, New York* **65**, 729–36.

FILNER, P. (1965). Regulation of nitrate reductase in cultured tobacco cells. *Biochimica et Biophysica Acta* **118**, 299–310.

GLASZIOU, K. T. (1969). Control of enzyme formation and inactivation in plants. *Annual Review of Plant Physiology* **20**, 63–88.

GRANT, B. R., ATKINS, C. A., & CANVIN, D. T. (1970). Intracellular location of nitrate reductase and nitrite reductase in spinach and sunflower leaves. *Planta* **94**, 60–72.

HAGEMAN, R. H., & FLESHER, D. (1960). Nitrate reductase activity in corn seedlings as affected by light and nitrate content of nutrient media. *Plant Physiology* **35**, 700–8.

HATCH, M. D., & SLACK, C. R. (1966). Photosynthesis by sugar cane leaves. A new carboxylation reaction and the pathway of sugar formation. *Biochemical Journal* **101**, 103–11.

HEBER, U. W., & SANTARIUS, K. A. (1965). Compartmentation and reduction of pyridine nucleotides in relation to photosynthesis. *Biochimica et Biophysica Acta* **109**, 390–408.

HEIMER, Y. M., & FILNER, P. (1970). Regulation of the nitrate assimilation pathway of cultured tobacco cells. II. Properties of a variant cell line. *Biochimica et Biophysica Acta* **215**, 152–65.

HELLEBUST, J. A., & BIDWELL, R. G. S. (1963). Sources of carbon for the synthesis of protein amino acids in attached photosynthesising wheat leaves. *Canadian Journal of Botany* **41**, 985–94.

HEWITT, E. J. (1966). *Sand and water culture methods used in the study of plant nutrition.* Technical Communication No. 22 (Revised 2nd edition), Commonwealth Bureau of Horticulture and Plantation Crops. Commonwealth Agricultural Bureau, Farnham Royal, Bucks.

INGLE, J. (1968). Nucleic acid and protein synthesis associated with the induction of nitrate reductase activity in radish cotyledons. *Biochemical Journal* **108**, 715–24.

JACKSON, W. A., & VOLK, R. J. (1970). Photorespiration. *Annual Review of Plant Physiology* **21**, 385–432.

JACOB, F., & MONOD, J. (1961). Genetic regulatory mechanisms in the synthesis of proteins. *Journal of Molecular Biology* **3**, 318–56.

JOY, K. W. (1969). Nitrogen metabolism of *Lemna minor*. I. Growth, nitrogen sources and amino acid inhibition. *Plant Physiology* **44**, 845–9.

JOY, K. W., & FOLKES, B. F. (1965). The uptake of amino acids and their incorporation into the proteins of excised barley embryos. *Journal of Experimental Botany* **16**, 646–66.

KANAZAWA, T., KANAZAWA, K., KIRK, M. R., & BASSHAM, J. A. (1970). Differences in nitrate reduction in 'light' and 'dark' stages of synchronously grown *Chlorella pyrenoidosa* and resultant metabolic changes. *Plant and Cell Physiology, Tokyo* **11**, 445–52.

KANNANGARA, C. G., & WOOLHOUSE, H. W. (1967). The role of carbon dioxide, light and nitrate in the synthesis and degradation of nitrate reductase in leaves of *Perilla frutescens*. *New Phytologist* **66**, 553–61.

KENNEDY, I. R. (1966). Primary products of symbiotic nitrogen fixation. II. Pulse labelling of *Serradella* nodules with ^{15}N. *Biochimica et Biophysica Acta* **130**, 295–303.

LEECH, R. M., & KIRK, P. R. (1968). An NADP-dependent L-glutamate dehydrogenase from chloroplasts of *Vicia faba*. *Biochemical and Biophysical Research Communications* **32**, 685–90.

LIPS, S. H. (1971). Photorespiration and nitrate reduction. *Second International Congress on Photosynthesis Research, Abstracts,* p. 121. Bologna, Co-operative Libraria Universitaria.

MARCUS, A. (1971). Enzyme induction in plants. *Annual Review of Plant Physiology* **22**, 313–36.

MIFLIN, B. J. (1969a). The inhibitory effects of various amino acids on the growth of barley seedlings. *Journal of Experimental Botany* **20**, 810–19.

MIFLIN, B. J. (1969b). Acetolactate synthetase from barley seedlings. *Phytochemistry* **8**, 2271–6.

MIFLIN, B. J. (1970). Studies on the sub-cellular location of particulate nitrate and nitrite reductase, glutamic dehydrogenase and other enzymes in barley roots. *Planta* **93**, 160–70.

MORRIS, I., & SYRETT, P. J. (1963). The development of nitrate reductase in *Chlorella* and its repression by ammonium ions. *Archiv für Mikrobiologie* **47**, 32–41.

NOGUCHI, M., & TAMAKI, E. (1962). Studies in nitrogen metabolism in tobacco plants. A. Part II. Diurnal variation in amino acid composition of tobacco leaves. *Archives of Biochemistry and Biophysics* **98**, 197–205.

NUTMAN, P. S. (1956). The influence of the legume in root nodule symbiosis. A comparative study of host determinants and functions. *Biological Reviews* **31**, 109–49.

OAKS, A. (1965). The effect of leucine on the biosynthesis of leucine in maize root tips. *Plant Physiology* **40**, 149–55.

OGREN, W. L., & KROGMANN, D. W. (1965). Studies on pyridine nucleotides in photosynthetic tissue. *Journal of Biological Chemistry* **240**, 4603–8.

ONGUN, A., & STOCKING, C. R. (1965). Effect of light on the incorporation of serine into the carbohydrates of chloroplasts and non-chloroplast fractions of tobacco leaves. *Plant Physiology* **40**, 819–24.

PATE, J. S. (1968). Physiological aspects of inorganic and intermediate nitrogen metabolism (with special reference to the legume *Pisum arvense* L.). In: Hewitt, E. J., & Cutting, C. V. (Eds.) *Recent Aspects of Nitrogen Metabolism in Plants*, 219–40. New York and London: Academic Press.

ROBERTS, G. R., KEYS, A. J., & WHITTINGHAM, C. P. (1970). The transport of photosynthetic products from the chloroplast of tobacco leaves. *Journal of Experimental Botany* **21**, 683–92.

SCHRADER, L. E., & HAGEMAN, R. H. (1967). Regulation of nitrate reductase activity in corn (*Zea mays* L.). *Plant Physiology* **42**, 1750–6.

SKOOG, F., & ARMSTRONG, D. J. (1970). Cytokinins. *Annual Review of Plant Physiology* **21**, 359–84.

STOCKING, C. R., & LARSON, S. (1969). A chloroplast cytoplasmic shuttle and the reduction of extraplastid NAD. *Biochemical and Biophysical Research Communications* **37**, 278–82.

TRAVIS, R. L., JORDAN, W. R., & HUFFAKER, R. C. (1970). Light and nitrate requirements for induction of nitrate reductase activity in *Hordeum vulgare*. *Physiologia Plantarum* **23**, 678–85.

UMBARGER, H. E. (1969). Regulation of amino acid metabolism. *Annual Review of Biochemistry* **38**, 323–70.

VIRTANEN, A. I., MOISIO, T., & BURRIS, R. H. (1955). Fixation of nitrogen by nodules excised from illuminated and darkened pea plants. *Acta Chemica Scandinavica* **9**, 184–6.

WALLACE, W., & PATE, J. S. (1965). Nitrate reductase in the field pea (*Pisum arvense* L.). *Annals of Botany* **29**, 655–71.

WEISSMAN, G. S. (1959). Influence of ammonium and nitrate on the protein and free amino acids in the shoots of wheat seedlings. *American Journal of Botany* **46**, 339–46.

WEISSMAN, G. S., & KEYS, A. J. (1969). Root growth with ammonium and nitrate in continuous flow nutrient solution. *Bulletin of the Torrey Botanical Club*, **96**, 149–55.

WHITTINGHAM, C. P., HILLER, R. G., & BERMINGHAM, M. (1963). The production of glycollate during photosynthesis. In: Committee on Photobiology. *Photosynthetic Mechanisms of Green Plants*, 675–83. Washington, DC: National Academy of Sciences–National Research Council, Publication 1145.

YEMM, E. W., & WILLIS, A. J. (1956). The respiration of barley plants. IX. The metabolism of roots during the assimilation of nitrogen. *New Phytologist* **55**, 229–52.

7

THE POTENTIAL OF CEREAL GRAIN CROPS FOR PROTEIN PRODUCTION

By R. N. H. WHITEHOUSE

Plant Breeding Institute, Cambridge

INTRODUCTION

Compared with some other crop products cereal grains tend to be low in protein content and are generally regarded as sources of metabolisable energy rather than of protein. It can be argued (Carpenter, 1970) that increasing the metabolisable energy concentration will more readily provide an economic gain than will improving the protein content or composition of cereals. However, Thielebein (1969) suggests that cereals provide 80 million tons of protein which is one half of the annual human requirement, although this is probably a good deal less than the total amount that the cereals actually produce. Thus even a small proportional increase could make a worthwhile contribution to world supplies. Furthermore the grain proteins of cereals have a low biological value and consequently at least 50 per cent more protein must be ingested by non-ruminant animals than would otherwise be necessary. The case for trying to improve cereal proteins, either by cultural or genetic means, rests on these two aspects, the large quantity and the low quality.

Precise and up-to-date figures for world protein production by cereals are not available but estimates can be made for the major crops: wheat, barley, oats, rye, rice, maize and sorghum. The yield figures given in Table 1 are necessarily very approximate and refer to the 1968 crop. The protein contents of all cereals vary widely according to growing conditions; consequently the use of a single average figure for each crop is subject to considerable error. Nevertheless the resultant figures for the production of cereal protein probably give a reasonable guide to the world situation. It will be seen that wheat is the major producer followed by maize and rice, with barley some further distance behind. If these estimates are correct then the total world production of protein by cereals is in the order of a hundred and twenty million metric tons per year. This is sufficient to supply a world population of 3000 million people with rather more than 100 g per day of protein each; unfortunately they do not all receive it.

The biological value of cereal grain protein varies from crop to crop and to a lesser extent from sample to sample because of differences in the nitrogen content of the grain. The value also depends on the nutritional requirements of the people or animals to whom it is fed. When assessing the relative merits of the various crops it is necessary to examine numerous samples of each grown in different environments and to compare these with the needs of the population consuming it. Considerable variation occurs in the reported proportions of amino acids in the cereal grain (Riley & Ewart, 1970); this stems from several causes, the natural variation between samples, the methods used for analysis and the way the results are commonly expressed in terms of total nitrogen content (g per 16 g nitrogen) rather than as a proportion of total amino acids.

Table 1. *World production of cereal grain and cereal protein* ($kg \times 10^9$)

	Wheat	Barley	Oats	Rye	Maize	Sorghum and millet	Paddy rice	All grain crops
Grain*	333	131	54	33	251	85	284	1180
Protein (% of fresh weight)†	12	10	10.5	12.5	10	11	8	10.2
Protein	40	13	6	4	25	9	23	120

* Food and Agriculture Organisation of the United Nations (1969).
† Adapted from Kent (1966).

The proportion of the essential amino acids for each of six grain crops are given in Table 2 and Figure 1. The values are taken from two published surveys for each crop and make no attempt to represent the full range of results which can occur; they serve as a reminder, however, that the variability of plant material must be taken into account and that many samples may have lower values for some amino acids than those quoted. The table also gives the composition of an ideal or reference protein for one of the most demanding classes of livestock, namely fast-growing chicks. Arginine and histidine, although not essential for human diets, are required by chicks and are therefore included in Table 2. The figures for methionine and cystine are combined, as also are those for phenylalanine and tyrosine, because although the first of each pair is essential the second can to some extent compensate for the deficiency of the first. The remaining amino acids are lysine, threonine, isoleucine, leucine, valine and tryptophan.

Comparing the reference protein with what is available in the crops shows that lysine is limiting in all the six crops quoted, namely rice, oats,

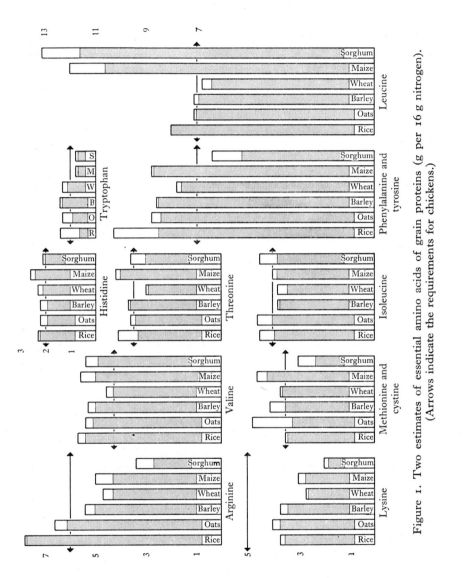

Figure 1. Two estimates of essential amino acids of grain proteins (g per 16 g nitrogen). (Arrows indicate the requirements for chickens.)

Table 2. *Approximate composition of grain proteins*
(*g per 16 g nitrogen*)

Amino acids essential in some diets	Ideal composition for chicks	Rice (1)	Rice (4)	Oats (1)	Oats (2)	Barley (1)	Barley (2)	Wheat (1)	Wheat (2)	Maize (1)	Maize (2)	Sorghum (1)	Sorghum (3)
Lysine	5.0	3.5	3.7	4.0	3.7	3.7	3.4	2.6	2.7	2.7	3.0	1.8	2.0
Methionine and cystine	3.5	3.4	3.5	4.8	3.2	4.1	3.5	3.6	3.7	4.6	4.2	3.0	2.3
Threonine	3.5	3.3	4.1	3.6	3.4	3.6	3.7	3.0	2.9	4.0	4.2	3.6	3.0
Isoleucine	4.0	4.5	3.9	4.0	4.6	3.7	3.8	3.4	3.8	3.8	4.0	4.5	3.8
Leucine	7.0	8.0	8.0	7.1	7.0	7.1	6.9	6.8	6.4	10.6	12.0	11.6	13.1
Valine	4.3	5.4	5.7	5.1	5.4	5.3	5.0	4.6	4.3	5.0	5.6	5.4	4.9
Phenylalanine and tyrosine	7.0	10.3	8.5	8.4	8.8	8.6	8.5	7.6	7.8	8.7	8.8	5.2	6.4
Tryptophan	1.0	0.6	1.4	0.9	1.3	1.3	1.4	1.1	1.3	0.7	0.8	0.8	0.7
Histidine	2.0	2.2	2.3	2.2	1.9	2.2	1.9	2.3	2.1	2.6	2.4	2.0	2.1
Arginine	6.0	7.8	7.7	6.1	6.6	5.4	5.0	4.7	4.3	4.3	5.0	3.4	2.7
Net protein utilisation by rats (%)		66		59		59		53		51		48	

(1) Data in first column for each crop from Eggum (1969).
(2) Data in second column for oats, barley, wheat and maize from Hughes (1960).
(3) Data in second column for sorghum (except for tryptophan) from Sykes (1970).
(4) Data in second column for brown rice (except for tryptophan) from Juliano, Bautista, Lugay & Reyes (1964).

barley, wheat, maize and sorghum. Oats, rice and barley have higher but still inadequate levels. Valine and histidine are unlikely to cause problems through shortage but each of the crops may be deficient in some other amino acid. Sorghum is outstanding because of its low level of methionine, phenylalanine, tryptophan and arginine. Maize is also short of tryptophan and arginine but both maize and sorghum have a large excess of leucine above requirement. Both these crops suffer from a considerable imbalance in their proteins and wheat is not much better having barely sufficient isoleucine, leucine, threonine, arginine and, of course, lysine. Rice, followed closely by oats, has the best spectrum with tryptophan and, possibly, methionine and threonine the only limiting amino acids apart from lysine. Barley is marginally worse particularly with regard to arginine.

Thus the gross deficiencies are for lysine in all these crops and for methionine, phenylalanine, arginine, threonine and tryptophan in certain crops, especially sorghum. The situation is summarised by figures for net protein utilisation in experiments with rats quoted at the bottom of Table 2; it will be seen that rice has the highest value at 66 per cent whereas sorghum is at the bottom of the list with 48 per cent. Thus compared with a completely digestible reference protein it would be necessary to feed 50 per cent more rice protein and 100 per cent more maize or sorghum protein in order to obtan the same food value. If the major defects in these proteins can be corrected then lesser ones which at present may be unimportant will become exposed.

RELATIONSHIPS OF YIELD WITH PROTEIN AND LYSINE CONTENTS

Before considering the possibilities of improving each crop it is necessary to be clear what are the component characters and how they are related, since selection for one character can nullify selection for another if they are negatively correlated. Total yield of grain per unit area of land is the most obvious measure of success or failure of any cereal crop. This is true whether it be for a peasant farmer or for one in an advanced agriculture whose crops are for sale in a market with no bonus for quality. The next character, protein content, is usually expressed as a percentage of the dry weight of the grain. Protein content is of direct concern to the users of the grain, for example, the manufacturers of compound animal feeding stuffs. This character usually refers to particular lots of grain such as the contents of the hold of a ship or of a silo without consideration of the number of acres which contributed to it. Thirdly, yield of protein per unit area of land is rarely of interest either to users or farmers; it is important to those concerned with national policies and the welfare of people dependent on the crops. Fourthly, the amount of lysine, or any other limiting amino acid, may be expressed in terms of production per unit area or as a percentage of the dry weight of the crop but is most conveniently represented as a percentage of the amount of protein in the crop.

Variations in the yield and in the protein content of the grain may be due either to the environmental conditions of the growing crop or to the variety grown, or both. There is a well established negative correlation between the yield of grain per unit area and its protein content. This applies to all the cereal crops and is tantamount to saying that conditions which raise the yield of grain increase the amount of carbohydrate more than they do the amount of protein, although the latter is normally higher when total yields are high. The highest yields of protein are thus often associated with low protein contents of the grain; this association can be demonstrated

(Whitehouse, 1969) when the variation is due either to changes in environment or to different genotypes. A common cause of high protein content is a restriction on growth, due to a shortage of water or some other adverse condition, during the later stages of grain filling. As a result there is a conflict between the need of farmers for high yield and the requirement of the nutritionist for high protein contents.

Correlations between grain yield and protein yield are usually positive. In a factorial experiment using barley varieties adapted to the area and imposed environmental treatments Whitehouse (1971) showed that this positive correlation was much stronger for genetic than for environmental differences. In contrast the correlations between grain yield and protein content, which were negative, were stronger for environmental comparisons. This shows that a change from a poor to a good environment results in a bigger increase in carbohydrate than in protein production whereas a change from a poor to a good variety produces similar increases in both components. Thus, if high yields of protein are required it would seem better to set about obtaining these through breeding rather than by cultural means.

A further correlation which is generally observed in cereal crops is a negative one between protein content of the grain and the amount of lysine in the protein (Hischke, Potter & Graham, 1968). This means that the increases in the amount of protein, particularly by cultural means, are not reflected by comparable increases in the amount of lysine (Whitehouse, 1970). Cereal albumins and globulins are relatively rich in lysine but occur in only small quantities in the grain (Swaminathan, Naik, Kaul & Austin, 1970; Sykes, 1970; Juliano, 1971); the glutelins have an intermediate amount and the prolamins very little lysine. The storage proteins in most of the cereals are prolamins, such as hordein, zein and kafirin; consequently as the amount of protein increases the proportion of lysine decreases. Table 3 indicates the proportions of these four fractions which may be found in cereal grain proteins and the proportions of lysine in each fraction. For some crops information is rather scarce and these figures should be taken for what they are, individual samples not necessarily representative of the crop as a whole. Rice is distinct from the other cereals in having a high proportion of glutelin, which must be largely responsible for the high biological value of its proteins.

Amongst some of the other crops, genotypes with improved lysine concentration have been obtained by an increase in glutelin at the expense of prolamin. Increasing the albumin content may prove difficult as much of this protein fraction is enzymic but raising the proportions of globulins and, to a lesser extent, glutelins would be valuable.

Table 3. *Protein fractions and their lysine content in some cereal grains*

| Protein fraction ... | Albumin | | Globulin | | Glutelin | | Prolamin | |
| Soluble in ... | Water | | Salt | | Alkali | | Alcohol | |
	(a)	(b)	(a)	(b)	(a)	(b)	(a)	(b)
Rice	5	4.9	10	2.6	80	3.5	5	0.5
Oats	1	—	78	—	5	—	16	—
Barley (normal)	13	7.9	12	6.3	23	4.8	52	0.8
Barley (Hiproly)	18	8.2	14	6.1	22	4.6	46	0.5
Wheat	5	—	10	—	16	1.9	69	0.6
Maize (normal)	4	3.8	2	6.1	39	3.4	55	0.2
Maize (*opaque*-2)	15	4.1	5	5.2	55	4.7	25	0.1
Sorghum (normal)	8	4.5	8	4.6	32	2.7	52	0.5
Sorghum (160-Cernum)	6	—	10	—	38	—	46	—

(a) Protein fraction as % of total protein; (b) Lysine as % of protein fraction. Data from: Bonner (1952), Breeze-Jones (1926), Brohult & Sandegren (1954), Folkes & Yemm (1956), Hischke et al. (1968), Jiménez (1966), Juliano (1971), Munck, Karlsson & Hagberg (1971), Nelson (1969), Skoch et al. (1970).

The structural proteins of a leaf in, for example, barley are relatively rich in lysine. Consequently during germination, breakdown and resynthesis of amino acids occur as storage proteins are converted to leaf proteins (Folkes & Yemm, 1956; Jones & Pierce, 1967). This has led to a suggestion that germinated grain, either from the malting process or deliberately produced, might have a higher nutritional value than raw grain. Samples produced at the Plant Breeding Institute by germinating two varieties of barley in the light for three different periods showed an increase in the lysine content of 25–40 per cent (Whitmore, 1971, personal communication), but when fed to chicks at the Rowett Research Institute there was little or no gain in nutritional value (Woodham, 1971, personal communication). Whether this is due to some other amino acid becoming limiting or to the production of an antimetabolite is still being investigated. It does indicate, however, the danger of relying on chemical tests when evaluating the potential of new cereal varieties.

IMPROVEMENT OF NUTRITIONAL VALUE

Attempts to improve the protein status of cereal crops may come about through changes in the cultural conditions of the crop, particularly manipulation of nitrogen and water supplies, or by breeding genotypes which have superior properties. The cultural methods for the cereal crops concerned are much too diverse to review here but the position can be summarised by saying that water and nitrogen supplied to the soil usually increase total yield of both grain and protein; nitrogen supplied late in the life of the crop, although difficult to apply, often results in an increase in the protein content of the grain but the extra protein is of low biological value. Probably the best cultural solution to the protein problem is, therefore, to grow the crops in such a way that the maximum yield of grain is obtained. This will also give the maximum yield of protein per unit area of land but in such a form that the protein content will be low although its biological value will be relatively high. It will then be up to the processor to put this extra protein to the best possible use.

Improvements brought about by breeding may be of two kinds. Great advances have been made in all the major cereal crops with regard to the characters which enable the farmer better to exploit the environment. Straw is stronger, disease resistance is greater and harvest losses are reduced by having earlier maturity. These and other characters allow increased yields and are already making a major contribution to protein production by cereal crops. There is no sign of this process either stopping or even slowing down at present although there must be some ultimate limit to yield. These indirect methods have only recently been supplemented by deliberate attempts to raise the content and biological value of protein in the cereal grains. Attempts to locate varieties which are capable of producing larger amounts of protein per unit area of land have been frustrated by the enormous environmental effect which conceals the genetic contribution. Similarly a belief that cereal protein composition was immutable discouraged the lengthy undertaking of growing thousands of varieties and determining their complete amino acid profiles. Progress has, however, been made and this will be surveyed for six of the most important cereal crops.

It is necessary to distinguish between those 'high-protein varieties' which are efficient at protein production and those which are inefficient at carbohydrate production, since the latter also have high protein contents (percentage of dry weight). There have been numerous claims for the production of 'high-protein varieties' by the application of mutagens to cereals; for example, Tanaka (1969) isolated rice mutants with increased protein contents but decreased grain size as a result of gamma irradiation. Unless these claims are supported by evidence that grain yields have been maintained there is no guarantee that the mutants would contribute to increased protein supplies.

SURVEY OF BREEDING WORK WITH CEREAL CROPS

Rice

During the milling of rice the aleurone and embryo are removed, thus reducing the dry weight by about 10 per cent, but the loss of both protein and lysine is relatively greater; even so rice remains the cereal with the highest biological value so that the efforts to improve its protein content are particularly worthwhile, especially since the normal levels are low even by cereal standards. Protein contents of milled rice vary greatly with growing conditions but 8 per cent (Juliano, Albano & Cagampang, 1964) and 9 per cent (Simpson et al. 1965) have been reported as mean values of commercial crops on a dry weight basis. In a survey (Juliano, 1971), however, of 7760 varieties in the International Rice Research Institute collection the mean protein content was about 12 per cent of dry weight with some varieties reaching 16 per cent. This may merely mean that these varieties were not well adapted to conditions at the test site and consequently gave low yields. Other data presented by Juliano (1971) also support this view. He found that thirty-five Asian varieties grown in their country of origin had protein contents ranging from 6.4 per cent to 10.8 per cent while the same varieties grown at the International Rice Research Institute in the Philippines, to which area some were presumably less well adapted, had protein contents ranging from 9.0 per cent to 13.9 per cent.

On the assumption that some of the variation observed can be ascribed to genes for high protein production, work has been started in an attempt to raise the protein level amongst high-yielding varieties by hybridisation with varieties such as IR8. Little attention has been paid in the past to this character so it is possible that considerable improvement will be made. About 80 per cent of rice protein consists of glutelin, a protein relatively rich in lysine compared with the prolamins which form the main storage protein of most of the other cereals. Varietal surveys are being undertaken at the International Rice Research Institute and also at the Indian Agricultural Research Institute with a view to isolating genotypes which are richer in lysine. Preliminary reports suggest that some variation occurs but it is too early yet to be sure that these will be confirmed. With work recently started in several places and the knowledge that success has been achieved in other crops, there is no reason to feel pessimistic about lack of positive results at this stage.

Sorghum

This crop starts with the disadvantage of having a low biological value resulting in a net protein utilisation of 48 per cent compared with 66 per cent for rice (Table 2). It is deficient in several essential amino acids, notably lysine and arginine, and, like maize, contains large amounts of leucine. Protein levels are quoted as having mean values of 11–12 per cent of the dry weight but the ranges are, as usual, large and depend on growing conditions. These protein figures are not unduly low for cereals so the amino acid composition is evidently the first nutritional character to which attention should be given. Surveys are now being undertaken in several countries but little information is yet available. Virupaksha & Sastry (1968) report that an Indian variety, 160-Cernum, has a markedly high level of lysine (Table 4) and high protein contents. If this characteristic could be combined with reasonable levels of yield a major advance would be made since the improvement does not seem to be at the expense of other essential amino acids. Indeed, arginine concentration is increased in 160-Cernum and Swaminathan (1971) reports that when lysine content is increased leucine content is reduced. This may be significant since it has been suggested that high values of the leucine to isoleucine ratio in sorghum are responsible for the incidence of pellagra.

Maize

The improvement of biological value has proceeded furthest in maize and it was the discovery of the high lysine content associated with the gene *opaque*-2 by Mertz, Bates & Nelson (1964) which stimulated much of the work with other cereal crops. Maize like sorghum is seriously deficient in several essential amino acids including lysine and tryptophan and has an excess of leucine which is sufficient to cause nutritional disturbances. Protein levels normally have means about 10–12 per cent but selection can alter this very greatly. After sixty-five generations of selection for high and low protein contents the values were 25 per cent and 4 per cent (Alexander, Lambert & Dudley, 1969) but the high protein selections were low in lysine, gave low yields and so were of little practical value.

Two genes affecting the biological value of the protein in maize are known; these are *opaque*-2 which raises the levels of lysine and tryptophan and *floury*-2 which results in a smaller increase in lysine but also raises the level of methionine (Cline & Krider, 1968). The increase in tryptophan is valuable since it compensates for a shortage of niacin, the pellagra-preventing factor. The *opaque*-2 gene is the more useful, being responsible for a 50 per cent rise in the lysine level (Table 4), and although the two genes can be combined there seems to be little benefit from doing so except for an improvement in grain texture. Unfortunately *opaque*-2 has some deleterious effects resulting in lower yields, softer and more moist grains which crack readily and seedlings which lack vigour. Breeders have, however, made good progress and yields of high lysine selections are now approaching those of normal types (Alexander, Dudley & Lambert, 1971), although some of the other problems remain. Breeding trials cannot be carried out with the best material until adequate stocks are available but Jensen (1968) showed that pigs made similar liveweight gains per day with similar feed conversion ratios when fed either a 12 per cent protein ration of normal maize and soya bean meal or a 9.5 per cent protein ration consisting of *opaque*-2 maize meal plus a supplement of 0.3 per cent lysine. The *opaque*-2 stock thus provided a gain equivalent to an extra 2.5 per cent of protein.

Table 4. *Amino acid composition of selected genotypes of sorghum, maize and barley (g per 16 g nitrogen)*

	Sorghum[1]		Maize[2]		Barley[3]	
	Normal[a]	160-Cernum	Normal	*opaque*-2	Normal[c]	High-lysine plants[c]
Lysine	1.9	3.1	3.0	5.0	3.4	4.2
Methionine and cystine	2.4	—	3.0[b]	4.2	2.3	2.4
Threonine	3.3	3.6	4.1	3.8	3.4	3.5
Isoleucine	4.8	3.8	4.2	3.4	3.7	3.8
Leucine	16.9	12.7	14.6	9.3	6.7	7.1
Valine	5.4	4.2	5.7	5.2	4.8	5.3
Phenylalanine and tyrosine	7.4	6.8	11.0	8.6	8.7	8.7
Tryptophan	—	—	0.7	1.3	—	—
Histidine	1.9	2.3	2.6	3.5	2.1	2.1
Arginine	3.3	4.9	4.9	7.2	4.6	4.8

[a] Mean of 8 Indian varieties.
[b] Low value possibly due to experimental error.
[c] Means of 35 normal and 27 high-lysine plants.
[1] Virupaksha & Sastry (1968).
[2] Nelson (1969).
[3] Munck (1971).

It has aready been demonstrated that children suffering from the protein deficiency disease, kwashiorkor, in Colombia can be cured by using high-lysine maize as the only source of protein which for human nutrition is said to be equivalent to skim milk (Pradilla & Harpstead, 1968). It is a considerable achievement when a crop can be changed by the use of a single gene from being the cause to becoming the cure for a deficiency disease of this kind.

Oats

Although the world production of oats is relatively low, and diminishing, this crop has the highest biological value of the temperate cereals except for rye and it exceeds rye considerably in its digestibility. The margin over barley is small but the lysine content is usually higher. Protein contents of the grain in its natural condition, that is with husk attached, are above average for cereals but when the husk is removed the protein content of the kernel may be 15 per cent or more of the dry weight. Some low-yielding wild oats (*Avena sterilis*) which have yet higher protein levels have been used in breeding work but the high-protein selections so far produced are low-yielding.

Oat varieties exist which shed their husk during threshing. There are some technical difficulties in their cultivation (Jenkins, 1968) but if these can be overcome such 'naked' varieties would provide a highly nutritious product containing high levels of protein with moderate lysine content. Little information is available concerning variations in amino acid content of oat grains but surveys are currently being undertaken to find out if there are differences which can be advantageously exploited. In one such survey Robbins, Pomeranz & Briggle (1971) tested single samples of 289 varieties and found some with unusually high contents of lysine even at protein levels above 15 per cent.

Wheat

The world acreage of wheat greatly exceeds that for any other cereal and although lower-yielding than maize it also heads the lists for total world production of cereals and for production of protein. In low-yielding areas its protein content may be very high, 16 per cent of dry weight, but in Western Europe where yields are high the protein content is usually much lower, about 10 per cent. Even so it makes a very considerable contribution to world protein supplies.

The possibilities of raising both the protein level and its biological value are being actively pursued in several places. Johnson, Whited, Mattern & Schmidt (1968) and Johnson, Mattern, Whited & Schmidt (1969) have shown that high protein content in some wheats has a high heritability in contrast to the situation observed in most of the cereals. They have manipulated this character so as to select plants which combine high protein content with high yield. Five of their selections produced a mean of 75 per cent more protein per unit area than did their low protein content parents. It has yet to be shown whether this genetic system will function in areas where high yields and low protein levels are normal.

The high protein levels are said not to be associated with the changes in amino acid balance commonly experienced and the lysine and methionine levels are maintained although threonine is reduced. Johnson *et al.* (1969) also report that there is sufficient variation in lysine content, after allowing for differences in protein, to provide a basis for trying to breed for improved lysine levels. Johnson, Mattern & Schmidt (1970) found that three old Australian varieties figured in the parentage of six out of eight high-lysine varieties chosen from amongst 9000 tested.

Wheat occupies a very special position among crop plants because of the advanced state of cytogenetic work with it. A range of new genotypes involving alien chromosomes from other genera is available for study; in particular, *Triticale*, the amphidiploid with rye, is now beginning to be grown as a crop. *Triticales* were first made from hexaploid wheats (*Triticum aestivum*) but more recently the tetraploid, *Triticum durum*, has been used with promising results. Villegas, McDonald & Gilles (1968, 1970) showed that rye proteins usually contained more lysine than did wheat proteins and that *Triticales* were intermediate. Technical difficulties have still to be overcome but this new crop may make a contribution to nutrition through rather higher protein levels than wheat, although Larter, Tsuchiya & Evans (1968) report that this has tended to decline as the performance of *Triticales* has improved. Very high protein levels in early reports should therefore be treated cautiously. The mean lysine and threonine levels of six *Triticales* were rather higher than those of wheat and similar to those of barley but methionine was lower. It is not yet clear, however, that these are essential characteristics of *Triticale*. The effects of individual rye chromosomes has been investigated by Riley & Ewart (1970) who found that three of the rye chromosomes when added separately to wheat increased the lysine content of the protein whereas the *Triticale* was similar in this respect to wheat. Variations in the relative lysine levels in wheat and *Triticale* will certainly depend on their protein contents and may be affected by the particular wheat-rye combinations used. When the genetic control of lysine content is better understood it may be possible to utilise the relatively high levels found in rye. The diploid wild wheat, *Triticum boeoticum*, also has high contents of lysine in the protein even at very high protein levels (Villegas *et al.*, 1970) and might provide another starting point for a breeding programme.

It is evident, then, that wheat is already responding to selection for improved nutritive value and that the flexibility of genetic control, provided by intergeneric hybrids and chromosome manipulation, is likely to result in yet further improvement.

Barley

Although the world production is relatively small, barley is important in areas where animals provide a major part of the protein ration and, like maize, it needs to be considered particularly in relation to the requirements of animal nutrition. It is superior to normal maize in its lysine content and even high-lysine maize is only slightly superior. The protein levels, as with other crops, are largely determined by environmental conditions and are often low because samples with less than 10 per cent protein are preferred for malting. These levels can be raised substantially when nitrogen is available to the crop late in its life. Surveys of the grain protein content of varieties grown under comparable conditions show a large range of values but it is difficult to determine how far these differences are under genetic control. Zoshke (1971) reported that differences in protein content were due to varying amounts of glutelin and prolamin while albumins and globulins remained more or less constant. He found considerable variation in protein content between varieties but it has yet to be shown whether this is due to difference in yield or to some other genetic factor. Breeding programmes for high protein production have been started in various places but results are not yet available.

The genotype with the most clearly defined nutritional advantages was located by Hagberg & Karlsson (1969) and named Hiproly. It originated in Ethiopia and contains larger amounts of lysine than normal varieties at a given protein level (Table 4). Although this variety suffers from numerous defects it is possible to transfer the lysine character to new selections by hybridisation. The increase in lysine content (Munck, Karlsson & Hagberg, 1971) is about 30 per cent, which is less than is obtained in maize using *opaque*-2 but sufficient to make high-lysine barley superior to high-lysine maize. Improved varieties are not yet available but work is in hand in many places and it is presumably only a matter of time until commercial high-lysine varieties are being grown.

CONCLUSION

Attempts to improve the composition of cereal grains with regard to their protein content and amino acid balance are relatively new compared with work on other characters. Changes in other quality components, for example bread-making quality in wheat and malting quality in barley, have been very successful even before the nature of the quality was well understood. The prerequisites for advances by breeding are a source of genetic variability and a means of recognising the desired phenotype. Searches for genotypes associated with grain of better nutritional value have already been successful in several crops but, although the total effort is as yet small, considerable progress has been made in combining the nutritional components with other characteristics essential for a modern agricultural variety. It seems probable that cereal breeders will be able to contribute significantly to world protein supplies; it is to be hoped that the world will be able to utilise the improvements.

REFERENCES

ALEXANDER, D. E., DUDLEY, J. W., & LAMBERT, R. J. (1971). The modification of protein quality of maize by breeding. In: Kovacs, I. (Ed.), *Proceedings of the Fifth Meeting of the Maize and Sorghum Section of Eucarpia*, 33–43. Budapest: Akademiai Kiado.

ALEXANDER, D. E., LAMBERT, R. J., & DUDLEY, J. W. (1969). Breeding problems and potentials of modified protein maize. In: Joint FAO/IAEA Division of Atomic Energy in Food and Agriculture. *New Approaches to Breeding for Improved Plant Protein. Panel Proceedings Series–STI/PUB/212*, 55–65. Vienna: International Atomic Energy Agency.

BONNER, J. (1952). *Plant Biochemistry*, 245–82. New York and London: Academic Press.

BREEZE-JONES, J. D. (1926). A new factor for converting the percentage of nitrogen into that of protein. *Cereal Chemistry* **3**, 194–8.

BROHULT, S., & SANDEGREN, E. (1954). Seed proteins. In: Newath, H., & Bailey, K. (Eds.), *The Proteins*, Vol. II, Part A, 487–512.

CARPENTER, K. J. (1970). Nutritional considerations in attempts to change the chemical composition of crops. *Proceedings of the Nutrition Society* **29**, 3–12.

CLINE, T. R., & KRIDER, J. L. (1968). Report of the Maryland Nutrition Conference. Quoted in *Trouw & Co. N.V. Bulletin* 1968, 205–6.

EGGUM, B. O. (1969). Evaluation of protein quality and the development of screening techniques. In: Joint FAO/IAEA Division of Atomic Energy in Food and Agriculture. *New Approaches to Breeding for Improved Plant Protein. Panel Proceedings Series–STI/PUB/212*, 125–35. Vienna: International Atomic Energy Agency.

FOLKES, B. F., & YEMM, E. W. (1956). The amino acid content of the proteins of barley grains. *Biochemical Journal* **62**, 4–11.

FOOD AND AGRICULTURE ORGANISATION OF THE UNITED NATIONS (1969). *Production Yearbook* **23**. Rome: FAO.

HAGBERG, G. A., & KARLSSON, K. E. (1969). Breeding for high protein content and quality in barley. In: Joint FAO/IAEA Division of Atomic Energy in Food and Agriculture. *New Approaches to Breeding for Improved Plant Protein. Panel Proceedings Series–STI/PUB/212*, 17–21. Vienna: International Atomic Energy Agency.

HISCHKE, H. H., POTTER, G. C., & GRAHAM, W. R. (1968). Nutritive value of oat protein. 1. Varietal differences as measured by amino acid analysis and rat growth responses. *Cereal Chemistry* **45**, 374–8.

HUGHES, B. P. (1960). *The Composition of Foods*. Medical Research Council Special Report Series No. 297. London: HMSO.

JENKINS, G. (1968). Naked oats. *NAAS Quarterly Review* **79**, 120–6.

JENSEN, A. H. (1968). Report of the Maryland Nutrition Conference. Quoted in *Trouw & Co. N.V. Bulletin* 1968, 206–7.
JIMÉNEZ, J. R. (1966). Protein fractionation studies of high-lysine corn. In: Mertz, E. T., & NELSON, O. E. (Eds.), *Proceedings of High-Lysine Corn Conference*, 74–9. Washington, DC: Corn Industries Research Foundation.
JOHNSON, V. A., MATTERN, P. J., & SCHMIDT, J. W. (1970). The breeding of wheat and maize with improved nutritional value. *Proceedings of the Nutrition Society* **29**, 20–31.
JOHNSON, V. A., MATTERN, P. J., WHITED, D. A., & SCHMIDT, J. W. (1969). Breeding for high protein content and quality in wheat. In: Joint FAO/IAEA Division of Atomic Energy in Food and Agriculture. *New Approaches to Breeding for Improved Plant Protein. Panel Proceedings Series–STI/PUB/212*, 29–40. Vienna: International Atomic Energy Agency.
JOHNSON, V. A., WHITED, D. A., MATTERN, P. J., & SCHMIDT, J. W. (1968). Nutritional improvement of wheat by breeding. *Proceedings of the Third International Wheat Genetics Symposium*, 457–61. Canberra: Australian Academy of Science.
JONES, M., & PIERCE, J. S. (1967). The role of proline in the amino acid metabolism of germinating barley. *Journal of the Institute of Brewing* **73**, 577–83.
JULIANO, B. O. (1971). Studies on protein quality and quantity of rice. Personal communication.
JULIANO, B. O., ALBANO, E. L., & CAGAMPANG, G. B. (1964). Variability in protein content, amylose content and alkali digestibility of rice varieties in Asia. *Philippine Agriculturist* **48**, 234–41.
JULIANO, B. O., BAUTISTA, G. M., LUGAY, J. C., & REYES, A. C. (1964). Rice quality studies on physiochemical properties of rice. *Journal of Agricultural and Food Chemistry* **12**, 131–8.
KENT, N. L. (1966). *Technology of Cereals, with Special Reference to Wheat.* London: Pergamon Press.
LARTER, E., TSUCHIYA, T., & EVANS, L. (1968). Breeding and cytology of *Triticale. Proceedings of the Third International Wheat Genetics Symposium*, 213–21. Canberra: Australian Academy of Science.
MERTZ, E. T., BATES, L. S., & NELSON, O. E. (1964). Mutant gene that changes protein composition and increases lysine content of maize endosperm. *Science* **145**, 279–80.
MUNCK, L. (1971). Aspects of the physiology of protein formation in cereal grains and its importance as a breeding objective. *Proceedings of Meeting of Sections Cereals and Physiology of Eucarpia, 20–22 October 1970*, 283–94. Dijon: INRA, Station d'Amélioration des Plantes.
MUNCK, L., KARLSSON, K. E., & HAGBERG, A. (1971). Selection and characterisation of a high-protein, high-lysine variety from the world barley collection. *Proceedings of the Second International Barley Genetics Symposium*, 544–58. Pullman: Washington State University Press.
NELSON, O. E. (1969). The modification by mutation of protein quality in maize. In: Joint FAO/IAEA Division of Atomic Energy in Food and Agriculture. *New Approaches to Breeding for Improved Plant Protein. Panel Proceedings Series–STI/PUB/212*, 41–54. Vienna: International Atomic Energy Agency.
PRADILLA, A., & HARPSTEAD, D. (1968). Unpublished data from Department of Paediatrics, Universidad del Volle, Cali, Colombia.
RILEY, R., & EWART, J. A. D. (1970). The effect of individual rye chromosomes on the amino acid content of wheat grains. *Genetical Research* **15**, 209–19.
ROBBINS, G. S., POMERANZ, Y., & BRIGGLE, L. W. (1971). Amino acid composition of oat groats. *Journal of Agricultural and Food Chemistry* **19**, 536–9.

SIMPSON, J. E., ADAIR, C. R., KOHLER, G. O., DAWSON, E. H., DEOBALD, H. J., KESTER, E. B., HOGAN, J. T., BATCHER, O. M., & HALICK, J. V. (1965). *Quality Evaluation Studies of Foreign and Domestic Rices.* USDA Agricultural Research Service Technical Bulletin No. 1331.
SKOCH, L. V., DEYOE, C. W., SHOUP, F. K., BATHURST, J., & LIANG, D. (1970). Protein fractionation of sorghum grain. *Cereal Chemistry* **47**, 472–81.
SWAMINATHAN, M. S. (1971). Genetic upgrading of nutritional quality in food plants. Personal communication.
SWAMINATHAN, M. S., NAIK, M. S., KAUL, A. K., & AUSTIN, A. (1970). Choice of strategy for the genetic upgrading of protein properties in cereals, millets and pulses. In: Joint IAEA/FAO. *Improving Plant Protein by Nuclear Techniques.* Proceedings Series–STI/PUB/258, 165–183. Vienna: International Atomic Energy Agency.
SYKES, A. H. (1970). *Grain Sorghum in Poultry Nutrition.* Technical publication. US Feed Grains Council, 28 Mount Street, London W1Y 5RB, England.
TANAKA, S. (1969). Some useful mutations induced by gamma irradiation in rice. In: Joint IAEA/FAO. *Induced Mutations in Plants. Proceedings Series–STI/PUB/231*, 517–27. Vienna: International Atomic Energy Agency.
THIELEBEIN, M. (1969). The world's protein situation and crop improvement. In: Joint FAO/IAEA Division of Atomic Energy in Food and Agriculture. *New Approaches to Breeding for Improved Plant Protein. Panel Proceedings Series–STI/PUB/212*, 3–6. Vienna: International Atomic Energy Agency.
VILLEGAS, E., MCDONALD, C. E., & GILLES, K. A. (1968). *Variability in the Lysine Content of Wheat, Rye and Triticale Proteins.* Research Bulletin No. 10, International Maize and Wheat Improvement Center, Mexico.
VILLEGAS, E., MCDONALD, C. E., & GILLES, K. A. (1970). Variability in the lysine content of wheat, rye and *Triticale* proteins. *Cereal Chemistry* **47**, 746–57.
VIRUPAKSHA, T. K., & SASTRY, L. V. S. (1968). Studies on the protein content and amino acid composition of some varieties of grain sorghum. *Journal of Agricultural and Food Chemistry* **16**, 199–203.
WHITEHOUSE, R. N. H. (1969). Barley breeding at Cambridge. *Report of the Plant Breeding Institute, Cambridge*, 1969. 6–29.
WHITEHOUSE, R. N. H. (1970). The prospects of breeding barley, wheat and oats to meet special requirements in human and animal nutrition. *Proceedings of the Nutrition Society* **29**, 31–9.
WHITEHOUSE, R. N. H. (1971). Variation in protein and amino acid levels in barley. *Proceedings of Meeting of Sections Cereals and Physiology of Eucarpia, 20–22 October 1970*, 307–16. Dijon: INRA, Station d'Amélioration des Plantes.
ZOSHKE, M. (1971). Effect of additional nitrogen nutrition, at later stages of growth, on protein content and protein quality in barley. *Proceedings of Meeting of Sections Cereals and Physiology of Eucarpia, 20–22 October 1970*, 317–18. Dijon: INRA, Station d'Amélioration des Plantes.

8

PLANTS AS SOURCES OF UNCONVENTIONAL PROTEIN FOODS

By N. W. PIRIE

Rothamsted Experimental Station, Harpenden, Herts.

We are confronted in this symposium with two words of uncertain meaning – *efficiency* in the general title, and *unconventional* in the theme assigned to me. Efficiency is not the same as rate; it should be expressed either as the ratio between experimentally observed alternatives, or as the ratio between what is achieved in practice and the maximum attainable according to currently accepted theory. Thus an ideal heat engine, working between fixed source and sump temperatures, may approach 100 per cent of the theoretical efficiency even although it is only getting mechanical work out of 40 per cent of the heat that passed through it. The practically important figure is the percentage by which any actual engine falls short of the performance of an ideal engine. In the present context, the relevant quantities are (1) the rate at which a crop accumulates dry matter compared with the theoretical maximum for photosynthesis in the conditions of the experiment, (2) the percentage of protein in that dry matter when different crops are used, and hence the protein yield in tons per hectare per year, and (3) the percentage of that protein that can be made into a form suitable for use as human food when the crop is treated in different ways.

Any food can be considered unconventional if it is being eaten by a community that has not hitherto eaten it, or if more of it is being eaten than is traditional. A few of the foods that will be considered are not at present eaten by anyone anywhere – the only unconventionality about the remainder is quantitative or geographical.

The potentialities of a piece of land depend on its characteristics. These can often be altered, e.g. by clearing stones off it into dykes in Scotland, by terracing, or by draining East Anglian fens. When we are unable or unwilling to alter the terrain it may have to be used for grazing which is not inefficient, in the strict sense, because no alternative use can at the moment be envisaged; steep, rough, or stony land is either used for grazing or it is not used for food production. Efficiency does however have a meaning when different grazing species or mixtures of species are compared, or different ages at slaughter. These points are made here lest it be thought that advocacy of plant proteins implies a general disparagement of animals, rather than the more limited disparagement of animals maintained on the product of arable land.

Marshes, shallow water and land that is seasonally flooded could, in principle, be drained and turned into arable land. When this is either impossible or inexpedient, the vegetation growing on these sites should not be wasted. There is growing awareness of the disadvantages of trying to control the growth with herbicides; it should instead be harvested and used, perhaps after pretreatment, as fodder, or used as the starting material for protein separation by the methods that will be outlined later. Apart from minor problems in collecting the material, there is no essential difference between the use of water and land plants. These points have been gone into in greater detail elsewhere (Boyd, 1971; Pirie, 1970).

PLANT PRODUCTS EDIBLE AFTER MINIMAL PRETREATMENT

In the present context, attention may be confined to plants grown on arable land, and in this connection the parts worthy of consideration are leaves, seeds, tubers, roots and other underground parts. The simplest policy is to grow plants of such conformation that part can be eaten directly with no more pretreatment than is involved in such traditional procedures as threshing, peeling and cooking. Somewhat arbitrarily, we may restrict the category 'protein foods' to foods containing more than 15 per cent protein. That figure is chosen because even the most ungenerous committees guess that people need 8–13 per cent protein in the dry matter (DM) of the diet – the smaller figure is for adult men and the larger for those who are pregnant, lactating or growing. If a diet is to be made palatable by including some fat and sugar, the part that is categorised as a protein food must contain a substantially greater percentage of protein than the minimum, and the organisation of a satisfactory diet will become easier the greater the percentage of protein in the protein food. Furthermore, although nitrogen balance can be maintained on diets with the meagre protein contents officially thought adequate, there is no evidence that a state of nitrogen balance is the same thing as optimal nutrition. It is therefore not unreasonable to restrict the term 'protein foods' to foods that contain more than 15 per cent protein in the DM.

Underground parts

Apart from the groundnuts, which are more conveniently dealt with among the seeds, no plants are known that yield an edible underground part containing 15 per cent protein in the DM. Thus an outstanding potato crop can yield a ton of protein per hectare, but it makes up only 8 per cent of the DM of the tubers. Some of the yams contain 2.5 per cent nitrogen (N) in the DM, i.e. more than 15 per cent protein if the N is all present as protein. This has not been demonstrated; the point merits detailed investigation.

Seeds

Although the cereal seeds supply more than half the protein eaten by the inhabitants of well-fed countries, and a still greater proportion in ill-fed countries, the seeds of conventional strains contain little protein. Furthermore, the mixture of proteins that they contain is not as favourable nutritionally as protein from most animal sources. The recent increase in interest in the amount and quality of the protein in food led quickly to the recognition that some strains of maize, which had been unesteemed for thirty years, were better sources of lysine than the conventional strains. Many institutes are now searching for cereal strains with all the merits of large yield, good quality protein, high percentage of protein, and disease resistance. The results are encouraging. There are sorghums with 3–4 per cent N in the grain DM, wheats with up to 3.8 per cent (Johnson, Schmidt & Mattern, 1968), a barley with 3.1 per cent (Munck et al. 1970) and an oat with 4.8 per cent (Murphy et al. 1968). Experience with other cereal seeds suggests that almost all the N will be protein N; it is too early to tell how well these strains satisfy the other requirements.

The protein content of cereal seeds is usually measured on a bulk sample because that is the type of material that an animal, or flour miller, will later use. Physiologically, the variation among individual grains is very interesting because this shows what potential the plant has that may be underexploited. In one strain of wheat, Levi & Anderson (1950) found individual grains ranging from 7 to 14 per cent protein in the DM; in another they ranged from 14 to 21 per cent. The smallest grains were not the ones with the most protein; there is no reason, therefore, to think that the effect was simply a failure to lay down starch. We made a casual observation (Pirie, 1966a) that may be connected; at the beginning of June, when flower initials were forming, wheat was given fertiliser in which 31 per cent of the N was ^{15}N, and this type of fertilisation was continued for a month. At harvest the ^{15}N enrichment of individual grains varied from 9 to 17 per cent. The mechanism of protein synthesis seems to be individually triggered in each grain.

Little need be said about the legume seeds; their use is traditional throughout the subtropics and those parts of the tropics where rainfall is so distributed that they are able to ripen. The only unconventionality envisaged, with regard to those that can be eaten with no more pretreatment than cooking, is in such aspects of husbandry as adequate use of manures, protection from disease and predation, and encouragement of those insects, including ants, that are necessary for fertilisation. The yields commonly attained are probably much smaller than those that are possible. Groundnuts (*Arachis* or *Voandzeia*) can yield 500 kg protein per hectare in four months; the more productive strains of *Phaseolus* can, in ideal conditions, yield as much as that but yields from *Cajanus*, *Vigna*, *Cicer* and *Pisum* are smaller. Two, or even three crops can, obviously, be grown in a year in some climates. Edible legume seeds can therefore supply about one ton of protein per hectare per year as a concentrate containing 25–30 per cent protein. The nature of the other material in the seed is not always known and it may sometimes limit use. Thus about half the weight of shelled groundnuts is useful oil, whereas some beans contain useless polysaccharides, some contain oligosaccharides that are fermented in the gut and produce unacceptable amounts of gas, and some take so long to cook that they are not welcome although they give good yields and have pleasant flavours. Advocacy of groundnuts as a food is sometimes condemned because of the risk of aflatoxin poisoning. This attitude is hysterical. Any food can develop aflatoxin if stored badly; the association with groundnuts arises because they have sometimes been eaten in larger quantities and stored in worse conditions than other crops.

It is now unconventional to eat seeds from plants other than cereals and legumes, though there is evidence that *Chenopodium* seeds were at one time a staple in Europe, and nuts are eaten on a small scale. Quinoa, and some other members of the *Chenopodium* family, are eaten in the highlands of South America (Simmonds, 1965). Although it has not been subjected to any up-to-date selection, and is probably suboptimally fertilised, it yields up to 3 tons of seed per hectare. The seed contains 15 per cent protein in the DM. It would seem to be a crop much more worthy of attention than buckwheat (*Fagopyrum*) which contains only 10 per cent protein.

Leafy vegetables

It is conventional to eat small amounts of green leafy vegetables but very unconventional to eat an amount that would contribute significantly to the protein supply. As Jelliffe tirelessly points out (in nearly every number of 'Cajanus', the newsletter of the Caribbean Food and Nutrition Institute, and elsewhere, e.g. Jelliffe, 1968) it is the dark green vegetables that are important, not 'anaemic' vegetables such as cabbage and lettuce. These contain little more than 1 per cent protein in the wet weight of the part usually eaten, whereas brussels sprouts, kale, spinach and broccoli flowers contain 3–6 per cent. Leafy vegetables and immature flowers vary more in composition than seeds and tubers, but the colour is a reasonable guide because the protein content, and, incidentally, the carotene content, is roughly proportional to the chlorophyll content. Unfortunately, the prestige of European and American technology gives an undeserved prestige to our vegetables and in many developing countries they are ousting superior local plants that were eaten traditionally but are now apt to be rated as food fit only for savages. It may be, as Li (1970) suggests, that the abundance of uncultivated edible leaves in the wet tropics inhibits the selection and breeding of cultivated strains, but even species that are cultivated, e.g. *Ipomoea aquatica*, *Basella*, and *Amaranthus*, are losing prestige. A useful and unconventional approach to the protein problem would be to encourage vegetable production in the tropics and, possibly, to introduce hardy strains of tropical leafy vegetables into the Temperate Zone. As Voltaire remarked in a somewhat different context, 'il faut cultiver notre jardin'.

Dark green leafy vegetables contain 3–5 per cent N in the DM. Specially selected leaves or parts of leaves can contain still more. Not all the N is in the form of protein but most of it is. Analyses of the edible parts of vegetables are given in many research publications and tables of food values. It is much more difficult to find reliable figures for the yield that can be expected. It is usually not clear whether what is weighed is the crop as harvested, as sold, or as eaten; sometimes it is even necessary to guess whether it is a wet or dry weight that is being recorded. It would be a welcome unconventionality if those interested in vegetable production would learn to publish their results in a manner comprehensible to others! Annual wet weight yields of cassava leaves can be 20 tons per hectare (Terra, 1964); *Ipomoea aquatica* can yield 90 tons (Edie & Ho, 1969). These yields correspond to 3–5 tons protein whereas a good crop of brussels sprouts in Britain yields only about 0.4 tons protein. FAO (1964) regards 400 kg per hectare in three to four months as a reasonable yield of protein from market garden crops in the tropics, and FAO (1969) regards 20 tons per annum as a reasonable yield of dry matter by crop rotations that include vegetables.

It is obvious from all this that leafy vegetables especially those derived by skilled selection from indigenous species, deserve much more attention than they get. They have limitations however. Terra (1964) reports the daily consumption of up to 500 g (wet weight) cassava leaves by people in the Congo and Java; consumption on that scale is unusual because of the (usually) limited capacity of the human gut to handle fibre. If that quantity of a leaf as dry as cassava were eaten it would supply 35 g protein.

It would probably be prudent to think of vegetables contributing only 2–4 g protein daily to the diet; this is very much more than is contributed by leafy vegetables in most countries now and it is about the amount contributed by fish. Oke (1966) estimates that in Nigeria 1–2 g protein per day comes from vegetables.

Fibre is not the only factor limiting the amount of leafy material that should prudently be eaten. Oxalates are often feared and this fear is probably justified with children because of interference with calcium absorption. Rhubarb leaves are poisonous; the amount of oxalic acid in them is not large enough to make it likely that it is the cause. Heavily manured crops, harvested when still growing vigorously, will contain nitrate and this has probably caused trouble because it is reduced to nitrite if bacteria are allowed to grow on the leaf. This is a matter of culinary technique; any food may become hazardous if bacteria are allowed to grow on it.

PROTEIN CONCENTRATES MADE BY FRACTIONATION

Many plants, or parts of plants, contain good quality protein that is not useable as human food because it is accompanied by other materials that are indigestible, unpalatable, or harmful; the other materials may be useful sources of energy but become troublesome diluents when it is a protein concentrate that is being sought. To get a protein concentrate, all that is necessary is fractionation. This procedure must be sharply distinguished from conversion in an animal, and from the use of a protein-deficient material, fortified with substances such as ammonium phosphate, as a substrate for the growth of micro-organisms. One or other of these processes is sometimes unavoidable if the material is not to be wasted, but conversion is a successful method for making a protein concentrate only because the animal excretes (*wastes* from our egocentric standpoint) even more of the carbon than the nitrogen in its fodder and so accumulates protein. For microbial growth, coal, or products made by electrochemical reduction of limestone, could be substituted for the plant material. A trivial entry into this form of food production has been made by the use of oil. These methods are dealt with by others at this symposium.

The palatability and digestibility of some foods, notably soya beans, have traditionally been increased in the Far East by germination, when the beans' own enzymes degrade unwanted material, and by fermentation, when the degradation is performed by enzymes from another organism. This is not, strictly speaking, a fractionation procedure because nothing is removed except some carbon dioxide and, possibly, water-soluble material. The products, bean sprouts, adamame, tofu, natto, tempeh, miso, ontjom, bongkrek, etc., are more digestible than the starting material; they keep well and have characteristic flavours that appeal to those accustomed to them, though not, at first, to visitors. There is no evidence that the quality of the protein, as opposed to its digestibility, is increased (Jelliffe, 1968; Hesseltine, 1965; Hesseltine & Wang, 1967; Standal, 1963; van Veen & Steinkraus, 1970; Wang, Ruttle & Hesseltine, 1968). In countries where raw milk is a hazardous food, it may be made safer and less perishable by fermentation. It would be unconventional to use these products in other parts of the world. Research is needed on the most suitable species and strains of micro-organisms so that deliberate seeding of a sterilised substrate could replace the rather primitive methods now used, for there is some reason to think that fermentation would be cheaper, and would introduce more variety, than the methods of heating and extraction that are now usually advocated.

Proteins have been extracted in the laboratory from many different underground parts of plants, notably potato tubers, horseradish, and legume root nodules (see Pirie, 1955). So far as I know, no one has used these methods on a technological scale as a means for making edible protein concentrates. The point deserves attention because one fifth of a potato is removed by mechanical methods of peeling, and the layer immediately under the skin is richer in protein than the remainder.

Seeds

The method of fractionation that can be most effectively used with a seed or nut depends on the manner in which protein is distributed within it. In rice, protein is concentrated in a layer immediately under the bran; it can be separated from the main body of the endosperm by suitably contrived abrasion and equipment for doing this is in use in various parts of the world. Starting with rice containing only 8 or 9 per cent protein in the DM, fractions containing up to 21 per cent protein can be separated (Milner, 1965). It is claimed that by combining abrasion with solvent extraction the yield of the protein-rich fraction can be increased; this process also concentrates the thiamine and riboflavin and the products are already being used commercially in baby foods. In a somewhat similar manner, coarse wheat flour is separated in an air stream into protein-rich and protein-depleted fractions. This is more convenient than the traditional wet process, developed in China many hundred years ago when gluten (*mien chin*) was separated, and introduced into Europe by Beccari in 1728. Fractions containing 30 per cent protein in the DM can be made (Stringfellow, Pfeifer & Griffin, 1965).

The problems raised by cottonseed are different from those raised by the cereal seeds. It is already a concentrate which, after defatting, contains 50–60 per cent protein in the DM but contaminated with various brown pigments and with gossypol, a phenolic aldehyde that combines with lysine in a manner that makes the lysine unavailable. There are strains of cotton free from these substances, but they are more subject than normal cotton to insect attack (Roux & Bui-Xuan-Nhuan, 1964). These harmful substances are contained in discrete glands which can be separated from judiciously milled cottonseed by flotation or classification in an air stream. The exploitation of these methods has elevated cottonseed from being a source of oil and cattle fodder to the status of a human food which is gaining acceptance in Central America largely through the enterprise of INCAP in Guatemala (Bressani, Elias & Braham, 1966; Shaw, 1967).

Soyabeans are a traditional source of oil in China, Japan and Manchuria. Most of the protein was wasted, but some was made edible by fermentation and some was turned into bean curd by grinding, extracting with water and coagulating with gypsum. Osborne & Campbell (1898) studied the extracted protein, and its potentialities were recognised in the USA (e.g. Horvath, 1937) as the plant gradually gained favour as a seed crop rather than as a forage. The extracted protein is now so widely used in the food industry that it can no longer be considered unconventional.

Recognition of the potentialities of coconut protein has come very much more slowly and it does not yet seem to be produced commercially. Dry copra contains 20–25 per cent protein in the DM and so, by the criteria used here, could be regarded as a protein concentrate, but the presence of 12–16 per cent fibre limits the amount that can be eaten. Early attempts to separate protein and fibre were partly frustrated by the use of dried copra, expeller residue, or autoclaved coconuts. All these pretreatments, to varying extents, coagulate the protein and so impede separation; this defect can be partly overcome by fermentation (e.g. Chandrasekaran & King, 1967; Puertollano, Banzon & Steinkraus, 1970). Technologists in Mysore, the Philippines and elsewhere have now stopped trying to be original and have reverted to the method that is traditional in the Pacific Islands; they disintegrate coconut meat with water warm enough to liquify the fat but not to coagulate the protein. The resulting emulsion is 'broken' and the protein coagulated from the aqueous layer. The concentrate contains 74 per cent protein (Rama Rao et al., 1967) and the oil made in this way is of better quality than oil made by the usual procedure of drying and screw expulsion. There seems to be no serious obstacle to the widespread adoption of this wet processing method; it would make available for human nutrition two-thirds of a million tons of protein that is at present either wasted or used as animal feed.

Residues left after expressing oil from these varied sources contain more than 20 million tons of protein. Less than 0.1 per cent of this is now used as human food and only about five times as much comes at present up to the quality standards that would make it acceptable as a food source. This very large potential protein source will not become useful until the producers realise the value of the residue and handle the crop, at all stages from harvest to processing, carefully and cleanly.

Cereal seeds are already being fractionated in Britain to make protein-rich breads and biscuits. The protein-depleted fraction is used, with added nitrogen, as a substrate for the growth of edible moulds. Field beans (*Vicia faba*) are a potential protein concentrate. The small-seeded variety, when harvested fully mature and dry, can yield 4–5 tons per hectare in a good year. This is about ten times the yield from the large-seeded garden variety harvested green, and this yield would contain about a ton of protein. However, even after prolonged cooking, these beans are usually considered inedible. By methods which are now well-known, the protein can be extracted from them and this is becoming a commercial process.

Leaves

In arid regions, or regions with seasonal rainfall, seed crops may yield better than leaf crops. In regions such as Britain, with well distributed rain, leaf crops yield better than seed crops, and in regions where it rains nearly every day, seed crops tend to rot rather than ripen. In Britain, therefore, leaves are more productive than any other crop; in still wetter tropical regions leaves and coconuts are probably the only source from which protein concentrates can be made reliably.

When the juice expressed from pulped fresh leaves is coagulated and the coagulum is filtered off, three products are made: protein coagulum, material soluble in water, and leaf fibre containing some unextracted protein (Table 1). Both for economy and to prevent local pollution each product must be used. At different times and places, each has indeed been regarded as the main product with the others as byproducts.

At Rothamsted we attach most importance to the extracted protein and we have designed equipment both for extracting crops on the 1–2 tons (wet weight) per hour scale (Davys & Pirie, 1960, 1965), and for making precise agronomic measurements on a laboratory scale (Davys & Pirie, 1969; Davys, Pirie & Street, 1969). More information about these pieces of equipment and about the equipment used in Hungary and the USA are given in Handbook No. 20 of the International Biological Program (Pirie, 1971*b*). That handbook also describes the separation and purification of the

Table 1. *The three products made by fractionating a fresh leafy crop*

Juice, which gives after coagulation	coagulum containing	proteins, fats, starch	FOOD for man and other non-ruminants
	fluid containing	amino acids, amides, sugars, salts, etc.	MEDIUM for the growth of micro-organisms
Fibrous residue containing	most of the	cellulose, hemicelluloses, lignins, pectin	Still a FODDER for ruminants and a substrate for microbial fermentation
	some of the	proteins, fats, starch	

protein and gives information about its composition, nutritional value and presentation on the table.

By using a succession of crops on the same land, the yield of extracted protein can, in a favourable year, exceed 2 tons per hectare; five cuts of cocksfoot yielded 1.67 tons in a less favourable year (Arkcoll & Festenstein, 1971). Figure 1 illustrates the improvement in yield of dry extracted protein during the last few years; this improvement is mainly the result of increasing skill in agronomy and in handling the extraction equipment. The actual yield in any year depends greatly on the weather – especially on rainfall in late spring and early summer. With inreasing skill and the use of irrigation water when necessary, we think it should be possible to get 3 tons per hectare in Britain regularly. In New Zealand and parts of India where cold weather does not stop growth, 3 tons per hectare has already been reached and 5 tons per hectare should be possible. Crops intended for leaf protein production are harvested young and so are less at risk to many pests, predators and diseases than crops harvested when more mature. As already pointed out, they can be grown in regions where seed crops do not ripen reliably. Leaf protein production has the added merit that it gives a greater yield of edible protein than any other method of land use.

These were the yields of protein from conventional varieties of such crops as winter wheat, rye, tares, fodder radish, lucerne and rape. They

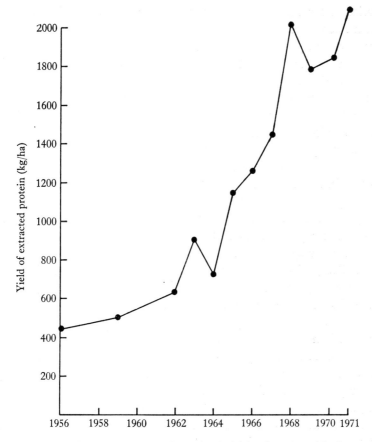

Figure 1. Improvements in the annual yields of extracted leaf protein.

should be surpassed when species and varieties are specially selected for this use. Joshi (1971) and Lexander et al. (1970) have embarked on the formidable task of finding more suitable plants. Joshi (1971) finds that *Tithonia tagetiflora* is particularly hopeful and Gunetileke (Pirie, 1971 a) estimates that the yield from *Basella alba* in Ceylon would be 7 tons dry extracted protein per hectare per year. This crop is used extensively as a vegetable, little would therefore be gained by extracting protein from it unless a community is already eating as much as is physiologically reasonable. Leaf protein should complement, rather than replace, fresh vegetables.

When forage has to be dried, either for pelleting or for storage as winter feed, it is inefficient to evaporate off all the water and discharge it as vapour because this entails the loss of the energy used to supply latent heat of evaporation. Attempts have been made to economise by using vapour compression and multiple-effect systems similar to those used in making fresh from salt water. Herbage is however an intractable material to feed into such a system and the expense of the necessary airlocks and so forth seems to outweigh the thermal gain. I pointed out (Pirie, 1951, 1953a, b) that one of the incidental advantages of making leaf protein is that the pressed fibre can be economically dried. Some attempts to press water out of leaves had been made before that; they failed because the leaves had not been pulped or in some other way robbed of their osmotic control. If the objective is simply to cheapen drying, the simplest procedure is to heat the fresh crop to 70 or 80 °C and then press it. The protein is thus coagulated and remains in the fibre (Pirie, 1966b). However, the protein content of a forage crop that is fertilised and irrigated so as to give the maximum yield, is greater than is needed in a ruminant feed. From such a crop we would expect to separate one-third of the protein by one extraction and half the remainder by a second. The residue then contains between 1 and 2 per cent nitrogen in the DM; it can be fortified, if necessary, with urea and has a feeding value (according to as yet unpublished experiments) greater than its nitrogen content suggests because it is relatively unlignified.

The presses that we use apply 1.5 to 2.0 kg cm^{-2} and produce fibrous residues containing 30–35 per cent DM. If the forage, as harvested, contained 12 per cent DM, only one-third as much water would have to be dried off from the pressed as from the fresh material to make a ton of dry forage. For protein extraction there is little advantage in pressing harder than this, but more intense pressing may be advantageous as a prelude to drying. DM in the fibrous residue can be increased to 45 per cent; at this point the residual water is largely 'water of hydration' and it amounts to only one-sixth of what was initially present. Whether this second pressing with powerful equipment would be worth while is a matter for experiment; it is unquestionable that the initial diminution of the drying load to one-third would be advantageous.

These points are now gaining attention. Thus Casselman *et al.* (1965) press crushed fodder solely for economical drying – they discard the protein-rich juice. Kohler & Bickoff (1971) and Hollo & Koch (1971) collect the extracted protein but are satisfied with what we would regard as very incomplete extraction because their main objective is economical drying. In my opinion, the value of extracted protein is so much greater than the value of residual protein in the fibrous residue, that it is false economy not to use equipment designed to make extraction as complete as practicable. Various organisations in Britain are now beginning to take an interest in the potentialities of fodder fractionation.

The water-soluble, non-protein components of the leaf were considered the primary products by the late E. C. Nilsson and by Jönsson (1962) who used them as microbial culture media. Shah *et al.* (Pirie, 1971*c*) is investigating the growth of yeasts on this 'whey'. Much of the nitrogen in the 'whey' is not metabolically useful in monogastric animals, and part of the carbohydrate is indigestible by them. The 'whey' is therefore probably most efficiently converted into a protein source when it is used as a microbial culture medium. Incompletely published results of experiments in Eire suggest that the uncoagulated leaf extract can replace a quarter of the conventional ration in pig feeding. In the USA the whole unfractionated pulp from relatively fibre-free leaves such as comfrey, and the laminae removed from sugar beet by rubber flails, are being used as pig feed. Unlike the original forage, the 'whey' has physical properties that make its economical drying easy. In spite of this, and in spite of the advocacy of Hartman, Akeson & Stahmann (1967), Hollo & Koch (1971) and Kohler & Bickoff (1971), it seems unlikely that it is worth drying in countries where the whey from milk is not fully used.

SOME OBSTACLES TO INNOVATION

Understandably, people are cautious about accepting a new procedure, especially one that will involve capital expenditure. Some of the more general obstacles to innovation are discussed elsewhere (Pirie, 1969*a*). Only two need to be discussed here – quality and price. There is a widespread illusion that all plant proteins are nutritionally inferior to animal proteins. Fortunately the old crystallisation of this illusion into the categories 'first' and 'second' class protein is disappearing. *A priori* such a generalisation is improbable. It is true that egg and milk proteins, the endproducts of 200 and 20 million years of adaptation to the feeding of young animals, are unrivalled, and that the proteins in many cereal seeds show amino acid imbalances. The proteins in other seeds tend to be better, but even in them the sulphur-containing amino acids are often deficient. This is not universally true; some varieties of maize are good methionine sources (Fowden & Wolfe, 1957). The mixture of protein in 'leaf protein' compares favourably with meat, fish, or FAO reference protein (Byers, 1971*a*, *b*) apart from a slight methionine deficiency. There is however no basis for the assumption that sulphur deficiency characterises all plant proteins. An outstanding example is the protein in the latex of the Upas tree (*Antiaris toxicaria*) with 7 per cent sulphur; 10.5 per cent cystine was isolated from it (Kotake & Knoop, 1911). A thorough study of the amino acid composition of the protein concentrates that can be made from cultivable plants would probably be rewarding.

Amino acid analysis is the first step in judging the merits of a protein. It establishes the potentialities of the protein either as a sole source, as with egg or milk, or as a component in a mixture in which, by judicious blending, proteins can supplement one another. A protein may however be indigestible, or some of its amino acids may have been rendered unavailable to monogastric animals by complexing with other substances. Complexes are often formed during extraction, processing or storage. Some preparations should therefore be made quickly and carefully, even if the methods used could not be easily adapted to large-scale practice, so as to establish the intrinsic qualities of the protein and demonstrate what should be aimed at. It is very easy to damage a protein, and some have been condemned for defects that turned out later to be the consequences of defects in processing.

The merits of novel protein sources have been discussed from several standpoints: how extensively could they be grown, and could they be produced on land that is now inadequately used, how does the yield per hectare and year compare with conventional sources and how do their nutritive values compare with those of other proteins? An obviously fundamental question is, how much would they cost? There is no direct answer, but a few relevant points may be made. First, too few people appreciate the cost of conventional protein supplements; they forget that only about 17 per cent of the wet weight of even carefully selected cuts of meat is protein. Secondly, some excellent concentrates, e.g. dried skim milk, are byproducts or are subsidised, so that it would not be possible to increase the supply greatly without an increase in price. Thirdly, and most important, intermittent production in a laboratory gives little or no basis for more than a guess at the costs of production in a continuously operated plant. Empirically minded scientists, and those with commercial experience, know that costs cannot be reliably assessed until production has started on a moderate scale. Economists and naive scientists are more gullible. The absurdity of making premature estimates was well illustrated at a recent Biochemical Society Symposium where an expert from one of the large oil companies said, but only when pressed, that yeast grown on oil would cost about as much as fishmeal. An expert from another oil company was incredulous. If, after many scientists have worked for these vast concerns for several years, there is still disagreement on a point of such direct commercial interest, caution is justified. Nevertheless, analogy permits some reasonable guesses. Thus the costs of growing soya are well known, and so are the costs of extracting protein from groundnuts. The costs of extracting protein from soya could therefore be estimated and, by further analogy, so could the costs of extracting protein from broad beans.

Similar reasoning, based on the observed costs of making dried forage in Britain, leads to the estimate (Pirie, 1969b) that leaf protein made here would cost 15 to 25p per kg (dry) if made from a crop grown specially for the purpose, and less if made from byproducts such as pea haulms or sugar beet leaves.

EFFICIENCY

When efficiency is assessed by comparing the yields of edible protein that can be got in comparable conditions by adopting different systems of husbandry, green vegetables are pre-eminent; next comes extracted leaf protein, then the legume seeds, and finally the meat, milk or eggs of animals fed on crops grown on arable land. This is unfortunate because most people prefer these animal products, but the facts are indisputable.

When efficiency is assessed in terms of the percentage of the solar energy that is used for the synthesis of dry matter and protein, there is more room for disagreement. In practical agricultural conditions, that is to say, when growth depends on sunlight and atmospheric carbon dioxide, is not limited by water or fertiliser deficiency, and is not seriously impeded by disease or predation, as much as 9 per cent of the light in the effective range (400 to 700 nm) can be used. Greater efficiencies are possible in feeble light but the yield is correspondingly less. Good farming in Britain uses sunlight with 1–2 per cent efficiency. This discrepancy arises because the plants showing the greatest efficiencies, e.g. corncockle (*Agrostemma githago*) (Blackman, 1962) and bulrush millet (*Pennisetum typhoides*) (Begg, 1965), do not yield an agriculturally useful product, the rate is not maintained throughout the year, and many species do not use strong light efficiently. Frost-sensitive crops, such as potatoes or sugar beet, do not cover the ground completely till Midsummer day and so waste much of the light during the first half of the year; cereals waste light during the second half of the year because they are then ripening rather than growing. To ensure the more efficient use of sunlight it is essential to maintain a complete green cover on the ground throughout the year, and to harvest that cover frequently, because photosynthetic efficiency tends to decline as plants mature. These conditions are not met by seed or root crops; only leafy crops meet them. Efficient production therefore depends on using leaves efficiently.

REFERENCES

ARKCOLL, D. B., & FESTENSTEIN, G. N. (1971). A preliminary study of the agronomic factors affecting the yield of extractable leaf protein. *Journal of the Science of Food and Agriculture* **22**, 49–56.

BEGG, J. E. (1965). High photosynthetic efficiency in a low-latitude environment. *Nature* **205**, 1025–6.

BLACKMAN, G. E. (1962). The limit of plant productivity. *Annual report of East Malling Research Station for 1961*, 39.

BOYD, C. E. (1971). Leaf protein from aquatic plants. In: Pirie, N.W. (Ed.), *Leaf Protein: its Agronomy, Preparation, Quality and Use*. IBP Handbook No. 20, 44–9. Oxford: Blackwell.

BRESSANI, R., ELIAS, L. G., & BRAHAM, E. (1966). Cottonseed protein in human foods. In: American Chemical Society, *World Protein Resources. Advances in Chemistry Series No. 57*, 75.

BYERS, M. (1971a). The amino acid composition of some leaf protein preparations. In: Pirie, N. W. (Ed.), *Leaf Protein: its Agronomy, Preparation, Quality and Use*. IBP Handbook No. 20, 95–114. Oxford: Blackwell.

BYERS, M. (1971b). Amino acid composition and *in vitro* digestibility of some protein fractions from three species of leaves of various ages. *Journal of the Science of Food and Agriculture* **22**, 242–51.

CASSELMAN, T. W., GREEN, V. E., ALLEN, R. J., & THOMAS, F. H. (1965). *Mechanical De-watering of Forage Crops*. Technical Bulletin 694. Agricultural Experiment Station, University of Florida, Gainesville.

CHANDRASEKARAN, A., & KING, K. W. (1967). Enzymic modification of the extractability of protein from coconuts (*Cocos nucifera*). *Journal of Agricultural and Food Chemistry* **15**, 305–9.

DAVYS, M. N. G., & PIRIE, N. W. (1960). Protein from leaves by bulk extraction. *Engineering* **190**, 274–5.

DAVYS, M. N. G., & PIRIE, N. W. (1965). A belt press for separating juices from fibrous pulps. *Journal of Agricultural Engineering Research* **10**, 142–5.

DAVYS, M. N. G., & PIRIE, N. W. (1969). A laboratory-scale pulper for leafy plant material. *Biotechnology and Bioengineering* **11**, 517–28.

DAVYS, M. N. G., PIRIE, N. W., & STREET, G. (1969). A laboratory-scale press for extracting juice from leaf pulp. *Biotechnology and Bioengineering* **11**, 529–38.

EDIE, H. H., & HO, B. W. C. (1969). *Ipomoea aquatica* as a vegetable crop in Hong Kong. *Economic Botany* **23**, 32–6.

FAO (1964). *The State of Food and Agriculture 1964*, p. 118. Rome: Food and Agriculture Organisation.

FAO (1969). *Provisional Indicative World Plan for Agricultural Development*, vol. 1, p. 132. Rome: Food and Agriculture Organisation.

FOWDEN, L., & WOLFE, M. (1957). The protein composition of some East African seeds. *East African Agricultural Journal* **22**, 207–12.

HARTMAN, G. H., JR, AKESON, W. R., & STAHMANN, M. A. (1967). Leaf protein concentrate prepared by spray-drying. *Journal of Agricultural and Food Chemistry* **15**, 74–9.

HESSELTINE, C. W. (1965). A millennium of fungi, food and fermentation. *Mycologia* **57**, 149–97.

HESSELTINE, C. W., & WANG, H. L. (1967). Traditional fermented foods. *Biotechnology and Bioengineering* **9**, 275–88.

HOLLO, J., & KOCH, L. (1971). Commercial production in Hungary. In: Pirie, N. W. (Ed.), *Leaf Protein: its Agronomy, Preparation, Quality and Use*. IBP Handbook No. 20, 63–8. Oxford: Blackwell.

HORVATH, A. A. (1937). The chemistry of soyabean protein extraction. *Chemistry and Industry (Journal of the Society of Chemical Industry)* **56**, 735–8.
JELLIFFE, D. B. (1968). *Child Nutrition in Developing Countries.* Public Health Service Publication No. 1822. Washington, DC: US Government Printing Office.
JOHNSON, V. A., SCHMIDT, J. W., & MATTERN, P. J. (1968). Cereal breeding for better protein impact. *Economic Botany* **22**, 16–25.
JÖNSSON, A. G. (1962). Studies in the utilisation of some agricultural wastes and byproducts by various microbial processes. *Kungliga Lantbrukshogskolans annaler* **28**, 235–60.
JOSHI, R. N. (1971). The yields of leaf protein that can be extracted from crops of Aurangabad. In: Pirie, N. W. (Ed.), *Leaf Protein: its Agronomy, Preparation, Quality and Use.* IBP Handbook No. 20, 19–28. Oxford: Blackwell.
KOHLER, G. O., & BICKOFF, E. M. (1971). Commercial production from alfalfa in USA. In: Pirie, N. W. (Ed.), *Leaf Protein: its Agronomy, Preparation, Quality and Use.* IBP Handbook No. 20, 69–77. Oxford: Blackwell.
KOTAKE, Y., & KNOOP, F. (1911). Ueber einen krystallisierten Eiweisskörper aus dem milchsäfte der Antiaris toxicaria. *Hoppe-Seyler's Zeitschrifte für physiologische Chemie* **75**, 488–98.
LEVI, I., & ANDERSON, J. A. (1950). Variations in protein contents of plants, heads, spikelets, and individual kernels, of wheat. *Canadian Journal of Research* **28**, F71–81.
LEXANDER, K., CARLSSON, R., SCHALEN, V., SIMONSSON, A., & LUNDBORG, T. (1970) Quantities and qualities of leaf protein concentrates from wild species and crop species grown under controlled conditions. *Annals of Applied Biology* **66**, 193–210.
LI, H. L. (1970). The origin of cultivated plants in Southeast Asia. *Economic Botany* **24**, 3–19.
MILNER, M. (1965). High protein fractions from white rice and simple methods for their production. *Protein Advisory Group Bulletin* **5**, 39. New York: United Nations.
MUNCK, L., KARLSSON, K. E., HAGBERG, A., & EGGUM, B. O. (1970). Gene for improved nutritional value in barley seed protein. *Science* **168**, 985–7.
MURPHY, H. C., SADANAGA, K., ZILLINSKY, F. J., TERRELL, E. E., & SMITH, R. T. (1968). Avena magna: an important new tetraploid species of oats. *Science* **159**, 103–4.
OKE, O. L. (1966). Chemical studies on some Nigerian vegetables. *Tropical Science* **8**, 128–32.
OSBORNE, T. B., & CAMPBELL, G. F. (1898). Proteids of the soybean (*Glycine hispida*). *Journal of the American Chemical Society* **20**, 419–28.
PIRIE, N. W. (1951). The circumvention of waste. In: LeGros Clark, F., & Pirie, N. W. (Eds.), *Four Thousand Million Mouths,* 180–99. Oxford University Press.
PIRIE, N. W. (1953a). Large-scale production of edible protein from fresh leaves. *Annual Report, Rothamsted Experimental Station, for 1952,* 173.
PIRIE, N. W. (1953b). Food and the future. Part 3(A). The efficient use of sunlight for food production. *Chemistry and Industry* **72**, 442–5.
PIRIE, N. W. (1955). Proteins. In: Paech, K., & Tracey, M. V. (Eds.), *Modern Methods of Plant Analysis,* vol. 4, p. 23. Heidelberg: Springer.
PIRIE, N. W. (1966a). *Annual Report, Rothamsted Experimental Station, for 1965,* 107.
PIRIE, N. W. (1966b). Fodder fractionation: an aspect of conservation. *Fertiliser and Feeding Stuffs Journal* **63**, 119–22.

PIRIE, N. W. (1969a). *Food Resources: Conventional and Novel.* Harmondsworth: Penguin Books.
PIRIE, N. W. (1969b). The production and use of leaf protein. *Proceedings of the Nutrition Society* **28**, 85–91.
PIRIE, N. W. (1970). Weeds are not all bad. *Ceres* **3**(4), 31–4.
PIRIE, N. W. (1971a). A survey of other experiments on protein production. In: Pirie, N. W. (Ed.), *Leaf Protein: its Agronomy, Preparation, Quality and Use.* IBP Handbook No. 20, 29–43. Oxford: Blackwell.
PIRIE, N. W. (1971b). Equipment and methods for extracting and separating protein. In: Pirie, N. W. (Ed.), *Leaf Protein: its Agronomy, Preparation, Quality and Use.* IBP Handbook No. 20, 53–62. Oxford: Blackwell.
PIRIE, N. W. (1971c). The use of the byproducts from leaf protein extraction. In: Pirie, N. W. (Ed.), *Leaf Protein: its Agronomy, Preparation, Quality and Use.* IBP Handbook No. 20, 135–7. Oxford: Blackwell.
PUERTOLLANO, C. L., BANZON, J., & STEINKRAUS, K. H. (1970). Separation of the oil and protein fractions in coconut (*Cocos nucifera* Linn.) by fermentation. *Journal of Agricultural and Food Chemistry* **18**, 579–84.
RAMA RAO, G., INDIRA, R. K., BHIMA RAO, U. S., CHANDRASEKHARA, M. R., CARPENTER, K. J., & BHATIA, D. S. (1967). Nutritive value of coconut protein concentrates obtained by wet processing. *Indian Journal of Experimental Biology* **5**, 114–17.
ROUX, J. B., & BUI-XUAN-NHUAN (1964). L'IRCT devant le problème de la sélection de cotonniers sans glandes et de l'utilisation des farines de coton sans gossypol. *Rapports du Premier Congrès International des Industries Agricoles et Alimentaires en Zones Tropicales et Subtropicales, Abidjan* **1**, 291.
SHAW, R. L. (1967). Incaparina gains acceptance. *Science* **156**, 168.
SIMMONDS, N. W. (1965). The grain Chenopods of the tropical American highlands. *Economic Botany* **19**, 223–35.
STANDAL, B. R. (1963). Nutritional value of proteins of oriental soybean foods. *Journal of Nutrition* **81**, 279–85.
STRINGFELLOW, A. C., PFEIFER, V. F., & GRIFFIN, E. L., JR (1965). Effect of fertiliser on air classification of wheat flours. *Journal of Agricultural and Food Chemistry* **13**, 262–5.
TERRA, G. J. (1964). The significance of leaf vegetables, especially cassava, in tropical nutrition. *Tropical and Geographical Medicine* **16**, 97.
VAN VEEN, A. G., & STEINKRAUS, K. H. (1970). Nutritive value and wholesomeness of fermented foods. *Journal of Agricultural and Food Chemistry* **18**, 576–8.
WANG, H. L., RUTTLE, D. I., & HESSELTINE, C. W. (1968). Protein quality of wheat and soybeans after *Rhizopus oligosphorus* fermentation. *Journal of Nutrition* **96**, 109–14.

9
POTENTIAL PROTEIN PRODUCTION OF TEMPERATE GRASSES

By TH. ALBERDA

Institute for Biological and Chemical Research on
Field Crops and Herbage (IBS), Wageningen,
The Netherlands

INTRODUCTION

Since the Second World War the amount of nitrogenous fertiliser applied to grassland has increased tremendously. For exclusively grassland farms mean values from about 50 kg nitrogen (N) per hectare in 1950 to 150 kg per hectare in 1965 have been recorded (Oostendorp, 1964) and for 1970 the amount is around 200 kg per hectare.

This increase led to a type of pasture without any legumes, often consisting of one or two highly productive grass species such as *Lolium perenne*, *Dactylis glomerata* and *Phleum pratense*. The higher amounts of nitrogenous fertiliser also led to a change in chemical composition in that the percentages of crude fibre and soluble sugars decreased and the crude protein content increased. As a general comparison the composition of high- and low-nitrogen plants might be as given in Table 1.

Table 1. *Approximate chemical composition of perennial ryegrass plants at low and high levels of nitrogen fertilisation (per cent of dry weight)*

Level of N fertilisation	Low	High
Organic N × 6.25	18	24
Crude fibre	27	24
Water-soluble carbohydrates	25	19
Ash	12	12
Not determined	18	20

Such high protein contents as are obtained with high levels of N application are not necessary for the maximum productivity of animals. The lactating cow, having the highest protein demand, does not need more than 15 per cent protein in her daily ration of 15 kg grass dry matter (DM) to produce 30 l milk (de Groot, 1966). On many pastures the animals nowadays eat considerably more organic N compounds than they need for milk production and maintenance; the consequence is that the extra N is excreted in the form of urea. Although this situation does not seem to be disadvantageous to the animal, it leads to a less efficient use of the fodder and to some losses of nitrogen by excretion.

If one wants to keep the utilisation as high as possible the situation can be improved in two ways, viz. (1) by feeding the animal a mixture of grass and other fodders with a higher carbohydrate and a lower protein content such as maize and other grain crops, or (2) by extracting part of the protein from the grass for other purposes and feeding the residual pressed cake to the animal. The latter solution has been until now a theoretical one and implies that proteins from grass can be extracted in a usable form and in a profitable way. Whether this has indeed real possibilities is not the concern of this paper. Here only the question of the maximum possible production of organic N compounds will be considered and not the possible use of these compounds.

Usually, the organic constituents of herbage are recorded in terms of crude protein (6.25 × Kjeldahl N) and crude fibre. What remains is usually referred to as carbonaceous material including sugars, some structural carbohydrates, and lipids. Occasionally, the amount of total water-soluble carbohydrates is separately determined. With the increase in the level of application of fertiliser the content of nitrate N in the herbage may be raised to higher levels, and a variable proportion of this nitrate is recovered as Kjeldahl N. To avoid this difficulty, and to obtain more accurate and reproducible results, the Kjeldahl procedure was used in the modification with salicylic acid to include all the nitrate, and its result indicated as total nitrogen. In another subsample nitrate was determined and, by difference, a value for organic N was obtained. Multiplication of the percentage of organic N by 6.25 then gave the percentage of organic nitrogenous compounds of which about 90 per cent was protein.

THE RELATIONSHIPS BETWEEN NITROGEN DOSAGE, NITROGEN UPTAKE AND DRY MATTER PRODUCTION

When *Lolium perenne* is grown on a series of nutrient solutions with increasing nitrate concentrations (Alberda, 1965) the results of such an experiment can well be represented in a form first used by Frankena & de Wit (1958). Figure 1 gives the results of such an experiment. In quadrant I the relation between nitrate concentration in the solution and N uptake is given for two successive harvests. The curves show the usual saturation pattern and the maximum rate of uptake is already reached at

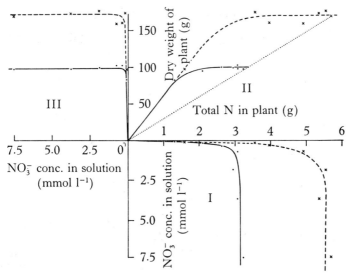

Figure 1. The influence of the nitrate concentration in the nutrient solution on nitrogen uptake and dry matter production of Lolium perenne at two successive harvests (from Alberda, 1965). The dotted line in quadrant II represents a constant N concentration in the DM of 3.4 per cent.

a concentration of 2 mmol nitrate per l. From quadrant III it can be seen that the maximum rate of dry matter production is already reached at a concentration below 1 mmol per l. Quadrant II gives the relation between N uptake and dry matter (DM) production and in this quadrant a straight line through the origin represents a constant N concentration in the plant DM. It appears from the data that at very low nitrate concentrations in the solution the N concentration in the plant does not get below a minimum value of 1.6 per cent of the dry weight. Apparently this is the minimum concentration needed to keep the plant alive but with increasing nitrate concentration the total N content of the plant shifts gradually to a maximum value of 3.4 per cent of the DM. This value holds for the whole plant; for the herbage the maximum value is about 4 per cent.

This way of representing results can also be used for field experiments. Figure 2 shows data of Mulder (1949) who carried out a series of experiments with different numbers of cuts and with N dressings of up to 420 kg N per hectare per cut, which means for six cuts a maximum of 2520 kg N per hectare per annum. The relation between N application and dry herbage production is given cumulatively in quadrant III for an experiment with six cuts per annum. The lowest curve gives the relationship for the first cut, the following one that for the first plus the second cuts and so

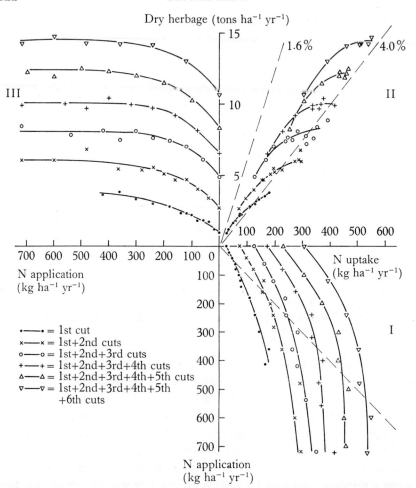

Figure 2. Nitrogen fertilisation, nitrogen uptake and herbage production of permanent pasture (after Mulder, 1949). The chain lines in quadrant II represent constant N concentrations in the DM of 1.6 and 4.0 per cent. The chain line in quadrant I represents 100 per cent uptake of applied N.

on. The first harvest still reacted positively to the highest amount of N applied but with subsequent cuts an increase above 100 kg N per hectare per cut had no positive effect and may even have depressed production to some extent. For this reason total amounts of fertiliser above 700 kg N per hectare per annum have been omitted as they do not further enhance N production. It even appeared in this experiment that the maximum annual herbage production with six cuts amounted to about 14 tons DM per hectare and had already been reached with 240 kg N per hectare per annum. In quadrant I the same relation between dosage and uptake was found as in the experiments with nutrient solution, except that there is no zero rate of application because the unfertilised plots contained a considerable amount of white clover which contributed an annual total N production of 300 kg per hectare. From quadrant II it can be seen that here also there was a minimum and a maximum N concentration in the plant DM, but these values were higher than in the nutrient solution experiments. For the minimum value this was undoubtedly caused by the presence of white clover. At high N dressings the clover was mainly replaced by quack grass (*Agropyron repens*), which may have caused the gradual shift towards the 4 per cent line when subsequent cuts were added. Whereas it was concluded that the maximum herbage production of 14 tons DM per hectare had already been reached at a fertilisation level of 240 kg N per hectare per annum (quadrant III), for the maximum total N production of 550 kg per hectare per annum an application of 600 kg was needed (quadrant I).

Mulder's experiments were carried out on permanent pasture with an initially high clover content and the different N dressings show different shifts in botanical composition of the sward which complicate the picture. Figure 3 shows the results of some of the author's experiments carried out on one-year-old perennial ryegrass (*Lolium perenne*) leys, and exhibiting no change in botanical composition. These experiments were designed to obtain the maximum possible dry herbage production with the minimum amount of fertiliser and not to obtain a high protein production. Two levels of N dressings were applied – a high level that was known by experience not to cause any N shortage and a lower one, called medium, that was thought to be just enough to reach the same herbage production as with the high level. The ratio of high to medium was about 4:3. Quadrant III shows that the maximum herbage production was around 20 tons per hectare per annum, and that a reduction of the fertiliser level caused a reduction in DM yield of about 4.5 per cent which was mainly caused by a difference in the first cut or the first two cuts. Quadrant I shows that a reduction in the N application has a pronounced effect on the rate of uptake. In the second quadrant it can be seen that the total N concentrations lie again between 1.6 and 4.0 per cent of the dry weight, that an application of 600 kg N per hectare per annum is needed for the production of 20 tons DM per hectare per annum, but that a higher level is necessary to reach the maximum N content of 4 per cent.

So far only the total N production has been considered. Mulder also determined protein by grinding the fresh grass with water and adding trichloracetic acid (TCA) after heating in a boiling waterbath. After filtering and washing repeatedly with TCA the residue was considered to be mainly protein. Figure 4 shows the result of such an experiment carried

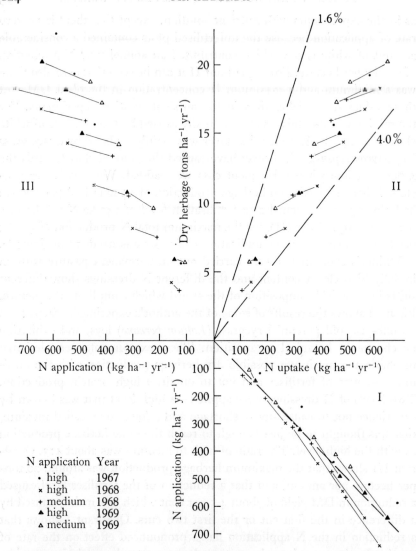

Figure 3. Nitrogen fertilisation, nitrogen uptake and herbage production of one year old perennial ryegrass (*Lolium perenne*) leys (from Alberda, 1972). See caption of Figure 2 for explanation of chain lines.

out one year later than that of Figure 2. The total DM production of six cuts reached a maximum value of nearly 16 tons per hectare per annum, 2 tons more than in the preceding year, at a level of N application between 600 and 700 kg per hectare. With higher amounts of N fertiliser a gradual decline in DM production was observed. The maximum amount of protein produced was around 2.5 tons per hectare. There was no decline with

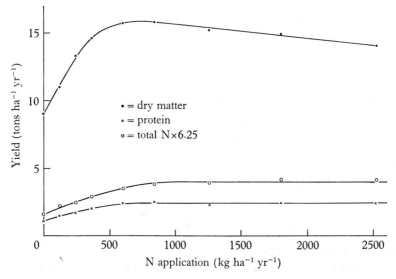

Figure 4. Nitrogen fertilisation of permanent pasture and yields of dry matter, protein and total organic nitrogen compounds (after Mulder, 1949).

increasing amounts of fertiliser since the gradually increasing protein content compensated for the decrease in DM production. The third line in the figure is the total N value multiplied by 6.25 to obtain the crude protein value. The difference between the two curves was mainly caused by the soluble organic nitrogenous compounds and inorganic N. Both curves run parallel, the maximum amount of total N also being reached at levels of application of 600–700 kg N per hectare.

In Figure 5 the data on total N of Mulder are compared with the data on organic N production with the high and medium N dressings described in Figure 3. Mulder's data for 1942 and the author's data suggest that maximum organic N production is obtained with amounts of nitrogenous fertiliser equivalent to at least 1000 kg N per hectare per annum. Similar results were obtained by Reid (1966) who found a maximum dry herbage yield of about 12 tons per hectare per annum with an N application of 400 kg per hectare, whereas at a dressing of 800 kg per hectare the maximum total N production had not yet been reached. However, unpublished results of the author suggest that such an amount of fertiliser cannot produce such high amounts of DM as 20 tons per hectare year after year on the same plot with applications of around 600 kg N per hectare. In the second year the sward becomes less dense and DM production decreases to 16 tons per hectare. It may be supposed that still heavier dressings will accentuate this effect.

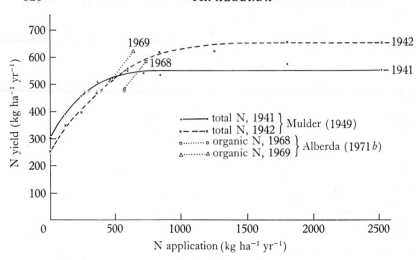

Figure 5. The influence of nitrogen fertilisation of swards on the yields of total nitrogen (after Mulder, 1949) and organic nitrogen (from Alberda, 1972).

There is still another reason why excessively high levels of fertiliser application should not be used to obtain high yields of protein. In Figures 2 and 3 the chain line in quadrant I indicates the situation in which just as much total N was harvested with the crop as had been applied. This means that the 600 kg per hectare needed to produce the maximum possible amount of DM can be recovered by the crop. Since this holds also for soils originally very poor in N, one may assume that such dressings do not lead to losses of N on any appreciable scale (Alberda, 1972). However, with dressings up to 1000 kg per hectare per annum distinct losses may occur of which a part may contribute to eutrophication of surface waters.

THE INFLUENCE OF THE CUTTING REGIME ON PROTEIN PRODUCTION

In Table 2 data from Mulder (1949) are given for the total N produced with 4 and 6 cuts per annum with different amounts of nitrogenous fertiliser. It can be seen that in 1941 6 cuts per annum gave a higher total N production than 4 cuts, but that there was hardly any difference between the two cutting regimes in the following year. In one experiment of the author 4 cuts per annum produced 605 kg per hectare organic N and 5 cuts 681 kg. However, the amounts of fertiliser N were 508 and 588 kg per hectare respectively so that 5 cuts produced less than 4 when expressed per kg N applied.

Also, other authors found no great influence of the number of cuts on the production of nitrogenous compounds. Cowling (1966) found no difference in N yields between 3, 6 and 62 cuts, although the DM yields decreased with the increase in number of cuts. Kadziulis & Tonkunas (1970) found that the efficiency of the N applied is independent of the number of cuts. There is only little evidence that there are marked optimum numbers of cuts per season to produce the largest amounts of nitrogenous compounds.

Table 2. *Amounts of fertiliser applied and total nitrogen recovered under two cutting regimes in two successive years (after Mulder, 1949)*

Year	1941		1942	
No. of cuts	4	6	4	6
Level of N application (kg per hectare per annum)	Total N recovered (kg per hectare per annum)			
0	282	307	208	179
120	351	395	282	272
240	409	462	333	—
360	457	513	422	451
600	505	552	555	555
840	493	539	609	631

AN ATTEMPT TO CALCULATE THE ORGANIC NITROGEN PRODUCTION OF A SWARD

In the early 1960s the author carried out a series of experiments to determine the rate of DM production of a closed green grassland sward under optimal conditions of water and nutrient supply in order to compare the results with the potential production calculated from a simulation programme devised by de Wit (1965). The results showed that the year to year variations were so small that for each month mean rates of herbage production could be calculated, which for the summer months came very close to the potential values. Furthermore it was established for each month how many days elapsed between cutting and the restoration of a closed green sward situation; this was called the recovery time. With a given herbage dry weight at the beginning of the growing season and a dry weight of 1500 kg each time the closed sward situation was reached, the total DM production could be calculated for any cutting regime (Alberda, 1971). As the total N content and the nitrate N content were also determined in these experiments, the organic N production could be

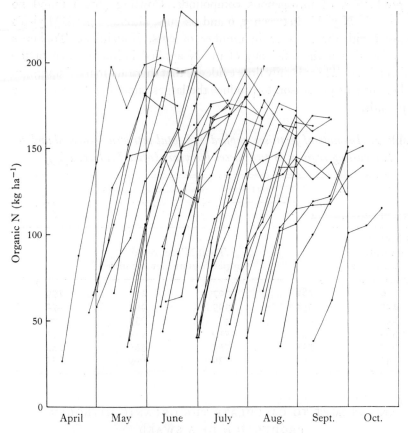

Figure 6. The production of organic nitrogen by swards at different times of the year.

calculated. The results are plotted in Figure 6. Although these results were less regular than those of DM production the general shape of the lines was the same; after a period of a fairly constant rate of production a ceiling value was reached where production stopped rather abruptly and sometimes became negative. These ceiling values dropped as the season progressed. From this figure also mean values were derived for mean monthly rates of organic N production and monthly ceiling yields. These figures are given in Table 3, together with the recovery times for each month.

Assuming that an amount of 75 kg per hectare of organic N is present when the crop is closed and an amount of 50 kg per hectare at 15 April, when the crop is also closed, the yearly organic N production can be calculated for any cutting regime. The results of such a calculation are given in Table 4. It can be seen that even with these rather rough data, there is

Table 3. *Ceiling yields and production rates for organic nitrogen by a closed green canopy, and recovery time to approximately 75 kg organic nitrogen per hectare during the months of April to September*

	April	May	June	July	Aug.	Sept.
Ceiling yield (kg per hectare)	200	185	165	145	130	110
Mean production rate (kg per hectare per day)	5.7	5.3	5.3	4.7	4.3	3.0
Recovery time (days)	25	20	22	25	27	31

Table 4. *Actual and calculated annual dry matter and organic nitrogen production by, and mean organic nitrogen concentration in, a sward under different cutting regimes*

	4 cuts	5 cuts	6 cuts	
DM production (tons per hectare)	20.9	22.3	—	actual
Organic N production (kg per hectare)	605	681	—	
Organic N content (%)	2.9	3.1	—	
DM production (tons per hectare)	18.6	21.1	18.4	calculated
Organic N production (kg per hectare)	570	680	630	
Organic N content (%)	3.1	3.8	4.1	

good agreement between the calculated amounts of organic N and those actually obtained. The calculations suggest an optimum of 5 cuts per season with a production of 680 kg organic N per hectare. These amounts were obtained with applications of around 600 kg N per hectare per annum.

When it is recalled that a nitrogen application higher than 600 kg per annum is likely to be unfavourable to the sward and that yearly DM yields on the same field will lie somewhat below the maximum of 20 tons per hectare it can reasonably be assumed that a sustained organic N production can amount to 600 kg per hectare per annum or, multiplied by 6.25, to 3700 kg organic N compounds.

REFERENCES

ALBERDA, TH. (1965). The influence of temperature, light intensity and nitrate concentration on dry-matter production and chemical composition of *Lolium perenne* L. *Netherlands Journal of Agricultural Science* 13, 335–60.

ALBERDA, TH. (1971). Potential production of grassland. In: Wareing, P. F., & Cooper, J. P. (Eds.), *Potential Crop Production. A Case Study*, 159–71. London: Heinemann.

ALBERDA, TH. (1972). Nitrogen fertilization of grassland and the quality of surface water. *Stikstof* 15 (Eng. ed.), 45–51.

COWLING, D. W. (1966). The response of grass swards to nitrogenous fertilizer. *Proceedings of the X International Grassland Congress*, 204–9.

FRANKENA, H. J., & DE WIT, C. T. (1958). Stikstofbemesting, stikstofopname en grasgroei in het voorjaar. *Landbouwkundig Tijdschrift* 70, 465–72.

DE GROOT, TH. (1966). De consequenties van de graslandintensivering voor de voorziening van het rundvee met organisch-chemische bestanddelen uit ruwvoer. *Stikstof* 51, 188–94.

KADZIULIS, L., & TONKUNAS, J. (1970). Nitrogen responses of legume-grass pastures in Lithuania. *Proceedings of the XI International Grassland Congress*, 397–400.

MULDER, E. G. (1949). Onderzoekingen over de stikstofvoeding van landbouwgewassen. I. Proeven met kalkammonsalpeter op grasland. *Verslagen van Landbouwkundige Fonderzoekingen in Nederland* 55, 7.

OOSTENDORP, D. (1964). Stikstofbemesting en bruto opbrengst van grasland. *Stikstof* 42, 192–202.

REID, D. (1966). The response of herbage yields and quality to a wide range of nitrogen application rates. *Proceedings of the X International Grassland Congress*, 209–13.

DE WIT, C. T. (1965). *Photosynthesis of Leaf Canopies*. Agricultural Research Reports No. 663. Wageningen: Centre for Agricultural Publications and Documentation.

DISCUSSION

By J. P. COOPER
Welsh Plant Breeding Station, Aberystwyth

AND P. F. WAREING
Department of Botany, University College of Wales, Aberystwyth

Protein production in plants can most usefully be considered in terms of the following input–output system.

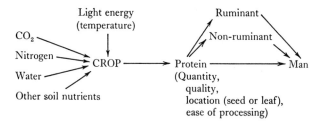

The efficiency of such a system can be defined in general terms as output/input, but, as pointed out by Spedding and Pirie, the most meaningful outputs and inputs to use will depend on the particular system. The appropriate output will usually be the quantity of total or utilisable protein but the relevant input may be the nitrogen (N) input, the land area or the energy input. For many purposes the output of utilisable protein per unit of land may be the most useful criterion.

Protein production by the crop can be limited not only by the input of N, as in much of British grassland, but also by the climatic inputs of light energy, temperature and water supply, as well as by other soil nutrients. In fact, in an intensive cropping system where ample N and water can be supplied the available light energy for photosynthesis and adequate temperatures for growth may well be the most important limitations, and the greatest return of protein per hectare for many field crops may be achieved by concentrating on maximum dry matter (DM) yield, rather than by stepping up the protein content. Even so Alberda has shown that in the grass crop, even when maximum DM production has been attained, further application of N can increase the N content of the crop, though it is not certain how much of this increase is in the form of protein.

The importance of the legume–*Rhizobium* association in providing the necessary N input was stressed by Donald, who pointed out that over large areas of the world this was the main source of both seed and leaf protein, while Braude mentioned the possible danger that intensive N inputs, particularly on grassland, could lead to the depletion of other elements important in animal nutrition, such as calcium and magnesium.

The exact output requirements of a cropping system will differ according to the purpose for which the protein is required. The protein yield per hectare will be determined by both the yield of DM and its protein content, while its quality in terms of amino acid composition, and the ease of harvesting, storing and processing can also be important. Seed or leaf proteins differ markedly in most of these characteristics.

The requirements of the following methods of processing must be considered:

1. Direct human consumption. Here a high protein content, as in seed proteins, is valuable, but so is adequate protein quality, in which certain seed proteins, such as cereals, may be deficient. As Pirie pointed out, the current intake in a European diet of leaf protein from vegetables is usually about 0.1 g per day, but this could be raised without difficulty to 1–2 g, while in Java, for instance, the intake may be 5–6 g.

2. Mechanical or chemical processing, as in the direct extraction of leaf protein, or the processing of seed protein to provide meat analogues.

3. Feeding to ruminants. Since most ruminants do not require more than about 10 per cent crude protein in the diet, a high protein content in the main feed is not necessary, nor is protein quality important, since the amino acids used by the ruminant are synthesised afresh in the rumen.

4. Feeding to non-ruminants, where in contrast to the ruminant, protein quality is important.

5. Processing through micro-organisms, which may be able to upgrade low quality protein.

These different methods of processing are not, of course, exclusive, as pointed out by Pirie and Alberda, in that a useful amount of leaf protein can be extracted from an intensively grown vegetable or grass crop and still leave a residue with adequate protein content for the ruminant, or for subsequent microbial processing.

Rather less is known of the characteristics of the plant which determine the efficiency of conversion of these inputs into the required outputs. These include the rates of the various stages of protein synthesis, the amino acid composition of the protein and also the partition of protein between leaf and seed. For instance, it has been suggested that under favourable light and temperature conditions and adequate N supply, the use of assimilates for protein synthesis may be limited by the activity of certain rate-limiting enzymes, e.g. nitrate reductase (Hageman, Lang & Dudley, 1967), by amino acid activation or indeed the transcription process.

In any particular environment, therefore, it should be possible to quantify the relevant inputs to the crop and, knowing the rate of conversion of these inputs by the plant, obtain an estimate of potential protein production. Several such estimates have been made for potential DM production and energy conversion (Holliday, 1967; Cooper, 1970) and, from a knowledge of the usual range of protein contents of the leaf or seed, these estimates can be extended to protein production. For most closed crop surfaces, for instance, the equivalent of 2.5–3.0 per cent conversion of incoming light energy into total DM during the growing season is often regarded as a useful prediction of potential DM production, when water and soil nutrients are not limiting. In a vegetative crop, such as the forage grasses, this corresponds to about 20 tons per hectare in western Europe and over 40 tons per hectare in the wet tropics. Assuming 15 per cent crude protein in the leaf, this would provide 3 tons per hectare leaf protein in western Europe and over 6 tons per hectare in the wet tropics, figures similar to those mentioned by Pirie for vegetable crops. Similar estimates can be made for cereal crops and for the grain legumes, and these estimates can then be compared with the values quoted by Pirie and Whitehouse for the output of seed or leaf protein in different environments.

The time scale over which both estimates and observations of protein production are obtained may also be important. As pointed out by Spedding, for most agricultural crops, a year or a growing season is the most appropriate unit, though year to year fluctuations should be taken into account. For some cropping systems, however, winter and summer crops should be considered separately, while in certain ecological situations, such as natural grasslands, continued production over a number of years is important, as stressed by Phillipson.

Many quantitative models of potential photosynthetic production in different environments have recently been developed (de Wit, Brouwer & Penning de Vries, 1971; Duncan *et al.* 1967; Idso & Baker, 1967) and these models could be extended to take account of protein production by the crop. Furthermore, as mentioned by Spedding and Boddington, once the appropriate models have been developed in biological terms, it should be possible to attach economic values to the various inputs and outputs.

Of course, the initial protein synthesis in the food chain does not inevitably have to be carried out by higher plants. As Blaxter pointed out an alternative chain can be added to the flow diagram, starting from the original N input and producing by direct chemical processes, either amino acids for human consumption, or urea for feeding to the ruminant.

Having considered the efficiency of current input–output systems and their potential protein production, the question arises, how can the efficiency of such systems be increased? This can be done firstly by choosing the most appropriate species for producing leaf or seed protein, possibly including currently uncultivated species such as water hyacinth (*Ipomoea aquatica*), as suggested by Pirie, and secondly, by improving existing crop species. The plant breeder can often alter the protein content of the crop, although as Whitehouse pointed out, in cereals the protein content of the grain is usually inversely correlated with yield. The earlier *Triticale* hybrids, for instance, which had an encouragingly high protein content, also had a low yield, and in fact useful varieties with both high grain yield and high protein content have not yet been obtained. Similarly, in many forage crops, the protein content has a high heritability, over 60 per cent in ryegrass, and can be readily selected by the plant breeder, but is usually inversely correlated with DM yield and, in particular, with the content of water-soluble carbohydrates which may be important in determining intake or silage quality. So that in forage crops an increase in protein production per unit area is often best achieved by increasing DM yield rather than protein content, particularly since protein contents in excess of 15 per cent are well above the requirements of the ruminant. In forage and vegetable crops, the breeder can also modify such features as lignin content or cellulose fractions, which influence both digestibility by the ruminant and the ease of mechanical processing (Cooper, 1973).

The breeder can also attempt to modify protein quality, in terms of amino acid composition, as has been done successfully in the high-lysine lines of maize and other cereals, discussed by Whitehouse. Such genetic modification is likely to be easier for seed storage proteins than for the functional proteins of the leaf.

More rapid progress in such a breeding programme is likely to be achieved if the rate-limiting physiological and biochemical steps in protein production can be identified. However, as Keys made clear, we are still ignorant of some of the most important steps in metabolism leading to protein synthesis. Thus, although our knowledge of nitrate reduction is reasonably extensive, we know little about the precise steps involved in the incorporation of ammonium ions into amino acids within the plant, viz. the extent to which they are formed in the root or the shoot, and, for amino acids synthesized in the shoot, whether they are formed predominantly in the mature leaves and exported to the young developing leaves (the main sites of protein synthesis), or whether they are mainly synthesized *in situ* in the latter. A fuller understanding of the physiological basis of the mobilisation and recirculation of leaf amino acids to the seeds is clearly important also for both grain and pulse crops.

In crops used for forage or for direct protein extraction it is necessary to consider not only the total leaf protein content, but also how this content is made up from different cellular components. Thus, leaf protein comprises cytoplasmic and chloroplast fractions, and each of these include both structural and enzyme proteins. The chloroplasts contain approximately 60–70 per cent of the total soluble proteins of the leaf and a certain proportion of this plastid protein ('Fraction I') appears to consist of the enzyme ribulose-1,5-diphosphate carboxylase ('carboxydismutase'), which is the enzyme involved in the primary fixation of carbon dioxide. These facts are clearly important for any attempts to increase the protein content of leaves by breeding. For example, protein content could probably be increased by increasing the number of chloroplasts per cell, but since chloroplasts are semi-autonomous organelles, this character is probably not inherited in an entirely Mendelian manner. A knowledge of how the total protein content is made up as between cytoplasmic and plastid fractions would seem to be relevant in selecting species for direct protein extraction, although Pirie stated that other characteristics, such as the pH of the extract, and the content of phenols and tannins, are also important.

Recent studies have revealed marked differences in the patterns of synthesis of cytoplasmic and plastid proteins during the life of the leaf. Thus, during the growth phase of the leaf, active synthesis of both fractions occurs, but as the leaf approaches its full size and plastid development is completed, synthesis of plastid protein effectively ceases, and in many plants there is a steady decline in the total plastid protein levels throughout the further life of the leaf (Smillie, 1969; Treharne et al. 1970; Woolhouse, 1968). On the other hand, the capacity for synthesis of cytoplasmic protein appears to be retained, even after full leaf expansion has been attained. The decline in the protein content of the leaf as it ages appears to be brought about by competition between the mature leaves and the new leaves continuing to be formed by the apical meristem, since if the shoot is decapitated and the axillary buds are removed, the decline in protein content of the mature leaves is arrested and indeed it may actually increase, a process referred to as 're-greening'. The mechanisms controlling protein breakdown and the export of amino acids from mature leaves are not fully understood, but it seems likely that they involve hormonal factors, such as cytokinins and gibberellins, which markedly affect protein synthesis in leaves.

Clearly, where leaves are to be used as a protein source for non-ruminants or for direct protein extraction, the percentage protein content (per unit of DM) is important, and if protein synthesis in the young leaves is achieved partly at the expense of the protein content of the older leaves, due to the recirculation and re-utilisation of amino acids, then it is not necessarily an advantage to have a high rate of new leaf production, since the overall protein content of the shoot will not keep pace with the total DM content. Thus, it would seem desirable to consider means of maintaining the protein content of the older leaves, and two possible methods suggest themselves:

1. To produce varieties in which the rate of new leaf production is in reasonable balance with the capacity of the plant to produce new organic N. Since this may result in a reduced rate of new leaf production and hence of DM production, different growth rates would seem to be required, depending upon whether the crop is to be used as food for ruminants or for non-ruminants.

2. Decapitation of the shoot (topping) when it has attained an optimum size. This treatment would not seem appropriate for most existing crop species (except possibly kale), but it may become important if new species are used for direct protein extraction. The use of 'chemical pruning' to destroy the apical meristems at the required stage of growth might be a useful alternative to mechanical topping. However, Alberda suggested that there might be other undesirable side-effects if the life of individual leaves was prolonged, such as the accumulation of lignin and polysaccharides.

In his paper, Keys suggests that one means by which the content of storage protein might be increased is to select and breed for genotypes in which there is modification of the control mechanisms, such as endproduct inhibition, which regulate various steps in protein synthesis. Wareing suggested that, in addition to such modification of the control mechanisms, consideration might also be given to the possibility of increasing the amount of DNA coding for the desired storage proteins, by 're-iteration' of the appropriate genome sequences. It is now well established that certain sections of the genome, such as those coding for ribosomal-RNA, are reiterated many times. Where the overall rate of protein synthesis is limited by the rate of transcription of DNA (synthesis of messenger-RNA) it may well be an advantage to re-iterate the relevant sequences. A crude method of such re-iteration might be achieved by reduplication of certain chromosome segments, but it would appear that natural re-iteration is probably achieved by other, less crude, means.

In conclusion, it seems clear that although useful genetic variation has been detected in many crop species for both protein content and protein quality, we are still a long way from understanding the physiological and biochemical basis of this variation, and from using such information in an advanced breeding programme (Hageman et al. 1967).

REFERENCES

COOPER, J. P. (1970). Potential production and energy conversion in temperate and tropical grasses. *Herbage Abstracts* **40**, 1–15.

COOPER, J. P. (1973). Genetic variation in herbage constituents. In: Butler, G. W., & Bailey, R. W. (Eds.), *The Chemistry and Biochemistry of Herbage* (in press). New York and London: Academic Press.

DUNCAN, W. G., LOOMIS, R. S., WILLIAMS, W. A., & HANAN, R. (1967). A model for simulating photosynthesis in plant communities. *Hilgardia* **38**, 181–205.

HAGEMAN, R. H., LANG, E. R., & DUDLEY, J. W. (1967). A biochemical approach to corn breeding. *Advances in Agronomy* **19**, 45–86.

HOLLIDAY, R. (1967). Solar energy consumption in relation to crop yield. *Agricultural Progress* **41**, 24–34.

IDSO, S. B., & BAKER, D. G. (1967). Method for calculating the photosynthetic response of a crop to light intensity and leaf temperature by an energy flow analysis of the meteorological parameters. *Agronomy Journal* **59**, 13–21.

SMILLIE, R. M. (1969). Synthesis of plastid and mitochondria proteins. *Proceedings of the Eleventh International Botanical Congress, Seattle*, 203.

TREHARNE, K. J., STODDART, J. L., PUGHE, J., PARANJOTHY, K., & WAREING, P. F. (1970). Effects of gibberellin and cytokinins on the activity of photosynthetic enzymes and plastid ribosomal-RNA synthesis in *Phaseolus vulgaris* L. *Nature* **228**, 129–31.

DE WIT, C. T., BROUWER, R., & PENNING DE VRIES, F. W. T. (1971). A dynamic model of plant and crop growth. In: Wareing, P. F., & Cooper, J. P. (Eds.), *Potential Crop Production. A Case Study*, 117–42. London: Heinemann.

WOOLHOUSE, H. W. (1968). Leaf age and mesophyll resistance as factors in the rate of photosynthesis. *Hilger Journal* **11**, 7–12.

PART III

THE BIOLOGICAL EFFICIENCY OF PROTEIN PRODUCTION BY ANIMALS

PART III

THE BIOLOGICAL EFFICIENCY OF PROTEIN PRODUCTION BY ANIMALS

10

CONSIDERATIONS OF THE EFFICIENCY OF AMINO ACID AND PROTEIN METABOLISM IN ANIMALS

By P. J. BUTTERY

University of Nottingham School of Agriculture, Sutton Bonington

AND E. F. ANNISON

Unilever Research Laboratory Colworth/Welwyn,
Colworth House, Sharnbrook, Bedford

The term 'biological efficiency of protein synthesis' may be interpreted in many ways. In its simplest form it might merely refer to the ratio: synthesised protein nitrogen/nitrogen supplied × 100, thereby ignoring the energy cost of synthesis. Alternatively, the term might refer to the energy cost of protein synthesis from dietary protein and be expressed in terms of the calorific value of the synthesised protein (Blaxter, 1967; Armstrong, 1969). We have chosen to consider the theoretical energy cost of protein synthesis from amino acids and an energy source using currently available information on the various steps involved in peptide bond formation. These calculations refer to ideal situations and provide some guide to the maximum possible efficiencies of protein synthesis. The wide variety of proteins synthesised by the mixture of cell types which comprise most animal tissues, however, implies that an ideal balance of individual amino acids for the synthesis of specific proteins is rarely encountered *in vivo*. Two of the many factors which impinge on 'biological efficiency', amino acid imbalance and the specific dynamic action (SDA) which accompanies protein and amino acid feeding, have therefore been considered at some length. As a further example of the multiplicity of factors which must be considered when the biological efficiency of protein synthesis is evaluated, we have briefly considered the microbial synthesis of protein in the rumen in relation to the overall economy of the ruminant.

MECHANISM OF PROTEIN SYNTHESIS

In the following discussion of some of the currently held views on the mechanism of protein synthesis in animal cells, we have made extensive use of the recent reviews of Munro (1970) and Baglioni & Colombo (1970). Special emphasis will be given to those steps in the process which require the expenditure of energy. It should be pointed out that, since micro-organisms make more suitable experimental subjects for protein synthesis studies than normal animal tissues, many of the currently accepted views of protein synthesis in the animal cell have been arrived at by extrapolation of results obtained from experiments with micro-organisms. In recent years several quite marked differences between the mechanism of protein synthesis in procaryotic (non-nucleated) and eucaryotic (nucleated) cells have been noted.

In animal cells the nucleus contains most of the total cellular DNA; significant quantities are also found in the mitochondria. In the nucleus the DNA is associated with a group of basic proteins called histones, a small amount of non-histone protein and a little RNA to form the substance chromatin of which the chromosomes are composed. The DNA directs the synthesis of RNA by the process of base pairing. In fact, if isolated chromosome preparations and ribonucleoside triphosphates are incubated together RNA is synthesised. The rate of synthesis can be greatly enhanced by the addition of the enzyme RNA polymerase (nucleoside triphosphate: RNA nucleotidyl transferase). The reaction can be summarised as:

$$n\,ATP + n\,UTP + n\,GTP + n\,CTP \underset{\longleftarrow}{\overset{Mg^{2+} + DNA}{\longrightarrow}} \begin{bmatrix} AMP \\ UMP \\ GMP \\ CMP \end{bmatrix}_n + n\,(P_i \sim P_i)$$

The pyrophosphate released is then broken down by the active enzyme pyrophosphate phosphohydrolase to inorganic phosphate; the energy of the pyrophosphate bond (6.5 kcal mol^{-1}, 28 kJ mol^{-1}; Atkinson & Morton, 1960) being lost as heat. It is worth noting that the hydrolysis of ATP to AMP and pyrophosphate yields nearly 20 per cent more energy than does the hydrolysis of ATP to ADP and P_i.

The rate of synthesis of RNA from chromatin is much greater, at least in *in vitro* systems, if the DNA is separated from chromosomal protein. Much speculation has been made as to the role of the histones as regulators of protein synthesis (Stellwagen & Cole, 1969). The suppression of DNA template activity varies from tissue to tissue.

Chromosomal preparations from tissues actively synthesising RNA, such as liver, have 20–30 per cent of the DNA template activity of the DNA extracted from the same tissue. In contrast, chromosomal preparations from nucleated duck erythrocytes, a tissue which makes virtually no RNA, have no DNA template activity. However, exact knowledge of the method of the regulation of DNA-directed RNA synthesis at specific cistrons is lacking. The histones so far isolated do not *appear* to be sufficiently different to account for the very specific role ascribed to them by several groups of workers. The synthesis of nuclear protein appears to take place in the cytoplasm and the synthesised protein is then translocated to the nucleus. Rapid movements of protein to and from the nucleus have been observed in *Amoeba proteus* (Goldstein & Prescott, 1968) and insect chromosomes have been seen to swell as a result of collecting protein from parts of the cell other than the nucleus (Beredes, 1968).

In normal animal cells the synthesis of all types of RNA is directed by the base sequence of the DNA from which the RNA is being transcribed. The RNA polymerase initiates the RNA synthesis at specific sites on the DNA. The RNA polymerase fraction, from *E. coli* at least, can be dissociated into several components including the so-called 'sigma factor'. This factor is released on the initiation of RNA synthesis and is then available for use at another initiation site (Geiduschek & Haselkorn, 1969). Other factors have been reported as being necessary for the termination of RNA synthesis, e.g. the 'rho factor' (Roberts, 1970), although certain codons on the DNA are thought to have the ability to induce the termination of RNA synthesis (Chamberlin, 1970).

There are three major types of RNA; messenger RNA (mRNA), ribosomal RNA (rRNA) and transfer RNA (tRNA). Ribosomal RNA, unlike the other types of RNA, is synthesised in the nucleolus and not the nucleoplasm.

Ribosomal RNA when it is first formed has a sedimentation coefficient of 45S and is subsequently broken down to form molecules of 18S and 28S which, when combined with protein, form the 40S and 60S ribosome components. The ribosomal proteins appear to originate from the cytosol. There are smaller quantities of another RNA species (5S) in ribosomes.

Transfer RNA accounts for about 10 per cent of total cellular RNA. There appears to be at least one specific tRNA for each amino acid. Several of these tRNAs have been isolated and all appear to have similar but not identical structures and to contain about 80 nucleotides. When drawn in two dimensions in such a way as to account for the base pairing that takes place they assume a shape which resembles that of a clover leaf (Figure 1). The actual shape of tRNAs *in vivo* is unknown but from thermodynamic

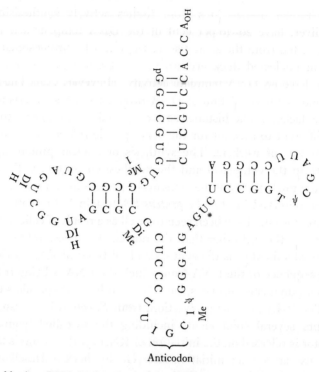

Figure 1. Alanine tRNA in the clover leaf arrangement. Abbreviations used were: A, adenosine or adenylic acid; C, cytidine or cytidylic acid; G, guanosine or guanylic acid; U, uridine or uridylic acid; p, phosphate residue (on the left of the nucleoside symbol it means a 5'-phosphate). The subscript OH is used to emphasise the presence of a 3'-hydroxyl group; Me, a methyl group whose position is as indicated; IMeG, 1-methylguanine, DiMeG, N^2-dimethylguanosine; I, inosine; ψ, pseudo-uridine; U*, a mixture of uridine and dihydro-uridine. (From Kit, 1970.)

considerations it would appear that the three loops seen in the two-dimensional representation are likely to occur (Ninio, Favre & Yaniv, 1969). The sequence of the three bases forming the OH-terminal end of the molecule, namely cytosine–cytosine–adenine–OH, is common to all tRNAs. The three bases which pair up with the codons on the mRNA are found in the so-called anticodon loop of the tRNA. The sequence of the three bases making the anticodon is peculiar to each type of tRNA.

Messenger RNA accounts for some 1–5 per cent of total cellular RNA. mRNAs differ in size, nucleotide composition and stability. They may either code for the synthesis of one protein or for several proteins (mono- or polycistronic).

Before amino acids can be incorporated into protein they must be activated and coupled to their specific tRNA. The reaction involved is as follows:

amino acid + enzyme + ATP $\rightleftharpoons$ enzyme $\sim$ amino-acyl-AMP + $P_i \sim P_i$

enzyme $\sim$ amino-acyl-AMP + tRNA $\rightleftharpoons$ enzyme + AMP + amino-acyl $\sim$ tRNA

$P_i \sim P_i \longrightarrow 2P_i$

It is interesting to note that the energy of the acyl bond in the amino-acyl tRNA is about 7 kcal mol^{-1} (29 kJ mol^{-1}), which is more than sufficient for the formation of the peptide bond which needs approximately 3 kcal mol^{-1} (13 kJ mol^{-1}) (Lengyel, 1969). There appears to be at least one specific enzyme (amino-acyl transfer RNA synthase) for the combination of each amino acid with its specific tRNA. The recognition sites for the enzyme on the tRNA are known in broad outline and do not appear to be the same as the anticodon site (Mirzabekov et al. 1971). Once the amino acids have been activated by combination with tRNA they are ready to be incorporated into proteins by the action of ribosomes following the directions held on the messenger RNA.

Eucaryotic ribosomes contain equal parts of RNA and protein and consist of two components – the so-called 40S component and the larger 60S component. The two components straddle the messenger RNA and the actual synthesis of the peptide bond takes place in the region of the cleft between the two components.

The intact ribosome has two binding sites for tRNA, namely an amino-acyl site and a peptidyl site. Initially the changed tRNA binds to the amino-acyl site and when it has acquired the growing peptide chain it moves to the peptidyl site.

The details of peptide bond synthesis in animal cells are lacking, although much information is available for bacterial cells.

In bacterial cells the initial step appears to be the assembly of the 30S ribosome component (equivalent to the 40S component in animal cells), mRNA, formyl-methionyl-acyl tRNA, GTP and several so-called initiation factors (Nomura & Lowry 1967). The formyl-methionyl-acyl tRNA binds with an initiator codon (adenine–uracil–guanine) followed by the addition of the 50S ribosome component (equivalent to the 60S ribosome component in animal cells). The formyl-methionyl-acyl tRNA is then transferred to the peptidyl site of the ribosome. At the same time GTP is hydrolysed to GDP and P_i. The next step involves the complexing of further protein factors with the amino-acyl tRNA that is coded by the next codon on the mRNA and GTP. The complex is bound to the ribosome in such a way that the new amino-acyl tRNA is inserted into the amino-acyl site of the ribosome. GTP is cleaved at this stage. A peptide bond is now formed between the carboxyl group of the formyl methionine and the amino group of the activated amino acid. The newly formed peptidyl tRNA is now transferred to the peptidyl site of the ribosome (Figure 2). At the same time as this transfer takes place GTP is hydrolysed. This process continues using the amino-acyl tRNAs coded for by the mRNA until the ribosome reaches the codon uracil–adenine–adenine, uracil–adenine–guanine or uracil–guanine–adenine, when polypeptide chain synthesis is terminated. The formyl-methionine is then removed from the N-terminal end of the released polypeptide chain (Livingston & Leder, 1969). The polypetide chain then folds up to form the protein. It is unlikely that this process requires the expenditure of large quantities of energy. The synthesis of the peptide bond in the ribosome would therefore appear to require the hydrolysis of two GTP β-phosphoanhydride bonds.

For the synthesis of the polypeptide chain it is of course necessary for the ribosome to move along the mRNA. Several suggestions have been put forward to explain in detail the movement of the ribosome along the message, e.g. the Inchworm hypothesis (Hardesty, Culp & McKeehan, 1969) where the mRNA folds inside the ribosome, and the sliding ribosomal subunit hypothesis (Bretscher, 1968). When the ribosomes have finished synthesising a polypeptide chain they break away from the mRNA and dissociate into their subunits.

In animals the mechanism of synthesis of the peptide bond appears to be similar to that in bacteria. It is not thought that formyl-methionyl-acyl tRNA acts as an initiator in animal systems, at least in cytoplasmic protein synthesis; mitochondrial protein synthesis appears to be identical to that of bacteria. In animal cells it has been suggested (Smith & Marcker, 1970) that methionyl-acyl tRNA may act as an initiator. The tRNA and the codon on the mRNA involved are thought not to be the same as would be required to insert methionine at any other point in the polypeptide chain. The enzymes involved in the synthesis of the peptide bond in animal cells have been partially characterised. It would appear that one enzyme is involved in the binding of the amino-acyl tRNA to the ribosome and another quite distinct enzyme is involved in the transfer of the growing polypeptide chain to the amino-acyl tRNA at the amino-acyl site of the ribosome (Hardesty et al. 1969). Both these enzymes will catalyse the hydrolysis of GTP although under slightly different conditions.

The mRNA can carry more than one ribosome at any one time. The message with its complement of ribosomes is termed a polysome. It has been found that some ribosomes are attached to the endoplasmic reticulum while others are found free in the cytoplasm. It would appear that the cytoplasmic ribosomes are responsible for the synthesis of proteins for use

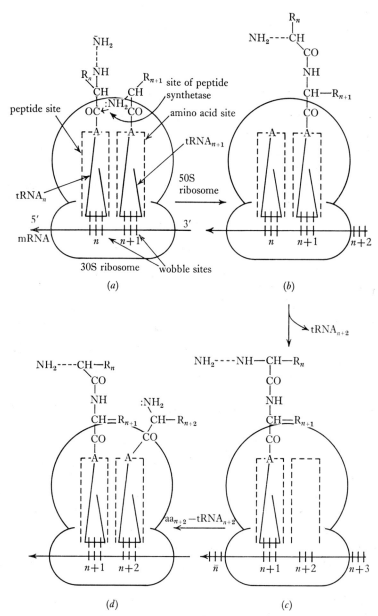

Figure 2. The scheme of peptide bond formation and translocation in protein synthesis. Peptide bond formation takes place in (*a*) and is followed by translocation of the peptidyl-tRNA in (*b*) and (*c*); binding of the incoming amino acid takes place in (*d*). (From Bretscher, 1968.)

within the cell while the membrane-bound ribosomes are responsible for the synthesis of proteins for export from the cell, e.g. serum albumen (Birbeck & Mercer, 1961).

ENERGETICS OF PROTEIN SYNTHESIS

From consideration of the current knowledge on the mechanism of protein synthesis it is possible to calculate the energy required for the synthesis of a peptide bond. The activation of an amino acid and its subsequent incorporation into a peptide bond would require the hydrolysis of four β-phosphoanhydride bonds of ribonucleoside triphosphates, the AMP generated in the activation step being assumed to be metabolised as follows:

$$AMP + 2ATP \rightleftharpoons 2ADP$$
$$2ADP + 2P_i \rightleftharpoons 2ATP.$$

If one assumes that the ΔG for the hydrolysis of the β-phosphoanhydride bond of adenosine and guanosine triphosphate to be -7 kcal mol^{-1} (-29 kJ mol^{-1}) and the ΔG for the hydrolysis of the peptide bond to be -3 kcal mol^{-1} (-13 kJ mol^{-1}), then the efficiency of incorporation of energy into the peptide bond is about 10 per cent $\left(\text{i.e. } \dfrac{3}{4 \times 7} \times 100\right)$.

Since, as mentioned previously, the energy of the amino-acyl bond is ample for the synthesis of the peptide bond it is interesting to speculate on the function of the two GTPs hydrolysed during the synthesis of the peptide bond. Lengyel (1969) speculates that the energy of these GTP molecules is used to drive the ribosome along the mRNA. For the present argument it is surely sufficient to say that the high energy cost of peptide bond synthesis is the price that has to be paid for the phenomenal specificity of the process. A complete evaluation of the energy cost of protein synthesis must also take into account the energy required to synthesise tRNA, RNA, mRNA and ribosomal protein. Adequate data on which to base this analysis is available for only a few proteins and we have chosen to consider myosin, a major component of the animal carcass.

Myosin, molecular weight about 500000, is made up of subunits of molecular weight about 200000 (Perry, 1967). Herrman (1968) estimated that a molecule of myosin could be synthesised every 1.44×10^2 s by the appropriate mRNA, i.e. each codon on the mRNA will be used every 1.44×10^2 s. If we assume that the half life of myosin mRNA is similar to that of the average mRNA of liver (1.44×10^4 s; Tominaga, 1971), then each codon could code for the synthesis of about 100 peptide bonds with the hydrolysis of 400 high-energy phosphate bonds. The synthesis of a codon on the mRNA would require the hydrolysis of six high-energy phosphate bonds. Thus the cost of synthesis of mRNA is minimal compared with the energy required for the synthesis of the peptide bonds which it directs. In bacterial systems it has been postulated that mRNA turnover accounts for less than 5 per cent or perhaps less than 1 per cent of the total energy required for protein synthesis (Salser, Gesteland & Ricard, 1969).

Only a small proportion of ribosome subunits are not involved in protein synthesis at any one time, 10 per cent in reticulocytes and 5 per cent in liver (Munro, 1970). Since 1 per cent of liver ribosomes are replaced each hour (Enwonwu & Munro, 1970) it can be concluded that the life expectancy of a liver ribosome subunit is about 3.6×10^5 s. During this time a chick embryo muscle ribosome would synthesise 2,500 myosin subunits containing 5×10^6 peptide bonds with the consumption of approximately 20×10^6 high-energy phosphate bonds. The molecular weight of ribosomal RNA is about 2.5×10^6 (see Munro, 1970) which is equivalent to about 7.5×10^3 bases. Since 50 per cent of a eucaryotic ribosome is protein it can be calculated that each ribosome contains 2.5×10^4 amino acids polymerised to form proteins. The total cost of synthesis of a ribosome is therefore approximately $(7.5 \times 2 \times 10^3) + (2.5 \times 10^4 \times 4) = 1.15 \times 10^5$ high-energy phosphate bonds plus any energy required for the synthesis of the ribosomes, mRNA and tRNA used for the synthesis of the ribosome. Even allowing for the turnover of the fragments of rRNA in the nucleolus it would appear, provided one accepts the assumptions made in the above calculations, that the energy required for the synthesis of a ribosome is about one order of magnitude less than the energy required for the synthesis of the peptide bonds that are promoted by the ribosome.

Little is known about the half life of tRNAs, but it is likely that they are able to be used many times before being degraded.

It would therefore appear that the energy required for the activation of amino acids and their condensation into peptide bonds is by far the greater part of the total energy required for the synthesis of proteins. To a first approximation, the number of high-energy phosphate bonds required for the synthesis of a protein is four times the total number of amino acids in the protein. It is not necessary to assume that in a protein the number of peptide bonds approximates to the number of amino acids (see, e.g. Blaxter, 1967), since each protein would appear to be initiated with an amino acid which is subsequently removed by the hydrolysis of a peptide bond.

Practical implications

The efficiency of incorporation of energy into the peptide bond from preformed high-energy phosphate bonds was shown earlier to be about

10 per cent. If the high-energy bonds are generated by the oxidative catabolism of glucose

(i.e. $C_6H_{12}O_6 + 6O_2 \longrightarrow 6CO_2 + 6H_2O$, $\Delta G \equiv -687$ kcal, $-2.9 + 10^3$ kJ),

then since the complete metabolic oxidation of glucose yields 38 ATP and the energy content of the peptide bond is about 3 kcal mol^{-1} (13 kJ mol^{-1}), the efficiency of protein synthesis is

$$\frac{(38/4) \times 3}{687} \times 100 = 4 \text{ per cent.}$$

When acetate serves as the main source of energy, as in ruminants, the overall efficiency of protein synthesis will be slightly reduced since the production of high-energy phosphate bonds from acetate is about 15 per cent less efficient than when glucose is the energy source (Blaxter, 1967).

The energy cost to animals for the synthesis of protein can be calculated. If it is assumed that the average molecular weight for amino acid is 100 then $(687/38) \times 4 \times 10^{-2} = 0.75$ kcal (3.14 kJ) is the energy required to synthesise 1 g of protein, excluding any consideration of the energy content of the amino acids incorporated into the protein. If the assumption is made that the metabolisable energy of protein is 5.12 kcal g^{-1} (21.42 kJ g^{-1}) then the total cost of protein synthesis is approximately 5.87 kcal g^{-1} (25 kJ g^{-1}) (Blaxter, 1967).

The efficiency of protein synthesis in animal production is conveniently expressed as follows (Blaxter, 1967; Armstrong, 1969):

$$\frac{\text{Calorific value of protein}}{\text{Calorific value of protein} + \text{calories required to synthesise ATP for peptide bond synthesis}} \times 100.$$

The value obtained for any particular protein is dependent on the value assigned for the moles of ATP/GTP required for the synthesis of each peptide bond and the source of energy for ATP synthesis. Armstrong (1969), on the basis of knowledge available when the review was written in 1964, calculated the efficiency of synthesis of casein assuming values for the number of high-energy bonds per peptide bond of 8, 10 and 12, which gave theoretical efficiencies of 81.4, 77.9 and 74.4 per cent respectively, when glucose was the energy source. When the best currently available estimate is used, i.e. 4 moles ATP/GTP per peptide bond, as discussed earlier, then an efficiency of about 90 per cent is obtained, using the same method of calculation. Blaxter (1967), using a similar method of calculation and a value of 3 high-energy bonds per peptide bond, calculated the theoretical efficiency of casein synthesis to be 91 per cent.

These calculations, however, neglect any consideration of protein turnover. In the young female rat (100 g body weight) 713 mg of muscle protein is synthesised per day while 564 mg is degraded (Millward, 1970). The net protein deposition is therefore 149 mg per day. Assuming, for the moment, that there is no loss of the amino acids released on the degradation of protein the total cost of synthesis of 1 g of muscle protein would be

$$(713/149 \times 0.75) + 5.12 = 8.64 \text{ kcal } (35.5 \text{ kJ}).$$

The relevance of data obtained with the young female rat to protein synthesis in farm animals is questionable, but information on the latter species was not available to us. If the data are applicable, however, protein turnover is obviously responsible for a major reduction in the overall efficiency of protein synthesis. These calculations would have to be modified if the evidence of Brostrom & Jeffay (1970) that protein degradation requires energy is substantiated. The effect of loss of amino acids released from proteins during protein turnover would depend on the fate of the lost amino acids. If they were catabolised within the tissue the energy generated would be available for protein synthesis; on the other hand, if the amino acids were catabolised to yield energy in another tissue there would be a reduction in the total efficiency of the protein synthesis within the former tissue but, presumably, the effect upon the efficiency of protein synthesis in the animal as a whole would be controlled by the efficiency of production of ATP during the catabolism of the amino acid. The production of ATP from amino acids is on average 20 per cent lower than from carbohydrate (Krebs, 1964).

Studies on whole animals confirm the prediction presented above that the cost of protein deposition is in excess of the cost of synthesis of the peptide bonds and the energy content of the amino acids in the deposited protein. Kielanowski (1965) measured the energy intake and deposition of fat and protein in young piglets and from a knowledge of the cost of fat deposition was able to extrapolate his data to obtain an estimate of the energy cost of protein deposition. Kielanowski suggested that 7.51 kcal (31.5 kJ) were required for the deposition of 1 g of protein. Kielanowski's value and that predicted above from theoretical considerations for the deposition of muscle protein in the young female rat (8.64 kcal) are in broad agreement, especially when allowance is made for the expected slower rate of protein turnover in the pig than the rat, at least as judged by albumin turnover (Munro, 1969) and for the fact that Kielanowski's value takes into account the total protein-synthetic activity of the animal.

THE EFFECTS OF AMINO ACID IMBALANCE

The amino acid composition of a protein is rigidly governed by the base sequence of that part of the DNA which codes for the particular protein. For the synthesis of a protein to proceed at its maximum rate all the amino acids to be incorporated into the protein must be freely available. During starvation there is a marked reduction in protein-synthetic activity, as revealed by changes in the polysome distribution of certain tissues (Figure 3). Not surprisingly the omission of one essential amino acid from the diet results in a reduction of protein synthesis at least as judged by liver polysome profiles. This is most easily illustrated for the amino acid tryptophan, since tryptophan pool sizes in animal cells are very small (Wunner, Bell & Munro, 1966). Under appropriate conditions similar effects can be seen on the omission of other essential amino acids (Pronczuk, Rogers & Munro, 1970).

Amino acids present in excess are degraded to provide energy (ATP). As stated previously, this process is less efficient than the conversion of the energy contained in carbohydrate or fat. In addition energy is required for the synthesis of urea from the ammonia produced.

Solberg, Buttery & Boorman (1971) examined the effects of feeding diets marginally deficient in methionine to chicks. Compared with methionine-supplemented diets the methionine-deficient diet caused increased food intake, less efficient food conversion, decreased nitrogen retention and increased uric acid excretion; measurement of plasma uric acid concentrations and liver xanthine dehydrogenase activity also indicated a more active state of uric acid synthesis in methionine-deficient chicks (Tables 1 and 2). The energy required for the synthesis of uric acid is difficult to assess from a consideration of the biosynthetic pathway of uric acid. Besides consisting of many reactions this pathway involves the utilisation of two one-carbon units which must be provided in a suitably reduced state, the withdrawal of glycine from the metabolic pool and the active secretion of uric acid into the kidney tubule. The energy requirements for the synthesis of urea are easier to calculate. Whole-animal experiments were in line with theoretical predictions that urea synthesis requires the hydrolysis of about two β-phosphoanhydride bonds for each nitrogen atom excreted (Martin & Blaxter, 1965).

The efficiency of amino acid utilisation in animals fed imbalanced diets has recently been reviewed by Harper, Benevenga & Wohlhueter (1970). After consideration of the large number of reports of work designed to investigate the biochemical effects of diets imbalanced in amino acid content they suggested that the evidence points to the following sequence of

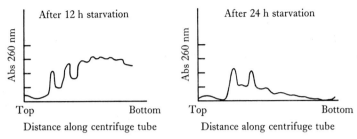

Figure 3. The effect of starvation on chicken liver polysome profiles (Wunner et al. 1966). Two-week-old chicks were starved for the length of time indicated and their liver polysomes were isolated. (Boorman & Buttery – unpublished observations.)

Table 1. *Weight gain, food intake and food conversion from 7 to 21 days of age by chicks fed a basal diet limiting in methionine and methionine-supplemented diets, and metabolisable energy contents of the diets (Mean values ± standard deviations of 6 replicates for treatment) (From Solberg et al. 1971)*

Diet	Basal	Basal + 0.15 per cent methionine	Basal + 0.30 per cent methionine
Weight gain (g/chick/day)	24.9 ± 1.5	25.3 ± 1.0	25.4 ± 2.1
Food intake (g/chick/day)	43.6 ± 2.2	42.3 ± 1.0	42.4 ± 3.2
Food intake (g/100 g body weight)[1]	15.1 ± 0.3	14.5 ± 0.2	14.4 ± 0.6
Food conversion[1]	1.75 ± 0.04	1.68 ± 0.05	1.67 ± 0.06
Metabolisable energy (kcal/kg)	3094 ± 26	3106 ± 34	3089 ± 21

[1] Difference between highest and lowest values significant ($P < 0.05$).

events: 'After ingestion of a meal of a diet with an imbalance the limiting amino acid is absorbed well, along with any surplus of other indispensable amino acids; the high concentration of the indispensable amino acids generally stimulates liver protein biosynthesis so that the limiting amino acid is used with high efficiency; efficient extraction of the limiting amino acid by the liver results in less remaining to circulate to other tissues but not enough to reduce the rate of protein synthesis in them.' The removal of the limiting amino acid from the plasma eventually results in a plasma amino acid pattern which resembles that obtained on feeding a severely deficient diet. The imbalanced plasma amino acid pattern induces a reduction in food intake with a subsequent further reduction in the availability of the limiting amino acid to the tissues. A depression in protein synthesis and growth rate follows.

Table 2. *Nitrogen retention, carcass water content, uric acid excretion, plasma uric acid concentration, liver xanthine dehydrogenase activity and liver weight of chicks fed a basal diet limiting in methionine and methionine-supplemented diets (From Solberg et al. 1971)*

Diet	Basal	Basal + 0.15 per cent methionine	Basal + 0.30 per cent methionine
Nitrogen retention[1],[3] (% ingested nitrogen)	53.9 ± 3.6	57.4 ± 2.2	58.9 ± 3.6
Carcass water (%)[2]	68.3 ± 1.4	68.6 ± 1.2	69.8 ± 1.7
Uric acid nitrogen excreted (% ingested nitrogen)[1],[3]	28.4 ± 1.0	23.2 ± 1.7	24.6 ± 4.0
Plasma uric acid concentration (mg/100 ml)[2]	5.5 ± 1.9	3.6 ± 1.9	3.6 ± 1.7
Liver xanthine dehydrogenase activity (μmol NADH produced/g body weight)[2]	5.0 ± 1.3	3.9 ± 0.8	4.1 ± 1.1

[1] Mean values ± SD of 6 replicates per treatment.
[2] Mean values ± SD of 9 chicks per treatment.
[3] Difference between highest and lowest values statistically significant ($P < 0.05$).

Excesses of a single amino acid can result in an inefficient conversion of amino acids into proteins. The responses to excesses of a single amino acid can be classified into toxic and antagonistic responses. The toxic responses of single amino acids are various. Excess tyrosine fed to a growing young rat fed a low-protein diet results in eye and paw lesions, retarded growth and, in severe cases, death (Schweizer, 1947). Threonine on the other hand can be tolerated even in great excess (Sauberlich, 1961).

Examples of antagonistic responses are the apparent valine and isoleucine deficiency in rats and chickens fed excess leucine and the apparent arginine deficiency on feeding excess lysine; Boorman (1971) reported that excess lysine in the plasma of cockerels resulted in a poor re-absorption of arginine in the kidney tubule. It is probable that the cationic amino acids share a common renal transport system in the fowl. Previously Boorman, Falconer & Lewis (1968) suggested that impaired re-absorption of arginine in the kidney could account for the characteristic antagonism between arginine and lysine.

The biochemical explanation of the antagonism between the branched-chain amino acids (leucine, isoleucine and valine) probably lies in the common degradative pathway of these amino acids. Phansalkar et al. (1970) demonstrated that excess leucine increased the oxidation of isoleucine by rats and Wohlheuter & Harper (1970) reported that excess leucine in the diet caused an increase in the ability of liver to dehydrogenate the α-keto acids derived from leucine, isoleucine and valine.

It is clear that the efficient use of amino acids by animal tissues requires that the profile of the amino acids presented to the tissue must closely resemble that demanded by the needs of protein synthesis.

HORMONAL REGULATION OF PROTEIN SYNTHESIS

Protein synthesis in animal tissues is subject to hormonal control. Tata (1968) considered the possible target sites of hormones in relation to protein synthesis. There is evidence that hormones either increase (androgens, growth hormones, insulin) or decrease (corticosteroids, thyroid hormones, estrogens) muscle production.

The anabolic steroids have a marked effect on muscle growth. Although there is no change in muscle DNA there is an increase in the thickness of muscle fibres. This group of steroids appears to act by increasing the activity of skeletal ribosomes and the production of mRNA from DNA.

The effects of the polypeptide growth hormone on mammals appear to be similar to that of the anabolic steroids. As with the anabolic steroids growth hormone appears to preferentially stimulate the deposition of muscle protein (Young, 1970). Combinations of growth hormone and testosterone had an additive effect on the weight increase and nitrogen retention of growing rats (Kochakian, 1960). The exact mechanism by which hormones regulate protein synthesis is unknown; current knowledge on the subject has been reviewed by Manchester (1970).

SPECIFIC DYNAMIC ACTION

The phenomenon of SDA of amino acids and proteins introduces a further loss of efficiency in protein metabolism. The SDA of food was a phrase coined by Rubner in 1902 (Borsook, 1936) to signify the increase in heat production and oxygen consumption that occurs after the ingestion of food by animals. The phenomenon is most pronounced with proteins or amino acids; relatively small effects are observed with fat or carbohydrate. Most experimental work on SDA has concentrated on dogs and man although SDA has been observed in other species, e.g. rats (Lewis & Luck, 1933; Munro, Clark & Goodlad, 1961; Buttery, Rowsell & Rowsell, 1973; Buttery, 1969).

The magnitude of the SDA effect promoted by an amino acid may be much greater than the energy required to incorporate the amino acid into protein. Wilhelmj & Bollman (1928) injected 9.3 g alanine into a dog and observed an SDA of 21.4 kcal (89 kJ) extra heat produced. To incorporate the alanine into protein would require approximately $4 \times 9.3/89 = 0.4$ high-energy phosphate bonds. Assuming that this energy was supplied by the metabolism of carbohydrate, the total heat loss would be approximately 7.4 kcal (31 kJ), i.e. only a third of the observed heat production.

It would appear that the liver is the major, if not the only site of SDA. What is certain is that SDA is not a function of intestinal movements, secretion, digestion and absorption induced by the injection of food, since the SDA of an amino acid is the same whether administered orally, intravenously, or subcutaneously (Nord & Deuel, 1928; Lundsgaard, 1931a, b; Weiss & Rapport, 1924; Wilhelmj, Bollman & Mann, 1931). Hepatectomy abolishes the SDA of glycine and alanine in the dog (Mann, Wilhelmj & Bollman, 1927; Wilhelmj, Bollman & Mann, 1928). Dock (1931) compared rats on a high-protein diet with control rats under different experimental conditions. When the splanchnic arteries of both groups were ligatured (i.e. the blood supply was cut off from the livers and intestines) the oxygen consumption of rats on the high-protein diet, which had been 35 per cent above that of the control rats, was nearly equal to that of the control rats. However, since a large difference in oxygen consumption persisted after ligaturing the renal arteries or after ligaturing the abdominal aorta, Dock concluded that at least 85 per cent of the heat production associated with SDA was coming from a region served by the splanchnic artery.

SDA has been observed in the isolated perfused dog liver but not in perfused hind quarters (Bornstein & Roese, 1930; Nothhaas & Never, 1930).

Lundsgaard (1942) studied the oxygen consumption of cats' livers when equimolar amounts of glycine, alanine or ammonium lactate were added to the perfused medium. The SDA observed in each case was about the same. These results confirmed earlier studies on the intact dog which demonstrated that the SDA observed with ammonium salts of lactate, acetate or chloride was comparable with that of equimolar quantities of alanine or glycine. SDA was not observed with sodium lactate, acetate or chloride (Lundsgaard, 1931a, b).

After considering the evidence available at the time Wilhelmj (1935) was of the opinion that SDA was a result of the stimulation of general metabolism which occurs as a consequence of the de-amination of amino acids.

In the first place he noted that the SDA measured for alanine and glycine often greatly exceeded the energy content of the amino acid de-aminated during the experimental period (Wilhelmj & Bollman, 1928). Secondly, when glycine and alanine were administered to phloridzinised dogs, the entire energy content of the amino acid was recovered in the urine, although a considerable SDA was observed (Lusk, 1930). Furthermore, studies on the respiratory quotient (RQ) of animals during the operation of SDA (Wilhelmj & Mann, 1930a, b) were indicative of a stimulation of general metabolism rather than the metabolism of the amino acid *per se*. Commenting on their observations, Wilhelmj & Mann stated that the RQ after the administration of an amino acid was unrelated to the theoretical RQ of the administered amino acid but was entirely dependent upon the nutritional state of the animal.

The SDA observed with ammonium salts (Lundsgaard, 1931a, b; Lundsgaard 1942) was consistent with a stimulation of metabolism. In comparatively recent work Hems et al. (1966) added ammonium chloride to the medium perfusing an isolated rat liver. The oxygen consumption of the liver was stimulated to an extent which could not be accounted for by the observed increase in urea production, although the authors did not interpret the results in relation to SDA.

Evidence for the stimulation of general metabolism during SDA was provided by Munro et al. (1961). They fed rats on a high-carbohydrate diet to obtain high liver glycogen levels. When these animals were fed casein, glycine, D,L-alanine or sodium L-glutamate there was a rapid loss of liver glycogen. Olive oil feeding was without effect. There was no change in muscle glycogen after the administration of casein or amino acids but blood sugar levels were reduced. As Munro et al. (1961) pointed out, the extent of loss of liver glycogen was consistent with the studies of SDA in the whole rat by Lewis & Luck (1933). Munro et al. calculated that the complete oxidative metabolism of the liver glycogen lost on giving 1 g glycine would yield 0.3 kcal (1.3 kJ), which was closely similar to the SDA effect of 0.32 kcal (1.3 kJ) observed when 1 g glycine was fed (Lewis & Luck, 1933).

Braunstein (1959) took the suggestion of Wilhelmj further and offered an explanation for SDA in terms of the known metabolic pathways of ammonia. Almost all of the ammonia produced when amino acids are de-aminated is either converted to urea or combined with glutamate to form glutamine. In both cases a hydrolysis of ATP results. Braunstein's hypothesis, however, is unable to account for the observed magnitude of SDA, i.e. assuming

(1) that there is tight coupling between electron transport and phosphorylation with a P:O ratio of 3 such that $3ADP + 3P_i \longrightarrow 3ATP \equiv \frac{1}{2}O_2$,

(2) urea synthesis involves a net change of $2ATP \longrightarrow 2ADP + P_i$ per mole of ammonia incorporated (with ammonium or aspartate as N donor), and

(3) glutamine synthesis involves a net change of ATP $\longrightarrow$ ADP+P_i per mole of ammonia incorporated,

then the conversion of 1 mole of N to urea, which involves the hydrolysis of 2 moles of ATP, is equivalent to $\frac{1}{3}$ mole of O_2. Similarly for glutamine synthesis, 1 mole of N is equivalent to $\frac{1}{6}$ mole of O_2.

Lundsgaard (1942) measured oxygen consumption and urea formation in the isolated perfused cat liver before and after adding glycine, alanine or ammonium carbonate. All three compounds promoted extra oxygen consumption of about 1.64 l per g N converted to urea or 0.90 l per g N added. The expected oxygen consumptions, on the basis of Braunstein's hypothesis, are 0.53 l per g N converted to urea, or 0.27 l per g N converted to glutamine.

Hems et al. (1966) have reported similar findings with the isolated rat liver. With the conversion of ammonia to urea they observed an increase in oxygen consumption that could not be accounted for by the measured urea synthesis. In the same series of experiments Hems et al. studied the increased oxygen consumption and glucose production on adding lactate to the perfusion medium. The observed increased oxygen consumption was found to be only slightly above the theoretical level predicted by assuming that lactate $\longrightarrow$ glucose requires 1 mole of oxygen per mole of glucose formed, confirming the effectiveness of the preparation.

The suggestion that SDA is due to the stimulation of oxidative metabolism has not gone unchallenged. Krebs (1964) has suggested that SDA is due to the metabolism of the amino acid *per se*. Krebs accounts for SDA by suggesting that the conversion of the energy of fat or carbohydrate into ATP is more efficient than the conversion of the energy contained in amino acids into ATP. Thus when fat or carbohydrate serves as a fuel less heat is released than when amino acids are catabolised. Krebs bases his suggestion on a detailed consideration of the known pathways of metabolism of the amino acids. In each case he was able to estimate the ATP yield per kcal heat liberated. This was found to differ markedly between different amino acids. By consideration of the amino acid composition of casein, rabbit myosin and albumen, Krebs calculated that the SDA of protein should be about 20 per cent. This value, Krebs suggests, is of the same order of magnitude as the 'most reliable estimates of SDA'.

The evidence against the Krebs hypothesis may be summarized as follows:

(1) A 20 per cent SDA would require that during the metabolism of an ingested protein total body ATP must originate only from the catabolism of amino acids, i.e. carbohydrate and fat metabolism must stop. If the total

metabolic activity of the liver which is the main site of SDA, were driven by amino acid catabolism, then since the liver only accounts for about 25 per cent of total body oxygen consumption (Brauer, 1963), the maximum attainable SDA would be only 5 per cent.

(2) It is difficult to explain the SDA of ammonia in terms of Krebs' hypothesis. In addition the observations that the extent of SDA appears to be related to the nitrogen content of an amino acid rather than its carbon skeleton is not compatible with Krebs' hypothesis.

(3) The loss of liver glycogen observed when amino acids were administered to rats (Munro et al. 1961). If Krebs' hypothesis were correct one would expect less depletion of liver glycogen following dosage with amino acids.

Grisolia & Kennedy (1966) have extended the hypothesis of Krebs to account for SDA effects greater than 20 per cent. They argued that amino acids, which are utilised for ATP generation less efficiently than carbohydrate or fat, also stimulate protein turnover. It was suggested that following the ingestion of protein there is an equivalent amount of protein synthesis and degradation and at the same time an equivalent amount of amino acid degradation. For the latter process they accepted Krebs' calculations but suggested that the value of the SDA effect can be increased from 21.8 to 39 per cent when the energy demands for synthesis of the peptide bond (activation and condensation of the amino acids) are considered. Grisolia and Kennedy also suggested that if the synthesis of mRNA requires large quantities of energy relative to the requirements for peptide bond formation, then the maximum obtainable SDA effect would be even higher. However, as pointed out earlier, the major energy demands for protein synthesis are made by the processes involved in the synthesis of the peptide bond rather than the synthesis of the mRNA.

Recently Rowsell and his co-workers have conducted experiments to test the validity of the two main hypotheses concerning the mechanism of SDA (Buttery et al. 1973; Buttery, 1969). The effects of amino acids administered intravenously to rats receiving a high carbohydrate diet were studied.

Rats were kept on a high-carbohydrate diet to induce high liver glycogen levels and were trained to meal feed. The glycogen reserves were labelled by feeding [U-^{14}C]glucose. The specific radioactivity of expired $^{14}CO_2$ was unchanged when 1800 μmol of unlabelled L-alanine (1.5 ml) was infused into the tail vein. Typical results are presented in Figure 4. It was concluded that there was no preferential oxidation of alanine.

To account for SDA Rowsell and his colleagues have suggested that the ammonia released from amino acids on de-amination catalyses the opera-

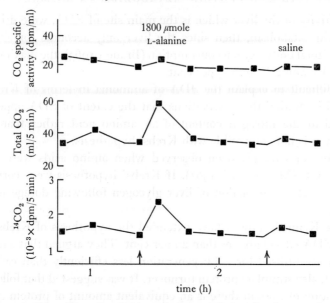

Figure 4. The effect of L-alanine administered via tail vein cannula on the rate of $^{14}CO_2$ expiration by a rat fed on a high-carbohydrate diet labelled with [U-^{14}C] glucose.

tion of a process which results in the hydrolysis of ATP. One possible cycle is presented in Figure 5. The proposed cycle uses enzymes which are all found in or in contact with the cytosol of liver. Such a cycle without exact stoichiometry between ammonia and ATP hydrolysis could account for the varied magnitudes of the SDA effect.

It is interesting to note that the majority of enzymes involved in the proposed cycle are all acutely body-size-dependent, i.e. they increase markedly with diminishing metabolic body size. It has been suggested that these enzymes are in some way involved in the maintenance of the higher basal metabolic rate in the small mammal (Rowsell, Buttery & Carnie, 1969). It is interesting to speculate that SDA is a transitory manifestation of a continuous process whose function is to generate heat in mammalian liver.

BIOLOGICAL EFFICIENCY OF MICROBIAL PROTEIN SYNTHESIS IN THE RUMEN

We propose to discuss the biological efficiency of microbial protein synthesis in the rumen in relation to the overall nitrogen metabolism of the host animal, a subject that has been extensively reviewed in recent years (Allison, 1970; Hogan & Weston, 1970; Purser, 1970).

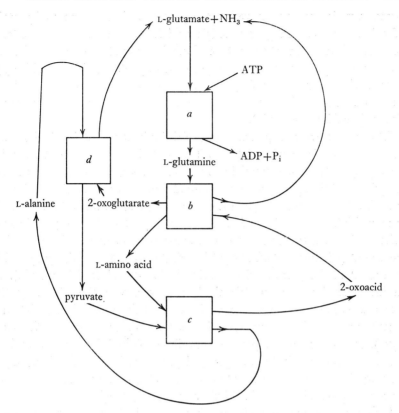

Figure 5. A metabolic cycle to account for specific dynamic action. Abbreviations used: a, glutamine synthetase; b, glutamine amino transferase; c, pyruvate-L-amino acid aminotransferase; d, L-alanine-2-oxoglutarate aminotransferase.

The amino acids established as metabolically essential for ruminants are similar to those for non-ruminants (Black et al. 1957; Downes, 1961). Quantitatively, the precise requirements for each individual essential amino acid are dependent on the physiological and nutritional status of the animal, which determines the nature and amount of protein synthesised. The extensive ruminal degradation of dietary protein (see Hungate, 1966) implies that the specific amino acid requirements cannot be directly determined, and that dietary supplementation with proteins or amino acids is usually not feasible.

A portion of most dietary protein survives ruminal degradation, and there is good evidence that as with other components of the diet, the proportion which reaches the intestine intact is dependent on the level of food intake (Ørskov, Fraser & McDonald, 1971). Quantitative data on the degradation of dietary protein in the rumen are scanty and limited to certain proteins which can be assayed in the presence of the mixture of proteins of microbial and endogenous origin which occur in duodenal contents. McDonald showed that the extent of ruminal digestion of zein and casein was about 50 per cent and 95–100 per cent respectively (McDonald, 1952; McDonald & Hall, 1957). In general, the extent of fermentation of dietary protein is largely dependent on the solubility of the material in rumen liquor (see Hungate, 1966).

The value to the nitrogen economy of the animal of protein which reaches the duodenum is closely related to its essential amino acid content and to its digestibility in the small intestine. Chalmers, Cuthbertson & Synge (1954) found that pregnant ewes retained more N when casein was infused into the duodenum than when the same supplement was infused into the rumen. This observation led to attempts to protect dietary protein from microbial attack in the rumen, and some success was achieved by heat treatment (Chalmers et al. 1954; Tagari, Ascarelli & Bondi, 1962) and by treatment of the protein with tannin (Delort-Laval & Zelter, 1968). An exciting development in this field, however, was the discovery that formaldehyde treatment substantially reduced the ruminal degradation of certain proteins (Ferguson, Hemsley & Reis, 1967; Hughes & Williams, 1971). An alternative approach skilfully exploited by Ørskov and his colleagues is to retain the oesophageal groove reflex in animals by liquid feeding after weaning in order to divert into the omasum certain nutrients such as high-quality protein (Ørskov & Benzie, 1969).

The peptides, amino acids and ammonia which arise as end products of protein fermentation act as nitrogen sources for the bacteria and protozoa which proliferate in the rumen. These microbial cells, which become available for digestion in the lower gut, constitute a major protein source for the ruminant. Clearly, the value to the ruminant of the conversion of dietary protein to microbial protein is largely dependent on the relative biological values of the two proteins. The essential amino acid profiles of bacterial and protozoal protein, and their digestibilities, imply that microbial protein is a moderately good protein source of relatively constant composition (see Purser, 1970).

A key feature of nitrogen metabolism in the ruminant, however, is the capacity of the micro-organisms in the rumen to utilise ammonia for cell synthesis. In effect, this constitutes a mechanism of upgrading ammonia into microbial protein. This property of the ruminant has been commercially exploited by incorporating into rations ammonium salts, or materials which give rise to ammonia in the rumen, such as urea or biuret.

MICROBIAL PROTEIN SYNTHESIS FROM AMMONIA

Allison (1970) has reviewed current knowledge on the assimilation of ammonia by rumen micro-organisms. There is good evidence that most strains of most rumen bacteria utilise ammonia as their main nitrogen source, and that although some strains accept either ammonia or amino acids, ammonia is essential for many organisms (Bryant & Robinson, 1962, 1963). Several groups of workers have shown that dietary protein may be completely replaced by non-protein nitrogen (NPN) (Oltjen, 1969), but growth rates and feed efficiency were invariably suboptimal. Supplementary amino acids did not improve performance, but dietary protein or peptides often led to marked improvements in nitrogen retention. These results are consistent with studies on whole rumen contents (Wright, 1967) and with a pure culture of *Bacteroides ruminicola* (Pittman & Bryant, 1964) which demonstrated the utilisation of peptides in preference to amino acids. A major difficulty in studies designed to explore the use of NPN as a sole nitrogen source is the need to use purified diets based on special ingredients possibly lacking other necessary food factors. The utilisation of urea and other sources of NPN in commercial practice has been fully reviewed by Chalupa (1968), Loosli & McDonald (1968) and Oltjen (1969).

An important determinant in the efficiency of utilisation of ammonia for microbial synthesis is the concentration of ammonia in the rumen. This reflects the relative rates of entry and exit, as summarised below:

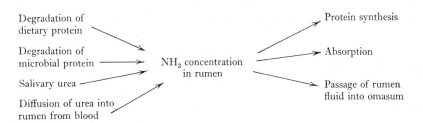

The entry of urea in saliva, and from blood by diffusion across the rumen wall, is not significant except at low nitrogen intakes, and the rate of degradation of microbial protein is probaby fairly constant. Changes in the rate of ammonia entry can be achieved by influencing the extent of degradation of dietary protein, or by supplementation with ammonia or materials which give rise to ammonia. The only practicable method of measuring the rate of ammonia removal is to increase protein synthesis, which is largely a function of cell numbers, and of the energy made available by the fermentation of carbohydrates and, to a much smaller extent, proteins and the glycerol moiety of fats.

The important question is the effect of ammonia concentration on the rate of microbial cell growth when energy is not limiting. The concentration at which ammonia becomes limiting for the growth of rumen bacteria has not been defined (Allison, 1970), although in a continuous culture system ammonia became limiting for the growth of *B. amylophilus* when the concentration in the uninoculated medium fell below 4.6×10^{-3} mol l^{-1} (Henderson, Hobson & Summers, 1969). This concentration of ammonia is probably close to the optimal level of rumen ammonia, since the cell crop of pure cultures of ruminal bacteria was proportional to the concentration in the medium between 0.5×10^{-3} and 4×10^{-3} mol l^{-1}. At levels about $4-5 \times 10^3$ mol l^{-1}, it is likely that the additional loss of nitrogen resulting from increased absorption of ammonia from the rumen is not made good by increased incorporation of ammonia into microbial protein.

A second factor which must be considered is the effect of increased available organic nitrogen on the microbial utilisation of ammonia. Pilgrim *et al.* (1970) and Mathison & Milligan (1971) have shown in studies with $^{15}NH_3$ that the percentage utilisation of ammonia by bacteria decreased as concentrations of rumen ammonia increased. A positive correlation between the concentrations of NPN, non-ammonia-N and ammonia-N in the rumen was demonstrated earlier (Blackburn & Hobson, 1960).

The relatively low concentration of ammonia for optimal utilisation, and the adverse effects of raised levels of organic N on ammonia utilisation explain the poor responses obtained when normal diets are supplemented with NPN (see Loosli & McDonald, 1968). A critical parameter is the level of rumen ammonia. Supplementary NPN in situations where the ammonia concentrations exceed $4-5 \times 10^{-3}$ mol l^{-1} is unlikely to improve nitrogen retention, and indeed the additional synthesis and excretion of urea imposes an additional energy cost (Martin & Blaxter, 1965).

ENERGY COST OF MICROBIAL PROTEIN SYNTHESIS

Hungate (1965, 1966) and Walker (1965) pointed out that the anaerobic nature of ruminal fermentation inevitably limits the conversion of digestible dry matter to microbial cells. Bauchop & Elsden (1960) in their much cited paper calculated that the yield of several micro-organisms was equivalent to 10–12 g (dry weight) cell material per mole ATP [termed Y (ATP)] made available by fermentation. Hogan & Weston (1970) have pointed out that Y (ATP) yield is not constant, but may vary with different micro-organisms, with different substrates and, in continuous culture, with variations in the dilution rate. Y (ATP) values of 10–20 g mol^{-1} have been reported in studies with pure cultures (see Hogan & Weston, 1970). Y (ATP) values may be calculated from the yield of volatile fatty acids (VFA) if the amounts of substrate fermented are not known (Hungate, 1966). The use of isotope dilution to measure ruminal VFA production rates (see Leng, 1970) has made the latter approach feasible.

Purser (1970) has considered the relationship between substrates fermented and microbial protein production in terms of digestible protein, i.e. protein available to the host animal. This approach was dependent on a number of reasonable assumptions concerning the cell yield per 100 g digestible dry matter (15 g), the protein content of rumen microbial cells (65.4 per cent) and the digestibility of microbial protein (80 per cent). A value of 18.3 g digestible protein per 1000 digestible kcal (4.2×10^3 kJ) was deduced. A survey of relevant data in the literature, however, revealed values of 2.3–54.2 g per 1000 g digestible kcal (4.2×10^3 kJ), using the same method of calculation (Purser, 1970). The possibility that cell yields on some occasions may be greater than considered theoretically feasible has been supported by the important studies of Mathison & Milligan (1971), who have used direct methods based on the use of $^{15}NH_3$ to show that microbial growth in the rumen resulted in the assimilation of 1.7–2.6 g N per 100 g digestible dry matter, or 19.8–30.1 g digestible protein per 1000 digestible kcal (4.2×10^3 kJ) using the method of calculation of Purser (1970). The obvious conclusion is that more precise information is required in defined dietary conditions. Baldwin, Lucas & Cabrera (1970), who have developed a sophisticated model of the fermentation system constituting the rumen which embodies a wealth of known data, reported theoretical microbial cell yields of 12–18 g per 100 g digestible organic matter, or 14.7–21.9 g digestible protein per 1000 digestible kcal (4.2×10^3 kJ).

The interrelationship of nitrogen metabolism in the rumen and in the host animal may be effectively studied using ^{15}N-labelled materials (Pilgrim et al. 1970; Mathison & Milligan, 1971), or by a double isotope technique based on ^{15}N and ^{14}C (Nolan & Leng, 1970). These procedures, when coupled with the use of animals surgically prepared to permit the collection and assay of digesta at all points in the digestive tract (see Hogan & Weston, 1970), are able to provide data which are essential for the construction of effective models of rumen and ruminant metabolism. The general case for planning and defining biological research objectives in relation to simulation analysis is well established (see Baker, 1969; Baldwin et al. 1970), and the model simulating rumen metabolism described by Baldwin et al. (1970) is a particularly fine achievement. It must be stressed, however, that simulation analysis in no way lessens the need for precise data, but indicates the data required and in complex systems allows the data to be fully exploited.

Practical implications

When the many factors which impinge on nitrogen metabolism in the ruminent are defined, it becomes apparent that the conditions which permit maximum efficiency of microbial protein synthesis in the rumen are not necessarily consistent with the maximum efficiency of usage of dietary nitrogen by the host animal. Purser (1970) has discussed this apparent paradox, which stems from the suggestion that maximum cell yield of rumen micro-organisms should occur when there is an excess of nitrogen (inorganic or organic) in the rumen. This situation, although favouring microbial protein synthesis, inevitably results in raised rumen ammonia levels leading to increased rates of ammonia absorption, higher levels of blood urea and a greater loss of urinary nitrogen. The disadvantages to the host animal, which must be offset against any increase in the availability of microbial protein, are a reduced retention of dietary nitrogen and the energy cost of the additional urea synthesis. The relationship between microbial cell synthesis and nitrogen levels in the rumen, when energy made available by fermentation is not limiting, clearly requires closer study. The scanty evidence summarised by Allison (1970) on the relationship between levels of rumen ammonia and microbial cell yield suggests that maximum cell yields are probably achieved at relatively low levels of rumen ammonia ($4-5 \times 10^{-3}$ mol l^{-1}). If this is so, then there is no advantage whatsoever to be gained from additional dietary nitrogen in excess of that necessary to provide this rumen ammonia concentration. This analysis highlights the significance of rumen ammonia as an index of the efficiency of utilisation of dietary nitrogen in the whole animal, and emphasises the potential advantages to be gained by diverting high quality dietary protein, or the necessary specific amino acids, into the lower gut.

REFERENCES

ALLISON, M. J. (1970). Nitrogen metabolism of rumen micro-organisms. In: Phillipson, A. T. (Ed.), *Physiology of Digestion and Metabolism in the Ruminant*, 456–73. Newcastle upon Tyne: Oriel Press.

ARMSTRONG, D. G. (1969). Cell bioenergetics and energy metabolism. *Handbuch der Tierernährung*, volume 1, 385–414. Hamburg and Berlin: Verlag Paul Parey.

ATKINSON, M. R., & MORTON, R. K. (1960). Free energy and the biosynthesis of phosphates. In: Florkin, M., & Mason, H. S. (Eds.), *Comparative Biochemistry*, volume 11, 1–95. New York and London: Academic Press.

BAGLIONI, C., & COLOMBO, B. (1970). Protein synthesis. In: Greenberg, D. M. (Ed.), *Metabolic Pathways*, volume 4, 277–351. New York and London: Academic Press.

BAKER, N. (1969). The use of computers to study rates of lipid metabolism. *Journal of Lipid Research* **10**, 1–24.

BALDWIN, R. L., LUCAS, H. L., & CABRERA, R. (1970). Energetic relationships in the formation and utilisation of fermentation endproducts. In: Phillipson, A. T. (Ed.), *Physiology of Digestion and Metabolism in the Ruminant*, 319–34. Newcastle upon Tyne: Oriel Press.

BAUCHOP, T., & ELSDEN, S. R. (1960). The growth of micro-organisms in relation to their energy supply. *Journal of General Microbiology* **23**, 457–69.

BEREDES, H. D. (1968). Factors involved in the expression of gene activity in polytene chromosomes. *Chromosoma* **24**, 418–37.

BIRBECK, M. S. C., & MERCER, E. H. (1961). Cytology of cells which synthesise protein. *Nature* **189**, 558–60.

BLACK, A. L., KLEIBER, M., SMITH, H. M., & STEWART, D. N. (1957). Acetate as a precursor of amino acids of casein in the intact dairy cow. *Biochimica et Biophysica Acta* **23**, 54–69.

BLACKBURN, T. H., & HOBSON, P. N. (1960). Proteolysis in the sheep rumen by whole and fractionated rumen contents. *Journal of General Microbiology* **22**, 272–81.

BLAXTER, K. L. (1967). *The Energy Metabolism of Ruminants*. 2nd impression, 263–79. London: Hutchinson.

BOORMAN, K. N. (1971). The renal reabsorption of arginine, lysine and ornithine in the young cockerel (*Gallus domesticus*). *Comparative Biochemistry and Physiology* **39A**, 29–38.

BOORMAN, K. N., FALCONER, I. R., & LEWIS, D. (1968). The effect of lysine infusion on the renal reabsorption of arginine in the cockerel. *Proceedings of the Nutrition Society* **27**, 61–2 A.

BORNSTEIN, A., & ROESE, H. F. (1930). Uber die Beeinflussung des Sauerstoffverbrauches überlebender Organe durch Glykokoll. *Pflügers Archiv für die gesamte Physiologie des Menschen und der Tiere* **223**, 498–508.

BORSOOK, H. (1936). The specific dynamic action of proteins and amino acids in animals. *Biological Reviews* **11**, 147–80.

BRAUER, R. W. (1963). Liver circulation and function. *Physiological Reviews* **43**, 115–213.

BRAUNSTEIN, A. E. (1959). Some aspects of the chemical integration of nitrogen metabolism. In: Anerswald, W., & Hoffmann-Ostenhof, O. (Eds.), *Proceedings of the Fourth International Congress of Biochemistry*, volume 14, 63–88. London: Pergamon Press.

BRETSCHER, M. S. (1968). Translocation in protein synthesis: a hybrid structure model. *Nature* **218**, 675–7.

BROSTROM, C. O., & JEFFAY, H. (1970). Protein catabolism in rat liver homogenates. *Journal of Biological Chemistry* **245**, 4001–8.

BRYANT, M. P., & ROBINSON, I. M. (1962). Some nutritional characteristics of predominant culturable ruminal bacteria. *Journal of Bacteriology* **84**, 605–14.

BRYANT, M. P., & ROBINSON, I. M. (1963). Apparent incorporation of ammonia and amino acid carbon during growth of selected species of ruminal bacteria. *Journal of Dairy Science* **46**, 150–4.

BUTTERY, P. J. (1969). Studies bearing upon the biochemical basis of basal heat production. Unpublished Ph.D. thesis, University of Manchester.

BUTTERY, P. J., ROWSELL, E. V., & ROWSELL, K. V. (1973). Manuscript in preparation.

CHALMERS, M. I., CUTHBERTSON, D. P., & SYNGE, R. L. M. (1954). Ruminal ammonia formation in relation to the protein requirement of sheep. I. Duodenal administration and heat processing as factors influencing fate of casein supplements. *Journal of Agricultural Science* **44**, 254–62.

CHALUPA, W. (1968). Problems of feeding urea to ruminants. *Journal of Animal Science* **27**, 207–19.

CHAMBERLIN, M. J. (1970). Transcription 1970: a summary. *Cold Spring Harbor Symposia on Quantitative Biology* **35**, 851–73.

DELORT-LAVAL, J., & ZELTER, S. Z. (1968). Improving the nutritive value of proteins by tanning process. *Proceedings of the Second World Conference on Animal Production, Maryland*, 457–8.

DOCK, W. (1931). Relative increase in the metabolism of the liver and of other tissues during protein metabolism in the rat. *American Journal of Physiology* **97**, 117–23.

DOWNES, A. M. (1961). Amino acid requirements of the sheep. *Australian Journal of Biological Sciences* **14**, 254–9.

ENWONWU, C. O., & MUNRO, H. N. (1970). Rate of RNA turnover in rat liver in relation to intake of protein. *Archives of Biochemistry and Biophysics* **138**, 532–9.

FERGUSON, K. A., HEMSLEY, J. A., & REIS, P. J. (1967). The effect of protecting dietary protein from microbial degradation in the rumen. *Australian Journal of Science* **30**, 215–17.

GEIDUSCHEK, E. P., & HASELKORN, R. (1969). Messenger RNA. *Annual Review of Biochemistry* **38**, 647–76.

GOLDSTEIN, L., & PRESCOTT, D. M. (1968). Proteins in nucleocytoplasmic interactions. II. Turnover and changes in nuclear protein distribution with time and growth. *Journal of Cell Biology* **36**, 53–61.

GRISOLIA, S., & KENNEDY, J. (1966). On specific dynamic action turnover and protein synthesis. *Perspectives in Biology and Medicine* **9**, 578–85.

HARDESTY, B., CULP, W., & McKEEHAN, W. (1969). The sequence of reactions leading to the synthesis of a peptide bond on reticulocyte ribosomes. *Cold Spring Harbor Symposia on Quantitative Biology* **34**, 331–45.

HARPER, A. E., BENEVENGA, N. J., & WOHLHUETER, R. M. (1970). Effects of injection of disproportionate amounts of amino acid. *Physiological Reviews* **50**, 428–558.

HEMS, R., ROSS, B. D., BERRY, M. N., & KREBS, H. A. (1966). Glucogenesis in the perfused rat liver. *Biochemical Journal* **101**, 284–92.

HENDERSON, C., HOBSON, P. N., & SUMMERS, R. (1969). The production of amylase, proteins and lipolytic enzymes in 2 years of anaerobic rumen bacteria. *Proceedings of the Fourth International Symposium on the Continuous Culture of Micro-organisms*, 189–204. Prague: Czechoslovak Academy of Sciences.

HERRMAN, H. (1968). Muscle differentiation and macromolecular synthesis. *Journal of Cellular Physiology* **72**, Supplement 1, 30.

HOGAN, J. P., & WESTON, R. H. (1970). Quantitative aspects of microbial protein synthesis in the rumen. In: Phillipson, A. T. (Ed.), *Physiology of Digestion and Metabolism in the Ruminant*, 474–85. Newcastle upon Tyne: Oriel Press.

HUGHES, J. G., & WILLIAMS, G. L. (1971). The utilization of formaldehyde-treated groundnut meal by sheep. *Animal Production* **13**, 396.

HUNGATE, R. E. (1965). Quantitative aspects of the rumen fermentation. In: Dougherty, R. W. (Ed.), *Physiology and Digestion in the Ruminant*, 311–21. London and Washington, D.C.: Butterworth.

HUNGATE, R. E. (1966). *The Rumen and its Microbes*. New York and London: Academic Press.

KIELANOWSKI, J. (1965). Estimates of the energy cost of protein deposition in growing animals. In: Blaxter, K. L. (Ed.), *Energy Metabolism*, 13–20. New York and London: Academic Press.

KIT, S. (1970). Nucleotides and nucleic acids. In: Greenberg, D. M. (Ed.) *Metabolic Pathways*, volume 4, 69–275. New York and London: Academic Press.

KOCHAKIAN, C. D. (1960). Summation of protein anabolic effects of testosterone propionate and growth hormone. *Proceedings of the Society of Experimental Biology and Medicine* **103**, 196–7.

KREBS, H. A. (1964). Metabolic fate of amino acids. In: Munro, H. N., & Alison, J. (Eds.), *Mammalian Protein Metabolism*, volume 1, 125–76. New York and London: Academic Press.

LENG, R. A. (1970). Formation and production of volatile fatty acids in the rumen. In: Phillipson, A. T. (Ed.), *Physiology of Digestion and Metabolism in the Ruminant*, 406–21. Newcastle upon Tyne: Oriel Press.

LENGYEL, P. (1969). The process of translation as seen in 1969. *Cold Spring Harbor Symposia on Quantitative Biology* **34**, 827–41.

LEWIS, H. G., & LUCK, J. M. (1933). The calorigenic action of glycine. *Journal of Biological Chemistry* **103**, 227–33.

LIVINGSTON, D. M., & LEDER, P. (1969). Deformylation and protein biosynthesis. *Biochemistry* **8**, 435–43.

LOOSLI, J. K., & MCDONALD, I. W. (1968). *Nonprotein Nitrogen in the Nutrition of Ruminants*. FAO Agricultural Studies No. 75. Rome: Food and Agriculture Organisation of the United Nations.

LUNDSGAARD, E. (1931 a). Ursachen der specifischen dynamischen Wirkung der Nahrungstoffe 1. *Skandinavisches Archiv für Physiologie* **62**, 223–42.

LUNDSGAARD, E. (1931 b). Ursachen der specifischen dynamischen Wirkung der Nahrungstoffe 2. *Skandinavisches Archiv für Physiologie* **62**, 243–81.

LUNDSGAARD, E. (1942). The specific dynamic action of amino acids and ammonium salts. *Acta Physiologica Scandinavica* **4**, 330–48.

LUSK, G. (1930). The specific dynamic action. *Journal of Nutrition* **3**, 519–30.

MCDONALD, I. W. (1952). The role of ammonia in ruminal digestion of protein. *Biochemical Journal* **51**, 86–90.

MCDONALD, I. W., & HALL, R. J. (1957). The conversion of casein into microbial proteins in the rumen. *Biochemical Journal* **67**, 400–3.

MANCHESTER, K. L. (1970). Sites of hormonal regulation of protein metabolism. In: Munro, H. N. (Ed.), *Mammalian Protein Metabolism*, volume 4, 229–98. New York and London: Academic Press.

MANN, F. C., WILHELMJ, C. M., & BOLLMAN, J. L. (1927). The specific dynamic action of glycocoll and alanine with specific reference to the fasting animal. *American Journal of Physiology* **81**, 496–7.

MARTIN, A. K., & BLAXTER, K. L. (1965). The energy cost of urea synthesis in sheep. In: Blaxter, K. L. (Ed.), *Energy Metabolism*, 83–91. New York and London: Academic Press.

MATHISON, G. W., & MILLIGAN, L. P. (1971). Nitrogen metabolism in sheep. *British Journal of Nutrition* **25**, 351–66.

MILLWARD, D. J. (1970). Unpublished results quoted by V. R. Young (1970).

MIRZABEKOV, A. D., LASTITY, D., LEVINA, E. S., & BAYER, A. A. (1971). Location of two recognition sites in yeast valine t-RNA 1. *Nature New Biology* **229**, 21–2.

MUNRO, H. N. (1969). Evolution of protein metabolism in mammals. In: Munro, H. N. (Ed.), *Mammalian Protein Metabolism*, volume 3, 133–82. New York and London: Academic Press.

MUNRO, H. N. (1970). A general survey of mechanisms regulating protein metabolism. In: Munro, H. N. (Ed.), *Mammalian Protein Metabolism*, volume 4, 3–130. New York and London: Academic Press.

MUNRO, H. N., CLARK, C. M., & GOODLAD, G. A. J. (1961). Loss of liver glycogen after administration of protein or amino acid. *Biochemical Journal* **80**, 453–8.

NINIO, J., FAVRE, A., & YANIV, M. (1969). Molecular model for transfer RNA. *Nature* **223**, 1333–5.

NOLAN, J. V., & LENG, R. A. (1970). Methods for the assessment of protein metabolism in sheep using (^{14}C) and (^{15}N) urea. In: Schurch, A., & Wenk, C. (Eds.), *Energy Metabolism in Farm Animals* 117–20. Zurich: Juris Druck & Verlag.

NOMURA, M., & LOWRY, C. V. (1967). Phage F2 RNA-directed binding of formylmethionyl-TRNA to ribosomes and the role of 30S in ribosomal subunits in initiation of protein synthesis. *Proceedings of the National Academy of Sciences of the USA* **58**, 946–53.

NORD, F., & DEUEL, H. J. (1928). The specific dynamic action of glycine given orally and intravenously to normal and adrenalectomized dogs. *Journal of Biological Chemistry* **80**, 115–24.

NOTHHAAS, R., & NEVER, H. E. (1930). Uber die specifische-dynamische Wirkung an der künstlich durchbluteten Leber. *Pflügers Archiv für die Gesamte Physiologie des Menchen und der Tiere* **224**, 527–34.

OLTJEN, R. R. (1969). Effects of feeding ruminants non-protein nitrogen as the only nitrogen source. *Journal of Animal Science* **28**, 673–82.

ØRSKOV, E. R., & BENZIE, D. (1969). Using the oesophageal groove reflex in ruminants as a means of bypassing rumen fermentation with high quality protein and other nutrients. *Proceedings of the Nutrition Society* **28**, 30–1 A.

ØRSKOV, E. R., FRASER, C., & McDONALD, I. (1971). Digestion of concentrates in sheep. 2. The effect of urea or fish-meal supplementation of barley diets on the apparent digestion of protein, fat, starch and ash in the rumen, the small intestine and the large intestine, and calculation of volatile fatty acid production. *British Journal of Nutrition* **25**, 243–52.

PERRY, S. V. (1967). The structure and interactions of myosin. *Progress in Biophysics and Molecular Biology* **17**, 325–81.

PHANSALKAR, S. V., NORTON, P. M., HOLT, JR, L. E., & SNYDERMAN, S. E. (1970). Amino acid interrelationships: the effect of a load of leucine on the metabolism of isoleucine. *Proceedings of the Society for Experimental Biology and Medicine* **134**, 262–3.

PILGRIM, A. F., GRAY, F. V., WELLER, R. A., & BELLING, C. B. (1970). Synthesis of microbial protein from ammonia in the sheep's rumen, and the proportion of dietary nitrogen converted into microbial nitrogen. *British Journal of Nutrition* **24**, 589–98.

PITTMAN, K. A., & BRYANT, M. P. (1964). Peptides and other nitrogen sources for the growth of *Bacteroides ruminicola*. *Journal of Bacteriology* **88**, 401–10.

PRONCZUK, A. W., ROGERS, Q. R., & MUNRO, H. N. (1970). Liver polysome patterns of rats fed amino acid imbalance diets. *Journal of Nutrition* **100**, 1249–58.

PURSER, D. B. (1970). Nitrogen metabolism in the rumen: micro-organisms as a source of protein for the ruminant animal. *Journal of Animal Science* **30**, 988–1001.

ROBERTS, J. N. (1970). The *p* factor: termination and anti-termination in Lambda. *Cold Spring Harbor Symposia on Quantitative Biology* **35**, 121–6.

ROWSELL, E. V., BUTTERY, P. J., & CARNIE, J. A. (1969). Species body size and liver glutamine synthetase, glutaminase and glutamine aminotransferase. *Biochemical Journal* **115**, 43 P.

SALSER, W., GESTELAND, R. F., & RICARD, B. (1969). Characterization of lysozyme messenger and lysozyme synthesized *in vitro*. *Cold Spring Harbor Symposia on Quantitative Biology* **34**, 771–80.

SAUBERLICH, H. E. (1961). Studies on the toxicity and antagonism of amino acids for weanling rats. *Journal of Nutrition* **75**, 61–72.

SCHWEIZER, W. (1947). Studies on the effect of l-tyrosine on the white rat. *Journal of Physiology* **106**, 167–76.
SMITH, A. E., & MARCKER, K. A. (1970). Cytoplasmic methionine transfer RNAs from Eukaryotes. *Nature* **226**, 607–12.
SOLBERG, G., BUTTERY, P. J., & BOORMAN, K. N. (1971). Effect of moderate methionine deficiency on food, protein and energy utilization in the chick. *British Poultry Science* **12**, 297–304.
STELLWAGEN, R. H., & COLE, R. D. (1969). Chromosomal proteins. *Annual Review of Biochemistry* **38**, 951–90.
TAGARI, H., ASCARELLI, I., & BONDI, A. (1962). The influence of heating on the nutritive value of soya-bean meal for ruminants. *British Journal of Nutrition* **16**, 237–43.
TATA, J. R. (1968). Hormonal regulation of growth and protein synthesis. *Nature* **219**, 331–7.
TOMINAGA, H. (1971). Metabolic turnover of messenger RNA in rat liver. *Biochimica et Biophysica Acta* **228**, 183–7.
WALKER, D. J. (1965). Energy metabolism and rumen micro-organisms. In: Dougherty, R. W. (Ed.), *Physiology and Digestion in the Ruminant*, 296–310. London and Washington, DC: Butterworths.
WEISS, R., & RAPPORT, D. (1924). The interrelations between certain amino acids and proteins with reference to their specific dynamic action. *Journal of Biological Chemistry* **60**, 513–44.
WILHELMJ, C. M. (1935). The specific dynamic action of food. *Physiological Reviews* **15**, 202–20.
WILHELMJ, C. M., & BOLLMAN, J. L. (1928). The specific dynamic action and nitrogen elimination following intravenous administration of various amino acids. *Journal of Biological Chemistry* **77**, 127–49.
WILHELMJ, C. M., & MANN, F. C. (1930a). The influence of nutrition on the response of certain amino acids. I. The effect of fasting. *American Journal of Physiology* **93**, 69–85.
WILHELMJ, C. M., & MANN, F. C. (1930b). The influence of nutrition on the response of certain amino acids. II. The effect of fasting followed by diets high in carbohydrate. *American Journal of Physiology* **93**, 258–66.
WILHELMJ, C. M., BOLLMAN, J. L., & MANN, F. C. (1928). Studies on the physiology of the liver. XVII. The effect of the removal of the liver on the specific dynamic action of amino acids administered intravenously. *American Journal of Physiology* **87**, 497–509.
WILHELMJ, C. M., BOLLMAN, J. L., & MANN, F. C. (1931). A study of certain factors concerned in the specific dynamic action of amino acids administered intravenously and a comparison with oral administration. *American Journal of Physiology* **98**, 1–17.
WOHLHEUTER, R. M., & HARPER, A. E. (1970). Coinduction of rat liver branched chain α-keto acid dehydrogenase activities. *Journal of Biological Chemistry* **245**, 2391–401.
WRIGHT, D. E. (1967). Metabolism of peptides by rumen micro-organisms. *Applied Microbiology* **15**, 547–50.
WUNNER, W. H., BELL, J., & MUNRO, H. N. (1966). The effect of feeding with a tryptophan-free amino acid mixture on rat liver polysomes and ribonucleic acid. *Biochemical Journal* **101**, 417–28.
YOUNG, V. R. (1970). The role of skeletal and cardiac muscle in the regulation of protein metabolism. In: Munro, H. N. (Ed.), *Mammalian Protein Metabolism*, volume 4, 585–674. New York and London: Academic Press.

11

POSSIBILITIES FOR CHANGING BY GENETIC MEANS THE BIOLOGICAL EFFICIENCY OF PROTEIN PRODUCTION BY WHOLE ANIMALS

By J. C. BOWMAN

Department of Agriculture, University of Reading,
Earley Gate, Reading, Berkshire, RG6 2AT

MEASUREMENT OF EFFICIENCY

Biological efficiency in whole animals involves the complex of relationships between input as feed on the one hand and the cost of maintenance plus the value of output on the other. The terms of the relationship can be in units of energy or of any or all nutrients involved in whole animal metabolism. It has become practice to refer to the ratio of output:input as efficiency, and its inverse, input:output, which is more commonly used in agricultural discussions, as utilisation.

The efficiency of food production by the animal industries has been discussed recently by Holmes (1971) who refers to several earlier studies. Holmes gives a generalised form of an equation to estimate the 'overall efficiency' (E) of an animal product. Thus

$$E = \frac{\text{Product in time } t \text{ (gain and/or milk, wool, eggs)}}{\dfrac{\text{Total maintenance for time } t\ (M)}{\text{Efficiency for maintenance}} + \dfrac{\text{Total milk, eggs, wool in time } t}{\text{Efficiency for milk Efficiency for eggs Efficiency for wool}} + \dfrac{\text{Total gain in time } t}{\text{Efficiency for gain}}}$$

or

$$E = \frac{P}{\dfrac{M}{K_m} + \dfrac{\text{milk}}{K_l} + \dfrac{\text{eggs}}{K_e} + \dfrac{\text{wool}}{K_w} + \dfrac{\text{total gain}}{K_f} - \dfrac{\text{total loss}}{K_m}}$$

where P is expressed as product or energy or protein,
E is expressed as a decimal,
K_m is expressed as a decimal, and is the efficiency of use of the nutrient for maintenance,
K_l is expressed as a decimal, and is the efficiency of use of the nutrient for lactation,

K_e is expressed as a decimal, and is the efficiency of use of the nutrient for eggs,

K_w is expressed as a decimal, and is the efficiency of use of the nutrient for wool,

K_f is expressed as a decimal, and is the efficiency of use of the nutrient for fattening.

Holmes indicates that all measures refer to the same time period (in some cases physiological time) and that the energy or nutrient cost of producing the young animal must be considered in some comparisons. It is worth discussing this equation in some detail because several parts of the equation refer to parameters which may be familiar to animal nutritionists under alternative names. If the equation is in units of energy, then E is 'gross efficiency' or the ratio Net Energy/Gross Energy. The efficiency terms in the denominator with the exception of those relating to maintenance and total loss are 'partial efficiencies' or 'net efficiencies' as defined by Brody (1945). The terms relating to maintenance and total loss may also be considered as net efficiencies but it is perhaps unwise in the present state of knowledge to use the same efficiency term (K_m) for maintenance and for weight loss.

If the equation is in units of protein or nitrogen (N) then it can be related to the parameter 'biological value' (BV) which is the proportion of absorbed N retained by the animal, viz.

$$BV = 100 \times \frac{I-(F-MFN)-(U-EUN)}{I-(F-MFN)}$$

where I   = dietary intake of N,
       F   = total faecal N,
       MFN = metabolic faecal N,
       U   = total urinary N,
       EUN = endogenous urinary N.

The gross efficiency of protein or N in the animal would be equal to

$$\frac{\text{true digestibility} \times BV}{100},$$

where true digestibility equals

$$\frac{100 \times I-(F-MFN)}{I}.$$

This gross efficiency is more familiarly known as net protein utilisation (NPU). The efficiency terms in the denominator are the net efficiencies of protein or N in the animal for the functions of maintenance, growth and production.

The distinction between gross and net efficiencies has been made by Morris (1971) when discussing the prospects for improving the efficiency of nutrient utilisation in poultry. He stresses the importance of considering the net efficiencies because it is possible to improve the gross efficiency of nutrient utilisation for growth and production merely by increasing output without altering the net efficiencies. It is equally important to investigate the possibilities for improving gross efficiency by an improvement in net efficiencies, and by modification of the requirements for maintenance.

FACTORS AFFECTING EFFICIENCY

Factors which have an effect on efficiency have been referred to by Holmes (1971) and in some greater detail by Blaxter (1964a). With respect to energy, and similar conclusions apply to the consideration of other nutrients, the sources of variation which can be expected to affect efficiency are:

(i) Intake – the ratio $\dfrac{\text{voluntary intake of feed}}{\text{maintenance feed requirement}}$ = relative feed level (RFL).

It is stated that efficiency is highest when relative feed level is highest.

(ii) Nutritive value of the feed – for an animal to remain in normal health and within the range of normal metabolism, it is necessary to feed nutrients in combination. The efficiency of use by the animal of a single nutrient is dependent upon the composition of the total intake. This is a particularly important factor in regard to the efficiency of protein or amino acid use.

(iii) Energy and nutrient requirements for maintenance relative to other requirements for productive purposes. This source of variation is itself dependent on the environment in which the animal is maintained. In particular, environmental temperature and the difficulty of obtaining feed will affect maintenance requirements.

(iv) The composition of weight gain and other productive output – it is clear that some modification of qualitative output can occur without change of quantitative output as a result of variations of input, nevertheless quantitative output is heavily dependent on input. Although under *ad libitum* feeding an animal may adjust its feed intake in relation to some aspects of its output capacity, it is clear that it cannot do this at all times for all nutrient requirements, depending on the feed available.

(v) The age of the animal and the age at which it reaches sexual maturity – discussion of the relationship between age and efficiency depends on the period of time involved and in particular whether the prepubertal (nonproductive) period is included in the calculation. Efficiency for growth and perhaps for maintenance declines with age; however, for other productive functions the relationship between net efficiencies and age is still a matter for investigation.

There are other possible sources of variation in efficiency but Blaxter (1964a) concluded that it was unlikely that much between-breed or individual variation existed with respect to some aspects of efficiency which included:

(vi) Digestive efficiency, though this comment was made in the context of comparisons between different ruminants.

(vii) Efficiency of the complex biochemical processes of converting absorbed nutrients to lipid and protein within the body.

(viii) The net efficiency of energy transfers within the body.

He also considered that 'if gains of weight are expressed in terms of their calorific value differences in efficiency due to changes in the composition of gains tend to disappear'.

Comparisons of feed intake between and within species have shown a close relationship between gross energy intake and metabolic body weight ($W^{0.73}$) although some exceptions within species have been reported (for instance Morris, 1968). Similarly, the energy maintenance requirements of different species are broadly related to the metabolic body weight. However much variation exists both between and within species (Blaxter & Wainman, 1966; Blaxter, Clapperton & Wainman, 1966; Agricultural Research Council, 1967; Taylor & Young, 1968). According to Taylor & Young (1968) little of the interspecies variation in efficiency of maintenance is caused by other than systematic covariation with mature weight. Between breeds or strains within species, between 40 and 87 per cent and within closed populations 90 per cent of the genetic variation in efficiency of maintenance is not accounted for by systematic covariation with mature weight.

From these considerations it seems clear that genetic variation in efficiency depends on one or more of three possible sources. These are variation of feed intake, variation in the partition of metabolic products in the animal between maintenance, growth and production, and variation in the relationship between efficiency of body metabolism and age. There are several reasons why these sources of variation have not been thoroughly investigated. For instance, many farm animals are grazing ruminants in which it has proved technically difficult to measure feed intake, many others are fed controlled quantities of conserved feed which has inhibited the investigation of variation in voluntary intake under an *ad libitum* feed regimen. It is also technically difficult to measure variation in the partition of nutrients between maintenance, growth and production, particularly on sufficient animals to be able to estimate genetic variation for this character. There is an urgent need to investigate these sources of variation in efficiency and their inter-relationships.

EFFICIENCY OF PROTEIN USE

We may turn now to a consideration of what is known of the variation in efficiency of protein and nitrogen use by the animal. The amino acid composition of muscle protein is very similar for a wide range of species (Mitchell, 1959) which include examples from mammals, birds, amphibians, fish and crustacea. There are close similarities in amino acid composition between liver, kidney and brain and between lung and stomach for a large number of species. Crawford, Paterson & Yardley (1968) have analysed the amino acid compositions of muscle protein in buffalo, domestic ox, warthog, domestic pig, hartebeest, topi, kob and elephant, and found them to be very similar. A comparison of the muscle protein figures with the amino acid composition of cow's milk and chicken egg proteins published by Ling, Kon & Porter (1961) indicates some marked differences of composition. There are large differences in the protein content and in the amino acid composition of milk of several species (Blaxter, 1964*b*).

It is interesting to consider the ways in which these similarities and differences affect the efficiency of feed protein use by the animal. Munro (1969) has presented evidence to support the view that several aspects of protein metabolism are related to metabolic body weight although the exponent of weight used to relate between species differences in some aspects of protein metabolism is not always 0.73. The biological value of dietary protein within a species varies between maintenance and growth (Allison, 1964) but is considered similar for lactation and juvenile growth (Blaxter, 1964*b*). Mitchell (1959) suggested that, because of the close relationship between amino acid composition of muscle protein in many species, the biological value of protein for growth varies little between species. This suggestion still requires to be substantiated.

Genetic differences in the net efficiency of protein and individual amino acids for egg production in the laying hen have been sought but not found (Morris, 1971). The only possible exception is some evidence from Tolan (1970) of individual bird variation in the net efficiency of methionine for maintenance.

MEASURES OF EFFICIENCY

Considerable effort has been devoted to estimating the feed and the constituent nutrient requirements, particularly energy and protein, of farm livestock. The results of much of this effort have recently been summarised in the United Kingdom in publications from the Agricultural Research Council (1963, 1965, 1967). Nutrient requirements for each of the farm species for maintenance, growth and, as appropriate, for pregnancy, lactation and egg laying have been suggested and these requirements are related largely to the weight or age of an animal within a species and to the level of production achieved. Considerable effort has been devoted also to assessing the energy and protein content of the productive output (body tissue, milk and eggs) of the farm species. From these two sets of data it is possible to estimate the comparative biological gross efficiency of conversion of feed energy into animal protein and of feed protein into animal protein for the farm species. The comparisons can be made most usefully on the basis of *an* individual of each species and in terms of daily performance or of productive lifetime performance. Such interspecies comparisons have been made by several people including Lodge (1970) and Holmes (1970, 1971). Some of their results are given in Tables 1 and 2. It is clear that gross efficiencies of weight gain in beef cattle and sheep are lower than similar production in pigs which in turn is worse than in broilers. Milk production appears to be the most efficient form of animal protein production per unit of protein intake although per unit of energy input, chickens, either for meat or eggs, are more efficient.

It is fair to suggest that the errors which attach to the estimates of energy and protein requirement of farm livestock are still so high that to attempt to compare net efficiencies of energy or protein for protein production is premature except in the case of poultry. As already mentioned no genetic variation in such net efficiencies has been found in poultry.

POSSIBILITIES FOR CHANGING EFFICIENCIES

It is clear that, if the possibility of changing, genetically, gross or net efficiencies is to be contemplated, it is necessary to obtain more accurate estimates of the efficiencies of our existing domesticated species, breeds and strains. This comment applies particularly to improvements in estimates of protein requirement in farm livestock.

There are several possible ways by which efficiencies of domesticated species might be improved genetically. Briefly these are (*a*) selection within populations for increased production, (*b*) selection within populations for

Table 1. *Efficiency of conversion of dietary protein to muscle protein (adapted from Lodge, 1970)*

	Muscle protein gain / Protein intake (%)
Cattle	
veal	13.3
beef	7.7
Sheep	
lamb	13.3
hogg	10.0
Pig	
pork	15.4
'heavy'	13.8
Chicken	16.6

Table 2. *Efficiency of conversion of gross energy to edible protein and of feed protein intake to edible protein in lifetime performance of farm species (from Holmes, 1970)*

	Edible protein / Feed protein intake (%)	Edible protein / Gross energy intake (g/Mcal)
Dairy herd	23	5.4
Dairy and beef herd	20	4.7
Beef herd	6	1.5
Sheep flock	3	0.8
Pig herd	12	4.0
Broiler flock	20	7.7
Egg flock	18	8.0

improved net efficiencies, (c) selection within domesticated species for reduced maintenance requirements whilst either holding production constant or increasing production, (d) selection for increased feed intake or, more precisely, for increased RFL and (e) selection for feed conversion efficiency. As has been mentioned earlier, restricted or controlled feed intake regimens are used extensively in pig, cattle, sheep and, to a lesser extent, in poultry production. Such regimens are practised in order to prevent below-optimum feed conversion efficiencies and the deposition of excessive and unwanted levels of fat in the carcass. There seem to be no reasons to support the argument for adopting similar feed regimens for animals under selection, but several support the view that selection programmes should be carried out on an *ad libitum* feed regimen even if the animals selected provide stock for production on controlled feed regimens. Animals fed on an *ad libitum* regimen generally have higher growth rates. This may enable more animals to be produced per unit of accommodation and per unit time and, provided there are no adverse concomitant effects on carcass quality, this would be economically advantageous.

In practice, for various reasons, and most importantly because of the difficulties of measuring feed or nutrient intake of individual animals, selection has been concentrated on increasing production and on decreasing maintenance requirements relative to the level of production. For example in the laying hen, breeders have selected principally for increased total weight of egg output and where possible for simultaneous reduction in body weight. Reduction in body weight has led to reduced maintenance requirement and the result has been a marked improvement in gross efficiency of egg production. There is still plenty of genetic variation for growth rate, body weight and production in domesticated species so that it should be possible to make progress to improve the gross efficiency of all forms of farm livestock production. No attempts have been made yet to select directly for changes in feed intake or net efficiencies in farm livestock and only now are estimates of genetic parameters of feed conversion efficiency becoming available. The estimates are sufficiently high to merit further investigation and direct selection for feed conversion efficiency. In cattle Taylor & Young (1968) reported a heritability of 0.89 ± 0.37 for energy efficiency for maintenance and suggested that direct selection for this trait should yield a rapid response.

However, an experimental study with mice by Sutherland *et al.* (1970), in which the same base population (previously selected for rate of gain) was subdivided and selected separately for feed conversion efficiency, feed intake or rate of gain, produced results for eleven generations of selection which are shown in Table 3. It is interesting to note that the selection line with the highest weight gain per generation and the highest improvement in efficiency per generation was the line directly selected for feed conversion efficiency. Though the line selected for feed intake had a very considerable increase in feed consumption this resulted in a very small increase in efficiency of conversion. As all the lines had previously been selected for rate of gain for ten generations, it is not surprising that the line directly selected for rate of gain showed only moderate response for all three characters. We should be encouraged to attempt direct selection for feed conversion efficiency in farm species.

Table 3. *Selection for rate of gain, appetite and efficiency of feed utilization in mice (from Sutherland et al. 1970)*

	Regression of mean weight gain on generation (g/generation)	Regression of mean feed consumption on generation (g/generation)	Regression of mean efficiency on generation (g gain/g feed/generation)
Selection for feed conversion efficiency	0.75 ± 0.163	1.99 ± 0.936	0.0031 ± 0.00067
Selection for feed intake	0.40 ± 0.100	3.11 ± 0.708	0.0006 ± 0.00057
Selection for rate of gain	0.38 ± 0.115	1.50 ± 0.918	0.0013 ± 0.00052
Control	0.08 ± 0.038	0.71 ± 0.642	0.0002 ± 0.00036

Finally we should consider whether the main farm species with which we currently work are as efficient converters as other species which might be used for food production. There are indications (such as in the eland and the oryx) in terms of rate of gain of muscle tissue and of physiological thermoregulatory mechanisms which may lead to nitrogen saving. These and many other species should be investigated to see if their ability to convert energy and feed protein and nitrogen into animal protein for human consumption is more efficient than the species we now use for this purpose.

REFERENCES

AGRICULTURAL RESEARCH COUNCIL (1963). *The Nutrient Requirements of Farm Livestock. No. 1. Poultry.* London: HMSO.
AGRICULTURAL RESEARCH COUNCIL (1965). *The Nutrient Requirements of Farm Livestock. No. 2. Ruminants.* London: HMSO.
AGRICULTURAL RESEARCH COUNCIL (1967). *The Nutrient Requirements of Farm Livestock. No. 3. Pigs.* London: HMSO.
ALLISON, J. B. (1964). The nutritive value of dairy proteins. In: Munro, H. N., & Allison, J. B. (Eds.), *Mammalian Protein Metabolism*, volume 2, 41–86. New York and London: Academic Press.
BLAXTER, K. L. (1964a). The efficiency of feed conversion by livestock. *Journal of the Royal Agricultural Society* **125**, 87–99.
BLAXTER, K. L. (1964b). Protein metabolism and requirements in pregnancy and lactation. In: Munro, H. N., & Allison, J. B. (Eds.), *Mammalian Protein Metabolism*, volume 2, 173–223. New York and London: Academic Press.
BLAXTER, K. L., & WAINMAN, F. W. (1966). The fasting metabolism of cattle. *British Journal of Nutrition* **20**, 103–111.
BLAXTER, K. L., CLAPPERTON, J. L., & WAINMAN, F. W. (1966). The extent of differences between six British breeds of sheep in their metabolism, feed intake and utilization, and resistance to climatic stress. *British Journal of Nutrition* **20**, 283–94.

BRODY, S. (1945). *Bioenergetics and Growth*. New York: Reinhold Publishing Corporation.
CRAWFORD, M. A., PATTERSON, J. M., & YARDLEY, L. (1968). Nitrogen utilization by the Cape buffalo (*Syncerus caffer*) and other large mammals. In: Crawford, M. A. (Ed.), *Comparative Nutrition of Wild Animals*, 367-79. New York and London: Academic Press.
HOLMES, W. (1970). Animals for food. *Proceedings of the Nutrition Society* 29, 237-44.
HOLMES, W. (1971). Efficiency of food production by the animal industries. In: Wareing, P. F., & Cooper, J. P. (Eds.), *Potential Crop Production. A Case Study*, 213-27. London: Heinemann.
LING, E. R., KON, S. K., & PORTER, J. W. G. (1961). The composition of milk and the nutritive value of its components. In: Kon, S. K., & Cowie, A. T. (Eds.), *Milk: the Mammary Gland and its Secretion*, volume 2, 195-263. New York and London: Academic Press.
LODGE, G. A. (1970). Quantitative and qualitative control of proteins in meat animals. In: Lawrie, R. A. (Ed.), *Proteins as Human Food*, 141-66. London: Butterworths.
MITCHELL, H. H. (1959). Some species and age differences in amino acid requirements. In: Albanese, A. A. (Ed.), *Protein and Amino Acid Nutrition*, 11-43. New York and London: Academic Press.
MORRIS, T. R. (1968). The effect of dietary energy level on the voluntary calorie intake of laying birds. *British Poultry Science* 9, 285-95.
MORRIS, T. R. (1971). Prospects for improving the efficiency of nutrient utilization. In: Freeman, B. M., & Lake, P. E. (Eds.), *Egg Formation and Production*. Edinburgh: Constable.
MUNRO, H. N. (1969). Evolution of protein metabolism in mammals. In: Munro, H. N. (Ed.), *Mammalian Protein Metabolism*, volume 3, 3-19. New York and London: Academic Press.
SUTHERLAND, T. M., BIONDINI, PATRICIA E., HAVERLAND, L. H., PETTUS, D., & OWEN, W. B. (1970). Selection for rate of gain, appetite and efficiency of feed utilization in mice. *Journal of Animal Science* 31, 1049-57.
TAYLOR, ST C. S., & YOUNG, G. B. (1968). Equilibrium weight in relation to food intake and genotype in twin cattle. *Animal Production* 10, 393-412.
TOLAN, A. (1970). A study of variations in nutrient requirements of domestic fowl. Unpublished Ph.D. thesis, University of Reading.

12

FACTORS AFFECTING THE EFFICIENCY OF PROTEIN PRODUCTION BY POPULATIONS OF ANIMALS

By R. V. LARGE

The Grassland Research Institute, Hurley, Maidenhead, Berkshire

A population consists of a group of individual animals which may differ from each other in age, size and sex. The input, for a calculation of efficiency on a population basis, is the total amount of protein (or whatever criterion is being used) consumed by all the individuals in the group. The output of a population is more difficult to define.

In the wild state, population size will be controlled by factors such as food supply, disease and predators; if the reproductive rate of the species matches the loss of individuals from disease and predation, then the population size will remain stable. In wild populations, there is no actual product comparable with that from domesticated animals, although, if the wild species is preyed upon, then the loss of individuals to the predator may be looked upon as a form of production in the same way that man may be considered as a predator on populations of domestic animals. A wild species that is not preyed upon may be in a period of increasing population size and thus will have no productive element. Efficiency, in this instance, may be measured as the rate of increase of population size. This will continue until such time as the population size is limited or reduced by a shortage of food or an outbreak of disease. There will be a basic wastage rate, in both wild and domestic situations, from death either by disease or 'natural' causes and in addition by the 'disposal' of domestic animals.

The essence of livestock farming is that man is able to exert a degree of control over the population structure and consequently over the level and type of production. Control of the population size is achieved by ensuring that enough replacement animals are produced and reared to breeding age. Animal products are obtained from two main sources – directly from the female as milk, wool or eggs and indirectly as meat from the growth of her progeny. Control of the type of product is achieved by ensuring that there are enough animals in the right physiological state to produce what is required (i.e. meat, milk, wool, eggs, etc.).

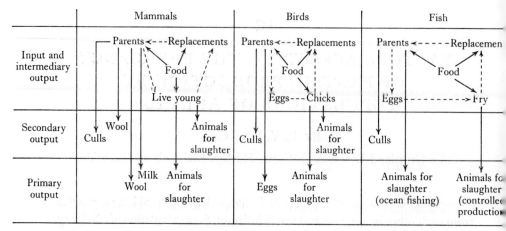

Figure 1. The structure and outputs of animal populations. (N.B. A basic 'wastage' rate from disease and mortality may occur at any stage; in the interests of simplicity this is not shown.)

THE STRUCTURE OF ANIMAL POPULATIONS AND THEIR COMPONENTS

Population structures, their food inputs and their primary and secondary outputs, are shown diagrammatically in Figure 1 for mammals, birds and fish. This figure shows that for production from mammals, the components of the population are the parents, the live young, and animals being reared as replacements for breeding stock. Over and above the replacements required to maintain the population size, the young are reared to slaughter weight to provide the primary output. A secondary output is obtained from disposing of animals at the end of their useful breeding life; these are known as culls. Breeding animals may also be disposed of at an earlier age as a result of selection in a breeding programme or, as in the case of cattle, by a deliberate policy of slaughtering young cows after they have produced one calf. Food inputs are shown as directed to the parents and to the progeny and replacements. In this instance it has been assumed that all the milk produced by the breeding females has been consumed by the progeny.

When milk is the primary output, it is not available (except perhaps in small quantities) for feeding the young; a milk substitute may then be provided as an additional food input. Progeny, over and above those required as replacements, may be considered as a secondary output, their value being taken as at birth or after their full potential has been realised by growing them on to slaughter weight. Secondary output will also arise from the disposal of breeding stock.

If sheep are used mainly for producing lambs for slaughter, secondary output of wool is obtained, both from the breeding stock and the replacements. In flocks where wool is the primary output it is usual to maintain as castrates larger numbers of males than would normally be required for breeding. In addition to the main sources of wool production some will also be provided by the replacement stock. Secondary output will arise from disposals and also from the surplus progeny not required for the replacement of breeding and wool producing stock.

The population structure for birds shows an extra component in the form of eggs. The primary output of eggs is represented by the total output of the laying hen, less a proportion required for providing replacements. Selection for replacement cannot be made until after hatching, resulting in a surplus of male chicks which could provide a secondary output if grown on to slaughter weight. Where broilers are produced as the primary output, all the chicks not required as replacements are grown on to slaughter weight.

For fish in controlled ponds and tanks, breeding stock and progeny will be kept separately and the fish will be harvested as a product without being used for breeding. In ocean fishing, however, it is likely that the harvest will include mature breeding males and females as well as immature fish.

THE IMPORTANCE OF THE POPULATION COMPONENTS AND THE FACTORS AFFECTING THEM

The relative sizes of the population components, i.e. the number of individuals contained in each of them, will largely determine the population efficiency. For a given number of breeding females the following factors will influence the size of the population components:

(1) The ratio of males to females. The cost of maintaining males may be negligible if artificial insemination can be used, but where 6–10 per cent of males are required the cost must be taken into account, particularly where the males may need replacing more frequently than the females.

(2) The number of females that fail to breed.

(3) The total number of progeny produced, their mortality rate and the number required as replacements.

(4) The number of replacement males and females required to maintain the breeding stock. This will be influenced by (a) the longevity of the animals and how well they sustain their level of performance, (b) the mortality rate, and (c) the need to replace animals at an early age in a breeding programme.

(5) The number of animals available for disposal; this will depend on the mortality rate of the breeding stock.

Blaxter (1968) has given values for some of these factors for cattle, sheep and pigs, and has produced a formula for the number of breeding animals required to maintain a given number of young animals for fattening. He has also included on the output side the value of disposable animals, i.e. as secondary output. Reid (1970a) has also made similar calculations for cattle, poultry and pigs, including assessments for the disposal value of the breeding stock, but, in the case of milk production, this does not include the value of the offspring. The calculations, made by Holmes (1970, 1971), do not appear to include the value of the secondary outputs. Comparisons, for different species of animal, are shown in Table 1 for the factors affecting the population structure, and in Table 2 for their reproductive rates.

THE EFFICIENCY OF PRODUCTION IN DIFFERENT SPECIES OF ANIMAL

Spedding (1973) has already pointed out that there are many ways of expressing efficiency; this paper is concerned with biological aspects of the efficiency of protein production by whole populations. The calculations have therefore been made for the nitrogen output of a population per hundred units of nitrogen input, over a period of one year. The input has been calculated as the total amount of nitrogen in the food consumed by the whole population in one year, and the output derived from the nitrogen content of the population production, whether as total body tissues of slaughtered animals, or as eggs, milk and wool. This makes possible comparisons of species, on a biological basis, independent of the different proportions of the total production that are normally regarded as being agriculturally useful. For example, the carcasses of cattle represent about 65 per cent of the whole body tissue, those of sheep about 55 per cent, whilst some 90 per cent of the egg is actually edible and milk is completely so.

Major biological objectives in increasing the efficiency of animal production would seem to be a high rate of reproduction, where the product is derived from the progeny and a high yield of products (i.e. milk, eggs and wool) per female, relative to body size (Dickerson, 1970). There are endless comparisons of factors which could be examined in both these types of production; a few examples have therefore been selected and worked out in some detail. It is not possible to include all the details of these calculations, but some of the more important are listed in the Appendix.

Table 1. *Factors affecting the population structure of domestic animals*
(*Average values derived from the literature*)

Species	Ratio of males to females	Age to first parity	Average length of breeding life	Mortality rate of breeding stock (%)	Replacement rate (%)
Cattle	A.I.	2 years	4 years	5	25
Sheep	1:30–40	2 years	5 years	5	20
Rabbits	1:15–20	4–5 months	2 years	20	50
Domestic fowl	1:10–20	24 weeks	1 year	10	100

Table 2. *Reproductive rate of domestic animals* (*Average values derived from the literature. Figures in parentheses are suggested potentials*)

Species	Gestation length	No. of parities per annum	Mean litter size	Total no. of progeny per annum
Cattle	9 months	1	1.0	1.0
Sheep	5 months	1 (2)	1.5 (3)	1.5 (6)
Rabbits	32 days	5 (10)	8 (10)	40 (100)
Domestic fowl	—	—	—	240

THE EFFICIENCY OF POPULATIONS OF RABBITS, DOMESTIC FOWL AND SHEEP, PRODUCING ANIMALS TO BE SLAUGHTERED FOR MEAT

The efficiencies of the populations have been calculated over a range of reproductive rates. Poultry are normally managed more intensively than ruminants and are therefore calculated up to a high reproductive rate. Management of rabbits is increasing in intensity and reproductive rates of up to 50 progeny per doe per year have been included; it is claimed that the latest hybrids will produce 80–100 progeny per year. Sheep normally have a relatively low reproductive rate but in these calculations values for ewes producing up to six lambs per year have been used.

Large (1970) has shown, for sheep, how efficiency calculated for the individual animal is reduced by adding the cost of keeping the breeding female for the whole year and the effect that reproductive rate has on this relationship. The upper curve in Figure 2 shows how the efficiency of rabbit production changes with increasing reproductive rate; this is in

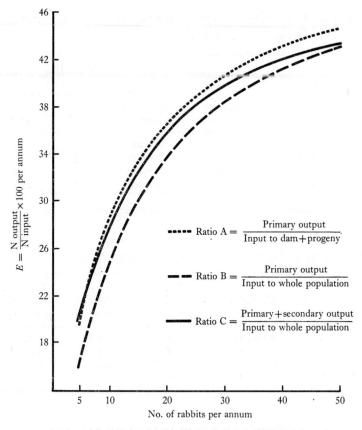

Figure 2. Population efficiency of rabbits.

effect the primary output per unit of input to the dam and progeny only (this will be referred to as Ratio *A*). To put this calculation on a population basis the output has to be divided by the input to the whole population, and the effect of doing this is shown by the lower curve (Ratio *B*). A complete appraisal of the efficiency of a population must include secondary outputs as well (Ratio *C*) and when the value of the culled breeding stock is added, the efficiency is raised to an intermediate level, except at the lowest level of reproduction where it gives the highest value for efficiency. Thus as the reproductive rate increases so the compensatory value of the secondary output has less effect.

Similar calculations for domestic fowl are shown in Figure 3. The relationship between the different ratios expressing efficiency are very similar to those for rabbits.

Results of calculations for sheep show a much greater reduction of

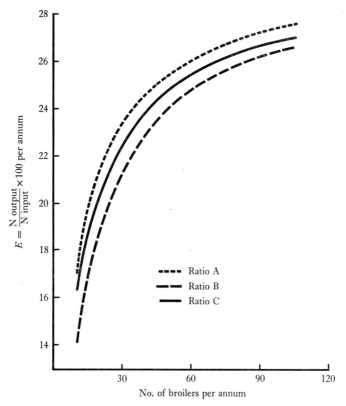

Figure 3. Population efficiency of domestic fowl (for broilers).

efficiency when the cost of maintaining the population is included (Figure 4). This is presumably because of the high maintenance requirements of the sheep population relative to a low reproductive rate. When the secondary output from the cull ewes is added, then an intermediate value for efficiency is obtained in a similar fashion to that for rabbits and poultry. With sheep an additional secondary output of wool must be accounted for. Wool has a high nitrogen content and its addition to the calculation results in the overall efficiency of the population exceeding that based on the primary product from the dam and progeny only.

From Table 1 it is clear that the rate of replacement of breeding stock will depend on their longevity and that this, in turn, will affect efficiency. The degree to which it does so will depend on the mortality rate, i.e. the number of breeding stock being replaced that survive to contribute to the secondary output. To give some idea of the magnitude of these effects calculations for sheep are shown in Figure 5 for three combinations of

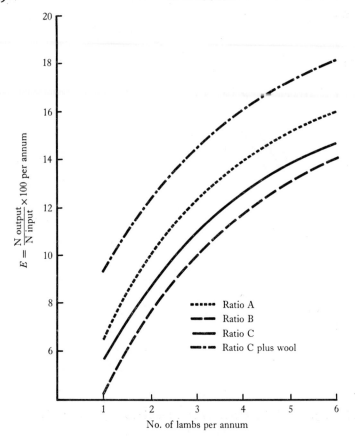

Figure 4. Population efficiency of sheep.

replacement and mortality rate; this has been worked out on the basis that if the replacement rate is higher, i.e. ewes are replaced at a younger age, then mortality will be lower. The result of this calculation shows that, on a population basis where the value of disposed animals is accounted for, the magnitude of the effects, for sheep at any rate, is small compared with the effect of the number of lambs produced per year.

Another major factor in determining the efficiency of sheep production is the relative size of the male and female (Large, 1970). Values for three sizes of ewe (80, 55 and 32 kg) mated to a large ram (100 kg) are shown in Figure 6 and confirm that, for a population, the size of the ewe is a factor of considerable importance, the greatest efficiency arising from small ewes mated to large rams.

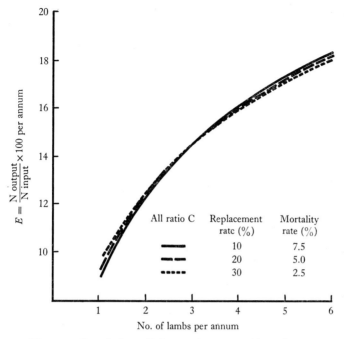

Figure 5. Population efficiency of sheep – effect of replacement and mortality rate of breeding ewes.

THE EFFICIENCY OF POPULATIONS OF DOMESTIC FOWL AND CATTLE, PRODUCING EGGS AND MILK RESPECTIVELY

The efficiency of domestic fowl for egg production is shown in Figure 7 for a range of yields. A curve is shown for the primary output per unit of input to the hen (Ratio A) and these values are reduced by adding the cost of maintaining a population (Ratio B). From the graph it appears, in absolute terms, that adding in the population costs has a greater effect on the efficiency ratio at higher yields, but when worked out on a percentage basis the reverse is true. The overall efficiency of the population (Ratio C) is shown in two ways. In both instances the value of replaced breeding stock, less those which have died, is included; in addition, the value of the surplus male chicks, hatched out in providing replacements, has been included either as day old chicks (i) or as the value of those chicks grown to slaughter weight (ii). The effect of this latter process is considerable at low egg yields but diminishes rapidly as yield increases.

The values for efficiency of milk production by cattle are shown over a range of yields in Figure 8. The relationship of the ratios is somewhat

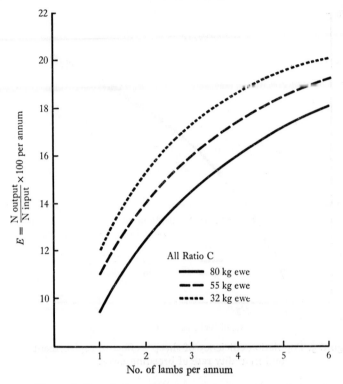

Figure 6. Population efficiency of sheep – effect of size of ewe.

similar to that for egg production. For the overall efficiency of the population the value of surplus calves has been taken either at birth (i) or grown to slaughter weight (ii). As with egg production, the value of taking the calf on to slaughter weight is relatively greater at low yields, but at higher yields it eventually falls below that where the calf is taken at birth because of the lower efficiency of increasing the body weight.

DISCUSSION

The comparative effects on efficiency of taking into account the cost of maintaining a population and then adding in the value of secondary output, are shown as percentage differences in Table 3. Values are given for the low and high ends of the production scale. It is clear that in all instances the greatest effect of adding in the cost of maintaining a population has been at low production levels.

This effect has been offset markedly by increasing the reproductive rate of the meat-producing species but less so by increasing the yield of milk

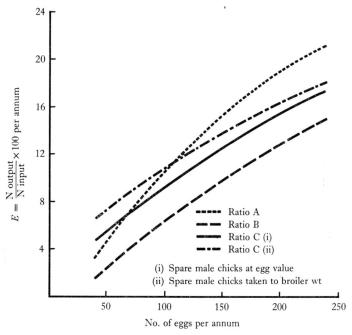

Figure 7. Population efficiency of domestic fowl (for eggs).

Table 3. *Percentage differences between protein efficiencies of animal production*

Ratio	Production level	Rabbits	Fowl (broilers)	Sheep		Fowl (eggs)		Cattle (milk)	
				(1)	(2)	(3)	(4)	(5)	(6)
$\dfrac{A-B}{A}$	Low	−19.7	−17.1	−36.4		−36.5		−19.0	
	High	−3.6	−3.6	−12.4		−30.9		−10.6	
$\dfrac{A-C}{A}$	Low	+1.3	−4.7	−13.3	+45.4	+32.0	+68.0	+16.1	+33.9
	High	−2.5	−1.8	−8.0	+13.4	−15.7	−12.6	−2.9	−8.4

Ratios

$A = \dfrac{\text{Primary output}}{\text{Input to dam (+progeny)}}$

$B = \dfrac{\text{Primary output}}{\text{Input to whole population}}$

$C = \dfrac{\text{Primary + secondary output}}{\text{Input to whole population}}$

(1) Sheep without wool
(2) Sheep with wool included
(3) Surplus male chicks at egg value
(4) Surplus male chicks taken to broiler weight
(5) Surplus calves at birth value
(6) Surplus calves taken to beef weight

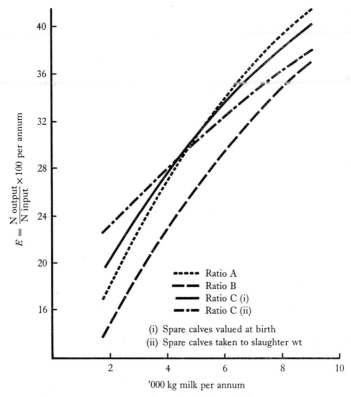

Figure 8. Population efficiency of cattle (for milk).

and eggs. When the value of the secondary products has been included the greatest effect has again been at the low levels of production where they are helping to balance the relatively higher cost of population maintenance. This is particularly so for egg and milk production where low yields are still linked to high replacement rates and the value of the disposable animals is high compared with the low yields of primary products.

All these efficiency ratios have been considered over a range of reproductive rates and yields. In Table 4 the percentage differences in population efficiency resulting from increases in reproductive rate or yield are shown for the low and high ends of the scale (e.g. an increase from 5 to 15 rabbits per doe per annum gives a 64.8 per cent increase in efficiency whereas increasing from 40 to 50 rabbits per doe only increases efficiency by 3.3 per cent; and so on for the other species). The effect is not so marked for eggs and milk because the efficiency curves show less tendency to level out than do those for meat production.

The effect of ewe size is shown in a similar way in Table 5, where the

Table 4. *Percentage differences in population efficiencies (Ratio C) due to increasing reproductive rate or yield, at different levels of production*

	Increase in reproductive rate (no. per female per annum)	Reproductive rate	
		L	H
Rabbits	10	64.8	3.3
Fowl (broilers)	24	41.7	0.7
Sheep	1	58.9	7.0
Sheep + wool	1	33.9	5.4
	Increase in yield	Yield	
		L	H
Fowl (eggs)	20	32.8	10.6
Cattle (milk)	1872 (kg)	19.6	8.9

Table 5. *Percentage differences in population efficiencies (Ratio C) for different sizes of sheep*

	Reproductive rate	
	L	H
Sheep (without wool)		
Medium–Large	10.3	3.0
Small–Large	25.7	6.5

advantage of using a medium or small ewe relative to a large one has considerably more impact at lower reproductive rates.

The comparative efficiencies of populations of rabbits, domestic fowl, sheep and cattle are shown over a range of reproductive rates and yields in Figure 9. The spacing of scales for the different products is very arbitrary and is based on the surmise that the higher end of the scales represents the limits of the productive capacity of the species used; these values may be fair for domestic fowl and cows but high for sheep and probably low for rabbits. Thus the scale of production for sheep has been projected forwards to a high potential; the other species have been projected backwards to rather lower yields than are at present being attained.

Values for sheep plus their wool have been given and also one value for single-suckled beef production. The efficiency of populations where animals are slaughtered for meat may seem surprisingly high when compared with that for eggs and milk, but an animal that is biologically

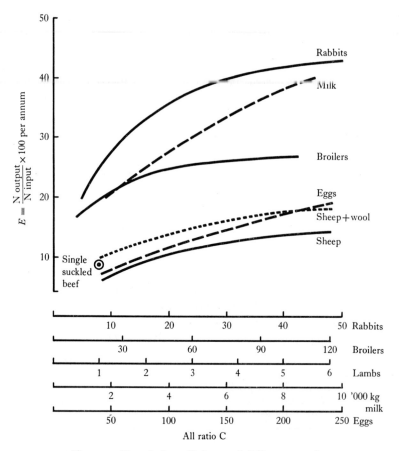

Figure 9. Population efficiency of different species.

efficient for nitrogen production, on a whole body basis, may be considered much less efficient agriculturally, because only a proportion of its body tissues are edible or marketable. This is evident from the calculations made by other workers (Blaxter, 1968; Coop, 1967; Holmes, 1970, 1971; Reid, 1970 a, b).

It appears from the graph that the curves for the meat-producing animals are levelling out, suggesting that they are nearing their maximum potential efficiency, whereas those for eggs and milk yield are not. Further efficiency in meat production must then depend on finding animals that are physiologically more efficient, or on utilising a higher proportion of the body tissues produced.

Efficiency of protein production as eggs and milk could also be improved by using physiologically different animals, but there is apparently still much scope for improvement by the production from the same kind of animals of higher yields such as are being achieved already by exceptional individuals.

In conclusion, it can be said that when the efficiency of protein production is measured for the whole population, then the cost of maintaining the population, in terms of replacement breeding stock, is largely balanced by the value of the secondary output. These calculations also show how important it is to consider combinations of factors and their interactions, rather than trying to make comparisons at only one level of production.

APPENDIX

Notes on calculations and sources of information

Rabbits

The calculations for rabbit production were based on the New Zealand White breed. The feed inputs were derived from data for commercial rabbit production, using a complete pelleted feed. Doe replacement rate was put at 50 per cent per annum and the mortality rate was calculated on a sliding scale, i.e. 0 per cent for 5 young per annum up to 20 per cent for 50 young per annum. Bucks were used at a ratio of 1:20 and replaced every three years.

Sources of information:
    Ministry of Agriculture, Fisheries and Food (1971). *Commercial Rabbit Production.* MAFF Bulletin No. 50. London: HMSO.
    Grassland Research Institute (1971). Unpublished data.
    Sandford, J. C. (1957). *The Domestic Rabbit.* London: Crosby Lockwood.

Domestic fowl

The calculations were based on the use of a 2.2 kg hen for egg production and a 3 kg hen for broiler production. The ratio of cocks to hens in broiler production was 1:10 and it was assumed that 60 per cent of the eggs were viable. Food inputs were based on standard commercial diets. The replacement rate of the hens was put at 100 per cent per annum and the mortality rate at 10 per cent.

Sources of information:
    Morris, T. R. (1971). Personal communication.
    Ministry of Agriculture, Fisheries and Food (1967). *Poultry Nutrition.* MAFF Bulletin No. 174. London: HMSO.

Sheep

The main calculations were made for Scottish Halfbred ewes mated to Suffolk rams; those comparing size of ewe also included Kerry Hill and

Welsh Mountain ewes mated to Suffolk rams. The food inputs were derived, for the rams, the ewes and the replacements, from data on the feeding of artificially dried herbage and those for the lambs from data on the artificial rearing of lambs on milk substitute and concentrates. The ewe replacement rate was put at 20 per cent per annum, with a 5 per cent mortality for the ewes for all numbers of lambs per annum: this was based on evidence (Newton, 1971) that increasing the reproductive rate does not increase the mortality rate of the ewes, provided that all the lambs are artificially reared. The rams were used at a ratio of 1:30 and replaced every three years. The number of lambs per annum is taken to mean the actual number of lambs that was reared.

Sources of information:

Grassland Research Institute (1971). Unpublished data.

Agricultural Research Council (1965). *The Nutrient Requirements of Farm Livestock. No. 2 Ruminants.* London: HMSO.

Newton, J. E. (1971). Personal communication.

Searle, T. W. (1970). Body composition in lambs and young sheep. *Journal of Agricultural Science,* **74**, 357–62.

Body composition in animals and man, National Academy of Sciences – Division of Biology and Agriculture – Agricultural Board (1968). Publication 1598 National Academy of Sciences.

Cattle

The calculations on milk production were for Friesian cattle. Food inputs were based entirely on conserved forages and concentrates, both for the lactating cows and the herd replacements. The replacement rate was put at 25 per cent per annum, with a mortality of 5 per cent. The cost of keeping males was not included. Calves, surplus to the herd replacements, were calculated as being reared on an intensive cereal/beef system.

Single suckled beef production was calculated from data on Welsh Black cows, managed indoors, and fed on conserved forages and concentrates.

Sources of information:

Grassland Research Institute (1971). Unpublished data.

Agricultural Research Council (1965). *The Nutrient Requirements of Farm Livestock No. 2. Ruminants.* London: HMSO.

Ministry of Agriculture, Fisheries and Food (1960). *Rations for Livestock.* MAFF Bulletin No. 48. London: HMSO.

Ministry of Agriculture, Fisheries and Food (1968). *Profitable Farm Enterprises. Booklet 1. Rearing Friesian Dairy Heifers.* London: HMSO.

Joint Beef Production Committee (1968). *The Intensive Cereal Beef*

System Using Calves from the Dairy Herd. Handbook No. 2. Bletchley, England: Meat & Livestock Commission.

Tayler, J. C., & Lonsdale, C. R. (1971). Personal communication.

Armsby, H. P., & Moulton, C. R. (1925). *The Animal as a Converter of Matter and Energy.* New York: Chemical Catalogue Company Incorporated.

The author readily acknowledges the assistance of Jean M. Walsingham, Miss A. Hoxey, Mr B. R. Bentley, Miss J. Bramley and Mrs B. Rodgers in making the calculations and drawing the figures.

REFERENCES

BLAXTER, K. L. (1968). Relative efficiencies of farm animals in using crops and by-products in production of foods. *Proceedings of the 2nd World Conference on Animal Production, Maryland,* 31–40.

COOP, I. E. (1967). The efficiency of feed utilisation. *Proceedings of the New Zealand Society of Animal Production* **27**, 154–65.

DICKERSON, G. (1970). Efficiency of animal production – molding the biological components. *Journal of Animal Science* **30**, 849–59.

HOLMES, W. (1970). Animals for food. *Proceedings of the Nutrition Society* **29**, 237–44.

HOLMES, W. (1971). Efficiency of food production by the animal industries. In: Wareing, P. F., & Cooper, J. P. (Eds.), *Potential Crop Production. A Case Study,* 213–27. London: Heinemann.

LARGE, R. V. (1970). The biological efficiency of meat production in sheep. *Animal Production* **12**, 393–401.

REID, J. T. (1970a). Will meat, milk and egg production be possible in the future? *Proceedings of 1970 Cornell Nutrition Conference,* 50–63.

REID, J. T. (1970b). The future role of ruminants in animal production. In: Phillipson, A. T. (Ed.), *Physiology of Digestion and Metabolism in the Ruminant,* 1–22. Newcastle upon Tyne: Oriel Press.

SPEDDING, C. R. W. (1973). The meaning of biological efficiency. In: Jones, J. G. W. (Ed.), *The Biological Efficiency of Protein Production,* 13–26. Cambridge University Press.

13

THE BIOLOGICAL EFFICIENCY OF PROTEIN PRODUCTION BY ANIMAL PRODUCTION ENTERPRISES

By P. N. WILSON

BOCM Silcock Ltd, Basing View, Basingstoke, Hampshire

The efficiency with which animals convert feed protein into protein food for man has received considerable attention in recent years and a large number of conflicting values have been calculated by different authors (e.g. Leitch & Godden, 1953; Coop, 1967; Blaxter, 1968). The difficulties in resolving the different viewpoints have been described by Holmes (1971) who emphasised that it is important to define both the numerator and the denominator of the fraction determining the efficiency index, and also to state clearly the time-scale upon which the standard of animal productivity is based.

The efficiency of converting feed into food protein necessitates the utilisation of nutrients other than crude protein *per se*, and for this reason certain authors, such as Holmes (1971), have employed parameters of efficiency which express the food protein produced as a function of the feed energy (in terms of calories of metabolisable energy) consumed. The use of such parameters reminds us that a single nutrient cannot properly be considered in isolation, but the correlation of such parameters to the basic parameter of feed protein consumed compared to food protein produced is very high, and for simplicity only this latter index is dealt with in this paper.

This paper will deal in turn with each of the principal types of animal production and to define for each the probable present level and range of feed conversion efficiency defined, after Holmes (1971), as 'that proportion of the specified nutrient which in the specified time is converted into a product for consumption or use by man'. It will then briefly discuss the various factors which could influence the present levels of efficiency and lead to their future improvement.

THE DAIRY COW

Assuming traditional and, probably, oversimplified standard protein requirements of 0.4 kg crude protein for the maintenance of a 450 kg cow and of 60 g crude protein per kg milk, the present feed conversion efficiency of an average cow calving at $2\frac{1}{2}$ years and yielding 4000 kg milk per lactation over an effective life of 4 lactations is about 24 per cent.

This figure can be markedly increased by raising milk yields over all lactations. If the yield can be elevated from the 'National Herd Average' of 4000 kg to the level achieved by the top 5 per cent of dairy herds, then the figure can be raised to over 34 per cent. This improvement is brought about by a marked widening of the ratio of the protein theoretically required for maintenance purposes and that required for milk production.

It has been argued (Wilson, 1968) that the theoretical 'genetic ceiling' for milk production in the dairy cow is of the order of 18 000 kg. Assuming that this figure could eventually be realised, then the feed conversion efficiency at this elevated level of milk yield would be raised to the highly respectable level of 45 per cent.

The efficiency of feed conversion is also influenced by longevity. If a cow only provides milk over four lactations then each lactation must bear one quarter of her protein requirements for rearing. Assuming that the protein demand over the two-year rearing period is about 270 kg crude protein, then the appropriate allocation is about 68 kg per lactation. If the longevity is increased to eight effective lactations the appropriate allocation is 34 kg per lactation, and if the longevity is further increased to 16 lactations then the figure is further reduced to 17 kg. The overall effect is to slightly increase the protein conversion efficiency from 28 per cent (four effective lactations) to 31 per cent (16 effective lactations).

Some workers, for instance Holmes (1971), discount improvement in protein conversion as a result of increased longevity since current breeding plans require the early replacement of a large proportion of cows within a herd with heifers of superior genotype. However, when milk yields are raised closer to their genetic ceiling, this argument will become less valid, and the chief reason for culling cows in favour of heifers will then be to replace animals which, through disease, accident or mismanagement, are suffering from a premature decline in yield. Again, under current systems of feeding and management, most dairy cows decline in yield after about the fifth or sixth lactation, with the result that the benefits of spreading the protein requirements of rearing the heifer over more years are offset by the declining protein yield from milk in later lactations. However, there are cases on record of milk yield in individual cows remaining at a high level for 10 or more lactations, and there are many case histories of cows living to over 20 years of age and producing 20 live calves in their lifetimes. There do not, therefore, appear to be any reasons why advances in our knowledge of the physiology of ageing should not enable the effective working life to be prolonged without consequent deleterious effect on milk yield.

Another possibility for improving protein conversion would be to decrease the protein required to produce a unit of milk. This can already be partially achieved by substituting non-protein nitrogen in one form or another for the more traditional sources of animal feed protein. The work of Virtanen (1966) has amply demonstrated that the whole diet of the successfully lactating and reproducing dairy cow can be supplied in this way, without noticeably adverse effect on calf health or on milk quality, but milk yields are not sustained at the higher levels which can be achieved when at least some normal protein sources are incorporated in the diet.

The last factor which could influence the protein conversion is the age at first calving. Current average age of heifers calving for the first time is several months in excess of two years, with the better farms regularly calving at or about 24 months. With selection for early maturity and rapid growth, there is a theoretical prospect of mating at about the yearling stage with first calves being dropped at 21 months. This would slightly reduce the protein input during rearing, possibly to about 50 kg protein, but the effect of this on overall protein conversion efficiency would be marginal, with an improvement of not more than 1 per cent for low-yielding cows, and appreciably less for high-yielding cows.

It has been shown that the milk yield of the cow is the prime determinant of protein efficiency, and a yield of 18 000 kg has been suggested as being theoretically obtainable. If this high level of yield is ever to be achieved several management factors will have to be radically altered. Firstly, the problem of ingestion of larger quantities of nutrients for milk production will have to be overcome, probably by the provision of high-nutrient density diets in liquid form, along the lines indicated by Clough and his co-workers at the National Institute for Research in Dairying. In addition, cows will have to be selected for Relative Feed Capacity since only those cows with very high maximum intakes of nutrients will be capable of having a sufficient surplus above maintenance requirements to sustain very high yields. Secondly, milking and feeding routines will have to be modified, probably to something approaching four times daily milking and *ad libitum* provision of feed throughout the 24 hours. Cows would be intensively housed and managed throughout their lactations and the elimination of mastitis and other diseases would be crucial to the maintenance of the udder in a healthy condition.

THE LAYING HEN

The protein inputs up to point-of-lay at five months are relatively constant at about 1.2 kg. Thereafter the efficiency of protein conversion is highly correlated to level of egg yield, being about 20 per cent at current average yields of 240 eggs per year and rising to about 30 per cent when yields are raised to the levels achieved in the top 5 per cent of the national flock, i.e. 320 eggs per year. It would be theoretically possible to raise the protein conversion efficiency to levels more closely approximate to that of the high-yielding dairy cow by extending the laying period for a further year, with or without an intervening moulting period. Indeed, there are some indications that such a trend may be developing in the industry at present. During the period 1965-70 it has been estimated that the proportion of flocks kept on into a 'second year' of lay was of the order of 20 per cent. Recent estimates put this figure significantly higher, at nearer to 25 per cent. If this trend continues and if the productivity during the 'second year' increases, either by higher yields in the same time or by extending the profitable laying period from the current length of about 30 weeks to 40 weeks or more, then the overall protein conversions will be raised, possibly up to a maximum of about 33 per cent.

However, it is the view of many that higher mean egg yields are more likely to be achieved by an elevation or extension of the peak laying period, or expressed another way, by achieving regular 24-hour ovulation cycles for longer periods. Prolonging the final phase of the laying period, when ovulation cycles have been increased from 24 to nearer 48 hours, is unlikely to prove an economic technique.

It would be theoretically possible to increase protein output, and hence improve protein conversion, by selecting for larger eggs of greater total protein content, and it might be theoretically possible to select strains capable of laying double-yolked eggs every 24 hours. The current economic trend, however, is to choose the lighter bird of lower maintenance requirement and replacement cost, and under these conditions the aim is to enable a larger percentage of the egg crop to achieve the minimum requirements for 'large' size rather than to produce 'very large' eggs at the expense of 'standard' eggs. However, it is clear that higher efficiencies of conversion could be achieved in this way if the market demand for eggs was to change more in favour of extra large egg size.

Lastly, slight increases in efficiency could be achieved by reducing the age at first egg. This has remained fairly constant at about five months for some years, and attempts to decrease the length of the pullet rearing period have been resisted on the grounds that the average egg weight, egg quality and the total egg yield usually suffer as a consequence. There appear to be no sound reasons why this should be so and it is therefore possible to contemplate a greater efficiency of protein conversion over a short four-month rearing period, with the result that lifetime efficiency would be increased by 3 per cent.

If all these effects are additive, one arrives at a maximum protein conversion efficiency of about 36 per cent for a laying hen coming into lay at 4 months of age and laying 350 eggs in the first 365 days, followed by a further 150 eggs in the second laying period after a short moult, only 15 per cent of which are below 'standard' size. At this high productive level, the efficiency is nearly equal to that of a milking cow first calving at two years and yielding 6800 kg milk in each of four lactations.

THE BROILER CHICKEN

Although different authors arrive at different figures for the protein conversion efficiency of the various livestock species, most are agreed that the broiler chicken and the laying hen are very close in rank order.

Current average standards of broiler production, in which birds of about 1.6 kg are raised in 56 days with a feed conversion ratio of 2.5, provide a protein conversion efficiency of about 20 per cent. By improving the feed conversion ratio to 2.0, a figure achieved by the top 5 per cent of producers, and by increasing the liveweight of the finished broiler to 2.0 kg, the efficiency is increased to just under 30 per cent.

In future we can envisage further improvements with feed conversion ratio being reduced to under 1.6 *either* with liveweight at eight weeks exceeding 2 kg *or* with 1.6 kg carcasses being produced at an earlier age of under 50 days. This improvement, either greater weight at the same age or equal weight at an earlier age, is most likely to be achieved by the use of diets with improved amino acid balance and with optimal relationships between the minor nutrient fractions. A direct consequence of the utilisation of new information on optimal amino acid levels will be a significant lowering of the overall crude protein levels. Further improvements are to be expected by treating male and female broilers as separate crops and formulating different diets for the two sexes (Filmer, 1970).

FUR-BEARING MAMMALS (RABBITS)

The most efficient meat-producing mammal is one with a short generation interval coupled to high growth rate and early maturity. The best-known example of such an animal is the rabbit but it is possible to speculate that other small mammals may also eventually achieve the required degree of gastronomic respectability. Current levels of productivity are low compared with the theoretical biological ceiling (Wilson, 1968), and this is primarily due to the current low levels of reproductive efficiency. At an average litter number of five produced three times a year over an effective three-year life the efficiency of protein conversion is of the order of 11 per cent. If this is increased to the easily attainable standard of four litters of 10 produced annually over a four-year life the protein conversion increases to about 17 per cent. If this could be further raised to four litters of 12 produced annually over a five-year life a top value of about 20 per cent may theoretically be obtained.

It follows that the production of meat from small mammals can never be as efficient a process as the production of chicken meat by the broiler, but it nevertheless has the potential of ranking higher than pigs and beef cattle in this respect, as will now be shown.

THE PIG

In any consideration of the protein conversion of the pig it is important to differentiate between the porker, the baconer and the heavy hog. The former type has been selected in this brief discussion, and different indices of efficiency will clearly relate to other production systems aiming at higher weights and consequently higher proportions of body fat in the resulting carcass. Greater absolute weights of lean meat will, naturally, be produced by the heavier animals, but the lean percentage of the carcass will be lower. World trends in pig meat production favour the medium weight porker rather than the baconer, and it is likely that this trend will continue as total world *per caput* consumption of pig meat increases.

Current levels of animal productivity in the pork industry are such that sows are maintained for an effective total life of about three years, during which about four litters are produced from which a total of about 30 porkers are grown to carcass weights of approximately 40 kg. At this productive level the protein conversion is approximately 12 per cent, depending very greatly on the breed or strain of pig and on its lean content at this carcass weight.

Earlier weaning techniques, and especially very early weaning immediately after colostrum feeding, would enable the farrowing index to be reduced and hence 2.25 litters to be born each year. Such early weaning systems would need to be coupled to efficient techniques to enable the sow to be successfully mated within one month of farrowing, possibly by endocrinological intervention to stimulate early *post-partum* ovulation.

The sow is capable of producing over 20 ova at each ovulation and the number of ova fertilised is usually greatly in excess of the number of live births, due to re-absorption of embryos in the Fallopian tubes or uterine horns. If more of the embryos could be carried to full term, litters of 20 piglets are theoretically possible and have been achieved in practice, but average litter sizes of 12 reared to maturity are already practical with the top 5 per cent of producers.

The two next most important factors determining protein conversion are the growth rate and feed utilisation of the growing and fattening pig. Current levels of weight gain average about 0.4 kg per day, and of feed conversion ratios about 2.9. These levels could be readily raised to 0.6 kg and 2.1, equal to the standards achieved on the top 5 per cent of pork-producing farms. Elevation to these levels is dependent upon both genetic and nutritional improvement. It is likely that extra lean breeds, such as the Pietrain, and the separate feeding of the two sexes with diets differing in both amino acid composition and nutrient density, will have a part to play. At these higher levels of productivity the protein conversion can be raised from 12 per cent to over 16 per cent. It is impossible to demonstrate that this figure could ever be raised to the 20 per cent level, even if all production parameters are raised to their respective theoretical maximum genetic ceilings.

THE LAMB

Current levels of sheep production are extremely low and are artificially depressed by the utilisation of a large proportion of the world flock, including much of the UK flock, for extensive grazing on low-quality hill and mountain forage. As has been demonstrated by Spedding (1969), the full productive potential of sheep cannot be realised until they are intensively managed in such a way that their potential fecundity can be exploited.

With ewes producing four crops of lambs over a five-year life, at an average lambing percentage of 150, and with lambs fattened to carcass weights of 16 kg in six months, protein conversion efficiencies are very low at about 4 per cent. If the ewes are kept for the same effective life but are mated out of season to enable them to produce two litters per year each of three lambs, and if the lambs are more intensively reared to produce 16 kg carcass weights at four months of age, then the protein efficiency may be raised to 9 per cent.

Extending the effective working life of the ewe beyond five years, and increasing the liveweight gain and feed utilisation of the lamb by early weaning and intensive rearing on low-roughage diets of high nutrient density, could further increase the efficiency index to 12 per cent.

However, it is unlikely that a significant proportion of the world sheep population will be intensively managed in this fashion. There are good agronomic reasons why breeding ewes should be maintained on less productive grassland, and it is unlikely that the market for veal-type lamb meat will merit a major change in management systems designed to produce such a product.

THE BEEF STEER

The chief factors limiting the protein conversion of beef animals are the gestation period of the cow, coupled to the production of a single calf at each parturition. Nothing can be done about the former constraint, which results in very high maintenance protein costs of the dam which must be spread over a limited number of calvings. The production of twinning strains of beef cattle is theoretically possible and from time to time it has been suggested that hormonal intervention should be used for this purpose. However, the danger of losing both calves at a difficult calving would offset the theoretical advantages of twinning and such a technique is therefore unlikely to be widely practised.

Assuming single calves and cows calving at 26 months of age, and assuming that carcass weights of 300 kg could be achieved in a 15-month fattening period, then protein conversion efficiencies of about 6 per cent may be achieved. Earlier calving of the beef cow, and more rapid growth of the calf so as to achieve the same carcass weight by 12 months of age, would increase this figure to about 8 per cent. The selection of breeds of cattle for lean content and for feed utilisation efficiency could elevate this figure still further but it is not feasible, adding all possible factors together, to exceed a protein conversion efficiency of 10 per cent.

All these calculations deal with each livestock class in isolation, and the figures can be altered by assuming, for instance, that beef calves can be produced as a byproduct of the dairy herd with the consequence that no provision need be made for the protein consumed by the milk-producing dam, or alternatively assuming that sheep, like goats, can be used as dual-purpose animals. Such considerations will be dealt with by subsequent speakers and in this paper each defined livestock class has been considered in isolation, from the standpoint of potential animal production and not from the standpoint of land utilisation and farming system.

REFERENCES

BLAXTER, K. L. (1968). The animal harvest. *Science Journal* **4** (5), 53–9.
COOP, I. E. (1967). The efficiency of feed utilization. *Proceedings of the New Zealand Society of Animal Production* **27**, 154–65.
FILMER, D. G. (1970). Segregating the sexes. *Poultry World* **121** (15), 23–5.
LEITCH, I., & GODDEN, W. (1953). *The Efficiency of Farm Animals in the Conversion of Feedingstuffs to Food for Man.* Commonwealth Bureau of Animal Nutrition Technical Communication No. 14. Farnham Royal, England: Commonwealth Agricultural Bureaux.
HOLMES, W. (1971). Efficiency of food production by the animal industries. In: Wareing, P. F., & Cooper, J. P. (Eds.), *Potential Crop Production. A Case Study*, 213–27. London: Heinemann.
SPEDDING, C. R. W. (1969). The agricultural ecology of grassland. *Agricultural Progress* **44**, 1–23.
VIRTANEN, A. I. (1966). Milk production of cows on protein-free feed. *Science* **153**, 1603–14.
WILSON, P. N. (1968). Biological ceilings and economic efficiencies for the production of animal protein, AD 2000. *Chemistry and Industry*, 899–902.

DISCUSSION

By V. R. FOWLER

Rowett Research Institute, Aberdeen

AND C. C. BALCH

National Institute for Research in Dairying, Shinfield, Reading

In a discussion on the significance of single-tissue and small animal studies in more practical situations, it was suggested that the magnitude of inefficiencies of protein production arising at the tissue level was of little practical significance at the whole animal level. It was stressed that, whilst in principle this was so, studies on the turnover rate of myosin indicated that protein turnover may be responsible for a high proportion of energy losses in the whole animal. This process may well be responsible for the considerable discrepancy which exists between theoretical values, computed on the basis of biochemical efficiencies, and values determined on whole animals.

The possible dangers of extrapolating findings made at the tissue level or with a small animal such as the rat to animals of agricultural importance were emphasised but, whilst extrapolation involves many assumptions, the importance of tissue studies in elucidating such mechanisms as the hormonal control of growth should not be overlooked.

The main area of discussion, which arose from the second paper on protein production in the whole animal, concerned the choice of objectives in breeding programmes; the contention that selection of animals should be based on a low daily energy expenditure for maintenance was disputed. It was suggested that maintenance *per se* was not a useful measurement but rather the maintenance cost per unit of protein produced; indeed in the growing animal 'maintenance' heat loss may well be positively correlated with the desirable characteristic of a high potential rate of protein synthesis. Conceding this point, Professor Bowman said that the classes of livestock which he had in mind when making his original statement were the lactating cow and the laying hen, in which maintenance costs represented a high proportion of the energy requirement.

Another topic which drew comment was the emphasis which was given to selection for increased appetite or relative feed level. The desirability of positive selection for such a trait was questioned, particularly in relation to the genetic improvement of the pig which has an almost legendary ability to eat in excess of its requirements. Doubts too were expressed about the reality of the advantages to be gained by breeding for increased intakes in some other classes of farm livestock on the grounds that in many cases increases could be more simply achieved by technological means, such as changing the energy concentration of the diet or the physical form in which the feed was presented. It was argued, however, that alterations in the energy content of the diet of the laying hen did not alter total daily energy intake and that, although the broiler chicken did not have the same fixity of voluntary intake of energy, much of the genetic improvement which has been made could be attributed to improvement in the potential for feed intake.

Questions were raised about what criteria should be considered in choosing the best population structure and size for obtaining maximal utilisation of a limited resource such as food. This problem, it was suggested, was analogous to that involved in choosing a stocking rate. Biological efficiency would be maximised if the highest output per animal were taken as the criterion of efficiency according to Mr Large but this view was disputed, particularly by the economists.

The usefulness of comparisons of biological efficiency made between different species and between different production systems using the same species was questioned. What, for example, should one conclude from the observation that improved biological efficiencies were associated with larger mature size? When extrapolated this observation would suggest that turkeys should replace broiler chicks and cattle replace sheep. One could not, however, dictate to the housewife that she should only eat meat which was produced by efficient animal systems but the comparisons did perhaps offer a clue to the way in which animal production systems might change in the future. One aspect of the subject which, it was suggested, had been treated only in passing was the potential for the improvement of the efficiency of protein production by the use of entire rather than castrated males. This was considered to be of considerable importance, particularly in the light of research carried out at Hurley on bulls and at the Rowett on boars. Attention was also drawn to the rather unexplored but exciting field of the use of exogenous anabolic steroids, which were already used in the production of poultry meat and beef.

The discussion was then widened to include issues of a more philosophical nature. It was suggested that, although the papers provided an excellent assessment of the situation in terms of *analysis*, there still remained the question of what should now be the strategy for a *re-synthesis*. Several proposals were made in response to this question and were of two main types. One group advocated a re-synthesis in purely economic terms; the other considered that the information should be regarded as providing a framework for producing improved models of biological systems and thereby highlighting suitable areas for future research. There was also ample recognition throughout the meeting of the fact that those who guide animal production policies in the wider sense will need increasingly to know how best to make use of the various systems by which inorganic nitrogen compounds can be converted to protein suitable for man, and that the biological efficiency of this conversion is one of the main factors to be considered. Insufficient attention was given at the symposium to the many virtues of the ruminant in effecting this conversion through the agency of the rumen microflora. All processes of protein synthesis required a source of energy in addition to a source of nitrogen. The ruminant had the advantage of being able to utilise a wide variety of sources of energy including some roughages so highly lignified that they are of little or no value to any other farm animal; ruminants were also able to harvest such unpromising material in highly unfavourable terrain. Insufficient attention had been given to the fact that in animals the amount of protein, or nitrogen in the case of ruminants, that could be converted into meat, milk, wool or eggs depended on the amount of energy consumed by the animal; the cost of producing protein must always have been considerably influenced by the cost of this energy. Ruminants presented a way of utilising very large sources of low-priced energy which would be unavailable for any other system of protein production. The response of ruminants to variations in the intake of energy and protein was somewhat reciprocal – the same response might be obtained with a diet high in energy and low in protein as with another diet lower in energy and higher in protein. This relationship permitted considerable adjustment to reduce feed costs and it might sometimes pay to accept an efficiency of protein utilisation which was lower than optimal.

PART IV

THE BIOLOGICAL
EFFICIENCY OF PROTEIN PRODUCTION
BY ECOSYSTEMS

14
THE BIOLOGICAL EFFICIENCY OF PROTEIN PRODUCTION BY GRAZING AND OTHER LAND-BASED SYSTEMS

By J. PHILLIPSON

Animal Ecology Research Group, Department of Zoology, University of Oxford

INTRODUCTION

To review the biological efficiency of protein production by land-based systems is not an easy task since the literature contains very few comprehensive data pertaining to whole ecosystems, be they agricultural or otherwise, and yet the present contribution is an attempt to:

1. Estimate the potential production of the global land ecosystem.
2. Compare the potential and realised production of non-woodland ecosystems.
3. Discuss the efficiency of production by woodland ecosystems.

In discussing the biological efficiency of protein production it must be realised that an efficiency is simply a ratio of input to output. Clearly, in the present context the output is protein, but protein in what form and for what purpose? At one extreme we can consider total protein production and at the other only that protein of direct use to man. A further complication arises in deciding the nature of the input. Is one interested in the efficiency of protein production in relation to the amounts of sunlight, water (including irrigation), or nutrients (including fertilisers) entering the system, or in relation to the energy expended by man to produce protein for his own ends? An efficiency is normally expressed as a percentage and thus it is necessary to adopt the same units for both input and output: clearly, however, the choice of units and the decision as to which efficiency is of the greatest interest is at the discretion of the investigator.

Just as the unqualified term 'efficiency' is conceptual so is the term 'ecosystem'. Indeed Tansley (1935) wrote: 'These ecosystems, as we may call them, are of the most various kinds and sizes.' In the terminology of thermodynamics natural, semi-natural and agricultural systems are 'non-equilibrium' systems and, apart from the virtually 'closed' biosphere, are to varying degrees 'open'. Ecosystems are thus characterised by rather fluid boundaries and the extent of any ecosystem is decided by the investigator according to the demands of the scientific enquiry being made.

Clearly, in the present context, there is a wide range of efficiencies and ecosystems from which to choose. It is my intention to deal with ecosystems of varying size and I shall select suitably qualified inputs and outputs to determine those efficiencies of protein production which, in my judgement, are the most interesting.

THE GLOBAL LAND ECOSYSTEM

It is generally agreed that the total surface area of the earth is 510×10^6 km²; however, estimates of the areas constituting land, fresh and salt water differ. For the total land surface Brown & Finsterbusch (1971) quote an area equivalent to 130×10^6 km², whilst Whittaker (1970) gives a figure of 149×10^6 km². The latter statistic includes 4×10^6 km² of lakes, streams, swamps and marshes; if these areas are considered as freshwater habitats then a corrected value of 145×10^6 km² is obtained for the total land surface. This figure is used in the present work.

Protein production in the global land ecosystem depends, as in most other ecosystems, on the ability of green plants to utilise solar radiation, nutrients and water to produce the plant tissues upon which heterotrophs depend. It is well known that in some areas of the world water or nutrients or both limit the amount of primary production but, for the purpose of calculating maximum potential production of protein, it is assumed that man will be technologically capable of solving problems associated with water and nutrients. It is implied therefore that the ultimate factor limiting primary production is the amount of solar radiation received at the surface of the earth. Different regions of the earth receive varying amounts of solar radiation but it has been estimated by Gates (1962) that the long-term mean value for visible light (0.4 to 0.7 μm) reaching the ground surface is 22 per cent of the total extraterrestial insolation (2.4×10^{15} J km^{-2} per annum). The total amount of visible light reaching the land each year is thus $(2.4 \times 10^{15})(1.45 \times 10^8) = 3.48 \times 10^{23}$ J, or 8.3×10^{22} cal. This figure is close to Wassink's (1968) estimate of 8.0×10^{22} cal.

Knowing the quantity of photosynthetically active radiation (PAR or visible light) reaching the land surface each year and given the maximum efficiency with which plants are capable of using this radiation on a sustained basis, it is possible to calculate the maximum potential production of plant material per annum. Bonner (1962) has calculated that the upper limit of plant yield (net production above ground) is between 2 and 5 per cent of the PAR received during the growing season, the exact figure depending on the average intensity of light. Accepting a mean growing season for the world of nine months a net photosynthetic efficiency

$$\left(\frac{\text{net primary production above ground}}{\text{incident PAR}} \times 100\right)$$

of 2.6 per cent can be calculated. Using this figure it can be estimated that the global land ecosystem has a potential maximum above-ground net primary production of 9.06×10^{21} J, which approximates to 540×10^{12} kg dry weight of plant matter (3.72×10^4 kg ha^{-1} per annum).

At present the mean realised net primary production above ground is much less than the calculated maximum potential. The results of studies on 46 woodland and 57 non-woodland ecosystems, man-made and natural, in all parts of the world were used to calculate the net photosynthetic efficiency of woodland and non-woodland areas; the efficiencies were 0.54 and 0.38 per cent respectively and do not differ markedly from the figures of Jordan (1971). The input of PAR to the land surface is 34.8×10^{22} J (8.3×10^{22} cal) per annum and the ratio of woodland to non-woodland areas on land is 4:6, hence woodland areas receive 13.94×10^{22} J (3.32×10^{22} cal) and the non-woodland areas 20.92×10^{22} J (4.98×10^{22} cal) of photosynthetically active radiation per annum. Using the net photosynthetic efficiencies of 0.54 and 0.38 per cent it can be shown that the net primary production above ground is 44.82×10^{12} kg dry weight in woodland areas and 47.31×10^{12} kg dry weight in non-woodland systems. The total of 92.13×10^{12} kg dry weight of plant matter is equivalent to 6.3×10^3 kg ha^{-1} per annum, or 630 g m^{-2} per annum. For comparative purposes this estimate is shown alongside those of other workers in Table 1.

The global land ecosystem can thus be subdivided into non-woodland and woodland systems and it is of some interest to explore the biological efficiency of protein production in each of these two categories.

NON-WOODLAND ECOSYSTEMS

Figure 1 summarises the results of a number of studies on above-ground net primary production in non-woodland ecosystems. The curve marked A indicates the maximum potential net primary production above ground when Bonner's (1962) net photosynthetic efficiency of 2.6 per cent is applied on the basis of 22 per cent of the extraterrestrial insolation reaching the ground as photosynthetically active radiation. However, different regions of the earth receive varying amounts of solar radiation and curve B shows the maximum potential above-ground net primary production when the mean annual prevalence of cloudiness for each latitude is taken

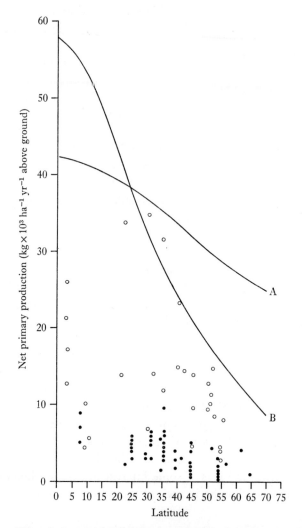

Figure 1. Net primary production in non-woodland ecosystems. Curve A is based on a net photosynthetic efficiency of 2.6 per cent and the assumption that 22 per cent of the extraterrestrial solar radiation reaches the ground. Curve B is also based on a net photosynthetic efficiency of 2.6 per cent but allows for the mean annual prevalence of cloudiness at different latitudes. Closed circles depict natural and semi-natural systems, whereas open circles depict man-made ecosystems.

into account. The closed circles are the results of studies on natural or semi-natural non-woodland systems and it is clear that these ecosystems do not reach the estimated maximum potential; in most cases they reach only 10–20 per cent of this value. On the other hand agricultural systems (indicated by open circles) in certain instances do reach the maximum potential value. Where the estimated maximum potential is not reached by either natural or man-made ecosystems it can be concluded that some factor, or factors, other than photosynthetically active radiation is limiting net primary production above ground.

Table 1. *Estimates of the net primary production of the earth's land surface ($kg \times 10^{12}$ dry weight per annum)*

Maximum potential (this study)	540
Realised (Wassink, 1968)	50
Realised (Whittaker, 1970)	109
Realised (this study)	92

So far I have concentrated on the biological efficiency of production of plant proteins in the form of net primary production above ground; there remains the biological efficiency of production of animal tissues.

If, with Hutchinson (1948) and Hairston, Smith & Slobodkin (1960), it is agreed that, over geological time, the accumulation of plant materials (fossil fuels) occurs at a rate that is negligible when compared with the rate of energy fixation through photosynthesis, then the equivalent of 5.44×10^3 kg ha^{-1} per annum dry weight of plant matter must be utilised by heterotrophs in non-woodland ecosystems. However, in natural and semi-natural non-woodland systems the mean value, as can be seen from Figure 1, is nearer 4.25×10^3 kg ha^{-1} per annum. The efficiency with which this autotroph protein is converted to heterotroph protein depends very much upon the biota which occur in different parts of the world. Unfortunately, very few data regarding heterotroph consumption in whole ecosystems are available. It is possible nevertheless to estimate the expected biomass, consumption and protein production by heterotrophs in non-woodland areas. This can be done by making use of the reasonably plentiful data on assimilation efficiency (food assimilated/food ingested $\times$ 100), production to respiration ratios, and production to biomass ratios. Clearly, all animal communities contain invertebrates and vertebrates in varying proportions but initially and for the sake of simplicity I shall consider (i) a purely invertebrate community, (ii) a purely small-mammal community, and (iii) a purely large-mammal community. In each case I shall assume that the above-ground net primary production is 4.25×10^3 kg ha^{-1} or 425 g m^{-2} per annum dry weight.

An invertebrate community

Terrestrial invertebrates which feed on plant matter either living or dead have a mean assimilation efficiency of the order of 30 per cent. Further, the work of McNeill & Lawton (1970) on population production to respiration relationships indicates that 43 per cent of the assimilated food is used for the production of new tissues and 57 per cent is used in respiration. Finally, the production to biomass ratio of terrestrial invertebrates varies between 2:1 and 6:1 but the mean value is 3:1, also given by Waters (1969) for freshwater invertebrates. Given the net primary production above ground, the assimilation efficiency, production as a proportion of assimilation, and the production to biomass ratio, it is possible to estimate production by this invertebrate community. Table 2 shows the various steps in arriving at an invertebrate herbivore biomass of 18.3 g m^{-2} with a production of 54.8 g m^{-2} per annum. The efficiency with which plant protein is converted to animal protein is 13 per cent.

Table 2. *Estimates of various production parameters for three hypothetical communities; a purely invertebrate community, a purely small mammal community, a purely large mammal community*

	Invertebrates only	Small mammals only	Large mammals only
Net primary production above ground (g m^{-2} per annum)	425.0	425.0	425.0
$\dfrac{\text{Individual assimilation}}{\text{Individual consumption}} \times 100$	30.0	75.0	40.0
Assimilation by plant feeders (g m^{-2} per annum)	127.5	318.8	170.0
$\dfrac{\text{Population production}}{\text{Population assimilation}} \times 100$	43.0	1.6	1.6
Production by plant feeders (g m^{-2} per annum)	54.8	5.1	2.7
$\dfrac{\text{Population production}}{\text{Population biomass}}$	3.0	2.0	0.2
Biomass of plant feeders (g m^{-2})	18.3	2.6	13.5
$\dfrac{\text{Trophic level production}}{\text{Trophic level consumption}}$	0.13	0.012	0.006

A small-mammal community

The mean assimilation efficiency shown by small mammals is approximately 75 per cent (Drozdz, 1968a, b), and from McNeill & Lawton (1970) it can be seen that on average only 1.6 per cent of the energy assimilated by homoiotherms appears as production. The production to biomass ratio can be as low as 1:3 (hares in Poland) but is most frequently of the order of 2:1. By adopting the procedure used for the invertebrate community it can be estimated (Table 2) that a purely small-mammalian herbivore community consuming 425 g m^{-2} per annum dry weight should have a biomass of 2.6 g m^{-2} and a production of 5.1 g m^{-2} per annum. The efficiency with which plant protein is converted to animal protein is 1.2 per cent.

Table 3. *Comparisons of biomass estimates for two hypothetical and three actual non-woodland ecosystems* (g m^{-2} *per annum dry weight*)

Ecosystem	Authority	Net primary production above ground	Biomass of plant-feeding invertebrates
Hypothetical meadow	Macfadyen (1963)	1000*	36.0†
Limestone grassland	Cragg (1961)	?	38.0†
Juncus	Cragg (1961)	?	15.6†
Ungrazed meadow	Oxford University students, 1969 and 1970	477†	17.9
Hypothetical invertebrate-dominated grassland	This study	425	18.3

* Assuming 4.19 kcal g^{-1}.
† Assuming water content of 80 per cent.

A large-mammal community

The mean assimilation efficiency of large herbivores on poor-quality forage is approximately 40 per cent (Hughes, Milner & Dale, 1964; Petrusewicz & Macfadyen, 1970). By using the homoiotherm value of 1.6 per cent for the proportion of assimilated energy appearing as production, and a production to biomass ratio of 1:5 (Wiegert & Evans, 1967) the annual production and biomass of a purely large-mammal community can be estimated. Table 2 indicates an annual production of 2.7 g m^{-2} per annum and a biomass value of 13.5 g m^{-2}. The efficiency with which plant protein is converted to animal protein is 0.6 per cent.

It may be inferred from these three hypothetical communities that the ratio of production per unit weight of food consumed by invertebrate, small mammalian, and large mammalian herbivores is of the order of 20:2:1. Invertebrates are thus twenty times more efficient than large mammals in converting plant material into animal protein.

No animal community is composed entirely of invertebrates, small mammals, or large mammals, but there are communities which can be said to be dominated by invertebrates or by large mammals. Complete data on net primary production, animal biomass and secondary production are not available for these land-based ecosystems, or indeed for any ecosystem. It is possible nevertheless to compare two of the hypothetical models with more realistic ecosystems by recourse to fragmentary information on invertebrate-dominated temperate grasslands and large-mammal-dominated East African plains.

Invertebrate-dominated grasslands

If the model of the hypothetical invertebrate community is to be considered acceptable then we would expect actual figures for total animal biomass per unit weight of net primary production above ground to approximate to the biomass estimates for the hypothetical invertebrate community given in Table 2. Table 3 lists those communities where total biomass figures are available and also gives the expected biomass as derived for the hypothetical community. The paucity of data is fully recognised, but on the information available the measure of agreement is acceptable. In view of the sparse data it is perhaps worthwhile to point out that in the ungrazed meadow studied by Oxford University students the herb layer animal biomass figure (0.637 g m^{-2}) agrees fairly well with herb layer biomass figures for grassland obtained by other workers; Balogh & Loksa (1948) calculated 0.2 to 0.6 g m^{-2} for Hungarian steppes, Odum (1959) gives 0.6 g m^{-2} for an old field in Georgia, Southwood & van Emden (1967) obtained a figure of 0.3 to 0.5 g m^{-2} for uncut grassland in Britain, and Gillon & Gillon (1967) quote 0.68 g m^{-2} for West African savannah. In contrast Menhinick (1967) working in a *Sericea lespedeza* stand in the USA estimated a herb layer invertebrate biomass of approximately 0.06 g m^{-2} but indicated that this figure is lower than those obtained by other workers.

Large-mammal-dominated grasslands

The open plains and savannah shrublands of Africa are the best known examples of near-natural ecosystems dominated by large mammals. Figures for the large mammal biomass supported by these systems have been published but little is known of the role of invertebrates and small mammals. Foster & Coe (1968) summarised a number of large-mammal studies and it is clear that their biomass varies between 1050 and 12600 kg km^{-2} live weight. Allowing a water content of 80 per cent these figures become 840 and 2520 kg km^{-2} with a mean value of 1200 kg km^{-2}. Hendrichs (1970), working in the Serengeti Plains, estimated a large mammal biomass of 5000 kg km^{-2} (1000 kg km^{-2} dry weight), and a small

Table 4. *Estimates of various production parameters for the Serengeti Plains ecosystem (biomass in g m^{-2} dry weight; all other figures as g m^{-2} per annum dry weight)*

		Percentage of net primary production above ground
Net primary production above ground	300.0	100.0
Large-mammal biomass	1.0	—
Production by large mammals (production:biomass = 1:5)	0.2	—
Forage assimilated by large mammals (using production as 1.6 % of assimilation)	12.5	—
Forage consumed by large mammals (assuming assimilation efficiency of 40 %)	31.25	10.42
Net primary production remaining for small mammals and invertebrates	268.75	89.58
Small-mammal biomass	0.08	—
Production by small mammals (production:biomass = 2:1)	0.16	—
Forage assimilated by small mammals (using production as 1.6 % of assimilation)	10.0	—
Forage consumed by small mammals (assuming assimilation efficiency of 75 %)	13.33	4.44
Net primary production remaining for and consumed by invertebrates	255.42	85.14
Forage assimilated by invertebrates (assuming assimilation efficiency of 30 %)	76.63	—
Production by invertebrates (using production as 43 % of assimilation)	30.65	—
Biomass of invertebrates (production:biomass = 3:1)	10.21	—

mammal biomass of 400 kg km^{-2} (80 kg km^{-2} dry weight). Braun (1969) has produced figures for the net primary production above ground in various regions of the Serengeti. Braun (1969) showed that the actual amount of primary production for any region depended upon rainfall and the range was from 600 to 6000 kg ha^{-1} dry weight with a mean value of 3000 kg ha^{-1} dry weight (300 g m^{-2}).

Using this mean value for net primary production above ground in conjunction with the biomass estimates of Hendrichs (1970), a tentative model for protein production in the Serengeti Plains has been constructed. Table 4 shows the various stages of the calculation. The conversion figures

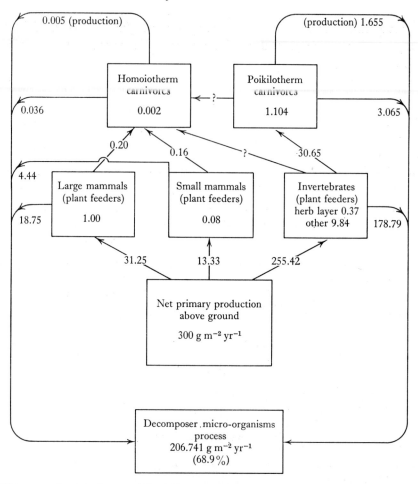

Figure 2. A tentative model of the Serengeti Plains ecosystem. Except where stated the figures in the boxes are for biomass and represent g m^{-2}, all other figures are in g m^{-2} per annum.

used were those employed in the earlier simplistic models when justification for their use was given.

Figure 2 shows the results in the form of a flow chart but, in addition, the role of carnivores and decomposers has been estimated. The only extra conversion figures used were an assimilation efficiency for carnivores of 90 per cent (a not unrealistic efficiency in the light of the many figures now published), a production/assimilation efficiency of 1.6 per cent for homoiotherm carnivores and a production/assimilation efficiency of 6.5 per cent for invertebrate carnivores (based on unpublished studies made with Phalangiida and Chilopoda by myself and colleagues). The invertebrate primary consumer biomass was divided into herb layer and other primary consumers on the basis of the ungrazed meadow community studied by Oxford University students where the herb layer component of the total invertebrate biomass equalled 3.7 per cent. The estimated 0.37 g m^{-2} dry weight compares favourably with the 0.24 g m^{-2} of herb layer primary consumers reported by Gillon & Gillon (1967) in West African savannah. Support for the homoiotherm carnivore level is to be found in Foster & Coe (1968), who estimated that the large carnivores in Nairobi National Park removed 0.14 g m^{-2} per annum dry weight, a figure not far removed from the 0.2 g m^{-2} per annum estimated in the Serengeti Plains model.

The model can be used to calculate the ecological efficiency of both primary and secondary consumer trophic levels. The ecological efficiencies (yield to trophic level $n+1$/consumption by trophic level $n \times 100$) of these two levels are 10.34 and 5.04 per cent respectively. It is of interest to note that these efficiencies are close to the 11.0 and 5.5 per cent calculated for the aquatic Silver Springs ecosystem by Odum (1957), and add terrestrial, as opposed to aquatic, support for the contention of Steele (1965) and Kozlovsky (1968) that ecological efficiency decreases above the primary consumer trophic level.

With regard to the efficiency of production of animal protein of direct use to man it can be seen that the 0.2 g m^{-2} per annum dry weight produced by large mammals represents a conversion efficiency (production/consumption $\times 100$) of 0.7 per cent, a figure some five to ten times lower than the 3.3 per cent of bullocks in Macfadyen's (1963) composite grazed meadow and the 6.0 per cent quoted by Petrides, Golley & Brisbin (1969) for beef cattle under idealised range conditions in western USA. It should be remembered however that the high figures for beef cattle relate to ecosystems where the quality of forage is high and assimilation efficiencies as high as 60 per cent or more are reached. A better comparison of production by wild herbivores and beef cattle is made possible by Ledger, Payne & Talbot (1961), who report the carrying capacity of European-managed ranches in East Africa to be between 3728 and 5600 kg km^{-2} live weight. Taking the highest value for the European-managed farms (a biomass equivalent to 1.32 g m^{-2} dry weight) and applying the growth/biomass efficiency of 11.5 per cent given by Petrides, Golley & Brisbin (1969) for beef cattle, a production of 0.15 g m^{-2} per annum dry weight results. The production by beef cattle on the best European ranches is therefore estimated as being approximately 75 per cent of the value attained by wild mammals on the same type of pasture.

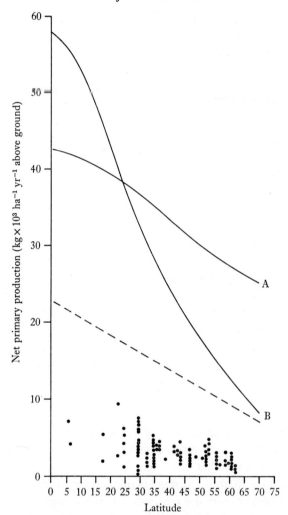

Figure 3. Net primary production in woodland ecosystems. Curve A is based on a net photosynthetic efficiency of 2.6 per cent and the assumption that 22 per cent of the extraterrestrial solar radiation reaches the ground. Curve B is also based on a net photosynthetic efficiency of 2.6 per cent but allows for the mean annual prevalence of cloudiness at different latitudes. Closed circles depict leaf and fruit (browse) production, whereas the broken line is based on Whittaker (1970) and represents total net primary production.

WOODLAND ECOSYSTEMS

Figure 3 summarises the results of a number of studies on above-ground net primary production in woodland ecosystems. As in Figure 1, curve A indicates the maximum potential production above ground when Bonner's

(1962) net photosynthetic efficiency of 2.6 per cent is applied on the basis of 22 per cent of the extraterrestrial insolation reaching the ground as photosynthetically active radiation. Curve B, again as in Figure 1, shows the maximum potential above-ground net primary production when the mean annual prevalence of cloudiness for each latitude is taken into account. The broken line is derived from Whittaker's (1970) estimates of the total net primary production of woodland ecosystems. It is of interest to note that this line approximates the mean values for the above-ground net primary production of agricultural systems in non-woodland areas. The closed circles are the results of studies on natural and man-managed woodland and show that part of woodland production which is potentially available to the leaf-eating secondary consumers (browsers). By comparing Figures 1 and 3 it can be seen that, at any given latitude, the mean amount of forage potentially available to grazers in natural non-woodland ecosystems is virtually the same as that potentially available to browsers in woodland ecosystems.

Unfortunately, information about the animal communities of woodland ecosystems is even less complete than that for non-woodland systems and any statements made about animal protein production from woodlands must, of necessity, be even more speculative than those already made about grasslands. By combining Varley's (1970) biomass estimates for the above-ground fauna and Macfadyen's (1963) biomass figures for the soil fauna in oak woodlands, I have produced a tentative model for this woodland system. Clearly, the animal biomass figures are from the literature and the transfer of materials per unit area per unit time have been calculated using conversion factors (assimilation efficiencies, production to assimilation, production to consumption, and production to biomass ratios) already published and used in the earlier non-woodland models.

Table 5 shows the order of application of the conversion factors to the primary consumer trophic level. Figure 4 shows the model in the form of a flow diagram and includes the secondary consumer trophic level. The transfer efficiency estimates for this level, where the biomass figures were known, were based on a carnivore assimilation efficiency of 90 per cent, a homoiotherm carnivore production over assimilation efficiency of 1.6 per cent, and a poikilotherm production over assimilation efficiency of 6.5 per cent. These efficiencies are the same as were applied to the secondary consumer level in the East African plains model.

The estimated net primary production of browse (413 g m^{-2} per annum dry weight) is very close to the mean value of *Quercus robur* and *Q. sessiliflora* litter production given in Bray & Gorham (1964). It may be inferred that the model is a reasonable approximation of the real situation.

Table 5. *Estimates of various production parameters for an oak woodland ecosystem (biomass in g m^{-2} dry weight; all other figures as g m^{-2} per annum dry weight)*

		Percentage of net primary production (estimated) above ground
Small-mammal biomass	0.033	—
Production by small mammals (production:biomass = 2:1)	0.066	—
Forage consumed by small mammals (production:consumption = 0.012)	3.000	0.73
Forage assimilated by small mammals (assuming assimilation efficiency of 75 %)	2.250	—
Biomass of feeding invertebrates	17.78	—
Production by plant-feeding invertebrates (production:biomass = 3:1)	53.34	—
Forage consumed by plant-feeding invertebrates (production: consumption = 0.13)	410.00	99.27
Food assimilated by plant-feeding invertebrates (assuming assimilation efficiency of 30 %)	123.00	—
Net primary production (browse) (410.0 + 3.0 g m^{-2} per annum)	413.00	100

This model, like the earlier East African plains one, can be used to calculate the ecological efficiencies of the primary and secondary consumer trophic levels; they are 12.95 and 5.84 per cent respectively and approximate the efficiencies estimated for the non-woodland ecosystem.

The oak woodland is an invertebrate-dominated system and there is little, if any, direct exploitation by man of its animal protein production.

DISCUSSION

The land surfaces of the earth are approximately 30 per cent of the world's total surface area and yet the estimated 92.13×10^{12} kg dry weight of net primary production per year in terrestrial situations is almost double the 50×10^{12} kg dry weight estimated by Ryther (1969) for the world's oceans. On a unit area basis it would appear that terrestrial ecosystems are four times more efficient than marine ones in producing plant matter. However,

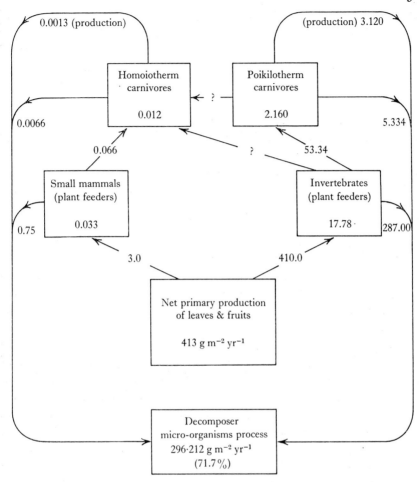

Figure 4. A tentative model of an oak woodland ecosystem. Except where stated the figures in the boxes are for biomass and represent g m^{-2}; all other figures are in g m^{-2} per annum.

most of man's fish protein currently comes from coastal and upwelling zones with a total surface area of 36×10^6 km^2 and a net primary production of 10×10^{12} kg per annum dry weight. Man's land-produced protein comes mainly from savannah, temperate grassland and tundra with a total surface area of 32×10^6 km^2 and an annual net primary production of 16×10^{12} kg per annum dry weight (Whittaker, 1970).

Ryther (1969) estimated the potential fish production from 10×10^{12} kg dry weight of net primary production to be 50×10^9 kg dry weight, a conversion efficiency of plant matter to fish of 0.5 per cent. Using the tentative model of the East African plains (net primary production above ground of

425 g m^{-2} per annum dry weight and a large-mammal production of 0.2 g m^{-2} per annum dry weight), it can be estimated that the 16×10^{12} kg dry weight of primary production in savannah, temperate grasslands and tundra will produce 7.5×10^9 kg dry weight of large mammal, a conversion efficiency of plant matter to mammal of less than 0.05 per cent. Clearly, in natural and semi-natural systems large mammals are ten times less efficient than fish in converting plant matter to edible protein; the reasons for this are that fish have assimilation efficiencies approximately twice that of large mammals and being long-lived poikilotherms have production to assimilation ratios (McNeill & Lawton, 1970) five times larger.

The estimated production of 50×10^9 kg dry weight of fish and the 7.5×10^9 kg dry weight of large herbivores is not potential harvest. Studies by workers such as Silliman & Gutsell (1958) suggest that the maximum sustainable yield for animal populations is of the order of 40 per cent of the annual production. Using current methods man might expect therefore an annual world fish yield of 20×10^9 kg dry weight and an annual large-mammal yield of 3×10^9 kg dry weight.

There are ways in which the animal protein yield to man from land-based systems could be increased. Using agricultural technology on non-woodland systems it has been shown (Figure 1) that man can already raise the level of net primary production per unit area by at least a factor of three; theoretically a factor of between five and ten is possible. Table 2 indicates that if man was willing and able to turn his attention to invertebrates a protein yield twenty times that considered possible for large mammals would follow. In practice invertebrates are not easy to harvest or process and man now removes much of the net primary production from the land and feeds it to domestic animals kept in environmental houses. In so doing, even though he replaces the nutrients removed by the application of fertilisers, he reduces the numbers and biomass of invertebrates, small mammals and micro-organisms which can be supported by the ecosystem. On the face of it a biomass reduction of this kind is not necessarily a bad thing, in that the ecological efficiencies of invertebrate-dominated systems at the primary and secondary consumer levels should remain at 13 and 6 per cent respectively. However, the end result of man's activities may not be just a proportionate reduction in the numbers and biomass of invertebrates, small mammals and decomposer micro-organisms; the danger is that a deleterious decrease in species diversity will occur. A decrease in species diversity could mean the loss of species essential to the condition of the soil which in turn will govern the long-term level of net primary production. As yet there is little or no information on the relationship between species diversity and soil condition; more information is essential and in the meantime the possibility of deterioration should be borne in mind and great care be exercised.

One source of animal protein not yet tapped to any extent appears to be woodland. Browsing animals, if kept at a proper stocking rate, or fed 'browse' in environmental houses, could prove an excellent source of animal protein. As was shown in Figure 2 woodlands have a world 'browse' production equivalent to that of 'graze' in non-woodlands, and hence there exists the possibility of doubling the land-produced animal protein for man's use. Clearly, much greater attention should be paid to browsing animals; nevertheless, the dangers of over-exploitation are just as great as those mentioned for non-woodland systems.

It has been suggested that excessive removal of net primary production in the form of 'browse' and 'graze' from natural situations could lead to land deterioration and reduced yields in the future. Is it not strange that man has failed to learn the lesson offered by the East African plains where grazers and browsers share the habitat and show an animal protein production per unit area higher than that achieved by beef cattle on similar habitats?

The tentative models of a so-called large-mammal-dominated East African plains ecosystem and an invertebrate-dominated oak woodland are, as far as I am aware, the only ones constructed for ecosystems of this type. It may be that further work will show them to be totally inadequate and yet it is of interest to note that despite the very different community structure of these two ecosystems the estimated ecological efficiencies are alike. The ecological efficiencies of the primary and secondary consumer trophic levels in the two systems are not only similar to each other but also to the few freshwater ecosystems studied so far. The models support the contention of Kozlovsky (1968) that the 10 per cent law does not apply at all trophic levels and that between 10 and 13 per cent is most likely for the primary consumer level and between 5 and 6 per cent for the secondary consumer level.

REFERENCES

BALOGH, J., & LOKSA, J. (1948). Quantitative-biozoologische Untersuchung der Arthropodenwelt ungarischer Sandgebiete. *Archiva biol. hung.* **18**, 65–100.

BONNER, J. (1962). The upper limit of crop yield. *Science* **137**, 11–15.

BRAUN, H. M. (1969). Grassland productivity. In *Annual Report 1969 of the Serengeti Research Institute*, 15–17. Arusha: Tanzania National Parks.

BRAY, J. R., & GORHAM, E. (1964). Litter production in forests of the world. In *Advances in Ecological Research*, vol. 2, ed. J. B. Cragg, 101–57. London: Academic Press.

BROWN, L., & FINSTERBUSCH, G. (1971). Man, food and environment. In *Environment: Resources, Pollution and Society*, ed. W. W. Murdoch, 53–69. Stamford, Connecticut: Sinauer Associates, Inc.

CRAGG, J. B. (1961). Some aspects of the ecology of moorland animals. *Journal of Ecology* **49**, 477–506.

DROZDZ, A. (1968a). Digestibility and assimilation of natural foods in small rodents. *Acta Theriologica* **13**, 289–367.

DROZDZ, A. (1968b). Studies on digestibility and assimilation of foods in rodents. *Ekologia Polska* **14**, 147–59.

FOSTER, J. B., & COE, M. J. (1968). The biomass of game animals in Nairobi National Park, 1960–66. *Journal of Zoology, London* **155**, 413–25.

GATES, D. M. (1962). *Energy Exchange in the Biosphere*. New York: Harper & Row.

GILLON, Y., & GILLON, D. (1967). Recherches écologiques dans la savane de Lamto (Côte d'Ivoire): cycle annuel des effectifs et des biomasses d'arthropodes de la strate herbacée. *La Terre et la Vie* **3**, 262–77.

HAIRSTON, N. G., SMITH, F. E., & SLOBODKIN, L. B. (1960). Community structure, population control, and competition. *The American Naturalist* **94**, 421–5.

HENDRICHS, H. (1970). Schätzungen der Huftierbiomasse in der Dornbuschsavanne nördlich und westlich der Serengetisteppe in Ostafrika nach einem neuen Verfahren und Bemerkungen zur Biomasse der anderen pflanzfressenden Tierarten. *Säugetierkundliche Mitteilungen* **18**, 237–55.

HUGHES, R. E., MILNER, C., & DALE, J. (1964). Selectivity in grazing. In *Grazing in Terrestrial and Marine Environments*, ed. D. J. Crisp, 189–202. Oxford: Blackwell Scientific Publications.

HUTCHINSON, G. E. (1948). Circular causal systems in ecology. *Annals of the New York Academy of Science* **50**, 221–46.

JORDAN, C. F. (1971). Productivity of a tropical forest and its relation to a world pattern of energy storage. *Journal of Ecology* **59**, 127–42.

KOZLOVSKY, D. G. (1968). A critical evaluation of the trophic level concept. *Ecology* **49**, 48–60.

LEDGER, H. P., PAYNE, W. J. A., & TALBOT, L. M. (1961). A preliminary investigation of the relationship between body composition and productive efficiency of meat producing animals in the dry tropics. Quoted in Foster & Coe (1968).

MACFADYEN, A. (1963). The contribution of the microfauna to total soil metabolism. In *Soil Organisms*, eds. J. Doeksen & J. van der Drift, 3–17. Amsterdam: North-Holland Publishing Company.

MCNEILL, S., & LAWTON, J. H. (1970). Annual production and respiration in animal populations. *Nature* **225**, 472–4.

MENHINICK, E. F. (1967). Structure, stability, and energy flow in plants and arthropods in a *Sericea lespedeza* stand. *Ecological Monographs* **37**, 255–72.

ODUM, E. P. (1959). *Fundamentals of Ecology*, second edition. Philadelphia and London: W. B. Saunders.

ODUM, H. T. (1957). Trophic structure and productivity of Silver Springs, Florida. *Ecological Monographs* **27**, 55–112.

PETRIDES, G. A., GOLLEY, F. B., & BRISBIN, I. L. (1969). Energy flow and secondary productivity. In *A Practical Guide to the Study of the Productivity of Large Herbivores*, eds. F. B. Golley & H. K. Buechner, 9–17. Oxford: Blackwell Scientific Publications.

PETRUSEWICZ, K., & MACFADYEN, A. (1970). *Productivity of Terrestrial Animals: Principles and Methods*. Oxford: Blackwell Scientific Publications.

RYTHER, J. H. (1969). Photosynthesis and fish production in the sea. *Science* **166**, 72–6.

SILLIMAN, R. P., & GUTSELL, J. S. (1958). Experimental exploitation of fish populations. *US Fish & Wildlife Service Fishery Bulletin* **58**, 214–52.

SOUTHWOOD, T. R. E., & VAN EMDEN, H. F. (1967). A comparison of the fauna of cut and uncut grasslands. *Zeitschrift für angewandte Entomologie* **60**, 188–98.

STEELE, J. (1965). Some problems in the study of marine resources. *Special Publication of the International Commission of the North-West Atlantic Fisheries No.* 6, 463–76,

TANSLEY, A. G. (1935). The use and abuse of vegetational concepts and terms. *Ecology* **16**, 284–307.

VARLEY, G. C. (1970). The concept of energy flow applied to a woodland community. In *Animal Populations in Relation to their Food Resources*, ed. A. Watson, 389–405. Oxford: Blackwell Scientific Publications.

WASSINK, E. C. (1968). Light energy conversion in photosynthesis and growth of plants. In *Functioning of Terrestrial Ecosystems at the Primary Production Level*, ed. F. E. Eckardt, 53–63. Paris: Unesco.

WATERS, T. F. (1969). The turnover ratio in production ecology of freshwater invertebrates. *The American Naturalist* **103**, 173–85.

WHITTAKER, R. H. (1970). *Communities and Ecosystems*. London: Collier-Macmillan.

WIEGERT, R. G., & EVANS, F. C. (1967). Investigations of secondary productivity in grasslands. In *Secondary Productivity of Terrestrial Ecosystems*, ed. K. Petrusewicz, vol. 2, 499–518.

15

THE BIOLOGICAL EFFICIENCY OF PROTEIN PRODUCTION BY STALL-FED RUMINANTS

By T. HOMB and D. C. JOSHI*

Institute of Animal Nutrition, Agricultural College of Norway,
Ås-NLH, Norway

One of the most important functions that ruminants subserve is the production of protein for human consumption. Efficiency of protein production in animals is defined as the edible protein in animal products as a percentage of the apparently digestible protein consumed (Byerly, 1967). This is the total or gross efficiency while the partial efficiency is the ability of an animal to deposit protein on the basis of the protein available for productive purposes, after the maintenance requirement is covered. Calculations based on the total protein content of the feed have also been used to some extent, but this does not seem to give a better picture. Since about 75–80 per cent of the protein in ruminant rations may be digestible, the efficiency figure arrived at, on the basis of total protein content, would be 25–33 per cent lower. In the present discussion the definition proposed by Byerly will be used to a great extent.

In the assessment of the efficiency of protein production only the edible protein in the animals is considered. According to Byerly this protein accounts for about one-half of the protein in the empty body. This is correct under American conditions, and probably also in many other regions though some variations of importance may exist, as will be shown later. It possibly costs the animal much the same to synthesise protein in the body, regardless of whether the protein is deposited in muscles, bones, skin or hair. Thus, if the efficiency is to be considered from a purely biological point of view, the figures arrived at, according to the accepted definition, should be doubled.

While considering the evaluation of the efficiency of protein production a question could be raised as to whether such evaluation should be done on the basis of the feed protein alone. A new light has been thrown on the energy expenditure of protein synthesis (Breirem & Homb, 1972). Protein

* Research Fellow, Norwegian Agency for International Development. On leave from the UP College of Veterinary Science and Animal Husbandry, Mathura, UP, India.

deposition in animal tissue is rather costly in comparison to fat deposition. Furthermore, a great number of balance trials have shown that the energy content of the ration up to a certain level, as well as the feed protein, affects the nitrogen retention of the animals (Kielanowski, 1971). The energy level in the feeds used should, therefore, also be considered in analysing the efficiency of protein production. This factor seems to be of greatest importance at an early age, as very young animals, particularly intact males, can realise their potential for protein deposition only at high energy intakes (Breirem & Homb, 1972). As an example it may be mentioned that Kielanowski (1962) calculated the net energy in the feed required per 100 kg of protein in edible animal products when he made comparisons of different production cycles. Reid (1970) has compared the protein production per Mcal in the ration.

THE PROTEIN CONTENT OF ANIMAL PRODUCTS

It is easy to assess the protein content of the milk produced by cows or goats. The protein content of cow's milk is correlated with the fat content, but varies relatively little from 3.5 per cent in most cases. A source of error is represented by the milk that is consumed by the animals. All milk is certainly edible for human consumption, but a certain quantity is needed for the replacement of the dairy herd. Furthermore, there is not always and everywhere a market for all the milk produced in different seasons, and this leads to a certain surplus of skim milk and whey that has to be fed to animals. In Norway about one fifth of the quantity of milk delivered to the creameries is usually returned to the farmers whether they need it for feeding purposes or not. There is, however, a declining trend in the feeding of milk protein to farm animals in this country as in others, because the spectrum of milk products for human consumption has steadily been widened, especially because of improvement in drying techniques.

Protein is distributed in most parts of the animal. Several reports show a negative correlation between the percentages of protein and fat. The protein percentage of beef carcasses declines by about 1 unit as the fat percentage increases by 5 units (Hopper, 1944; Callow, 1945, 1947; Hankins, 1946, 1947; Dahl, 1953; Homb & Offergaard, 1957). On a fat-free basis the empty body seems to be under a homeostatic regulation, the contents of essential nutrients being remarkably constant as stated by Bailey & Zobrisky (1968). They arrived at the following figures for the percentage distribution of total protein in the empty body of liberally fed cattle 1–2 years of age: blood 4–5, organs 7–8, hair and hide 14–18, skeleton 16–17 and fat and lean tissue 52–57. Watson (1943) concluded on the basis of an analysis of the old Missouri material that 8.5 per cent of the empty weight is made up of edible protein in the carcass. Some fat steers, however, did not fit into this system, with about 6 per cent edible carcass protein.

Even though protein is the predominant nutrient in meat and meat byproducts, it is primarily the meat the public requires, provided there is a sufficiency of fat to promote tenderness and to prevent drying out in cooking (Cooper, 1955). The protein of livestock is of excellent quality and most people like it in the many forms available from animal sources (Byerly, 1966). Other statements indicate that 'quality of meat is that which the consumer likes best' (Hammond, 1936), or 'the value of beef depends on the psychological effect of the pleasure derived from eating it, and on the protein and fat it contains' (Watson, 1943). The tedious work that has led to the discovery of different chemical compounds behind the meat taste (Wismer-Pedersen, 1968) is of importance but is hardly a factor of such magnitude that protein in meat should be neglected.

In contrast to milk it is a difficult task to evaluate the edible protein in different animal tissues. Though most of us consider hides as inedible, this is not always correct as African people very often disagree on this point (Nordrum, 1971). After drying and thoroughly cooking, hides with hair seem to be a suitable dish. The same could be said about the rumen wall and other protein-rich organs that usually are used as animal feeds. Of the various offals blood is one of the best protein sources, amounting to 5 per cent of the total protein in the empty body, or more than 10 per cent of the protein in the edible carcass (Moulton, 1923; Bailey & Zobrisky, 1968). Though edible, only a small part of blood is used as food for human beings. A significant part of the liver is wasted because of parasites. Leitch & Godden (1941) calculated the protein content of edible offal to be one-seventh to one-sixth of the edible carcass protein. Under Scandinavian conditions the authors have calculated the total edible protein in offals (organs that may be used) to be 2.5 per cent of the chilled carcass weight, or about 17 per cent of the edible carcass protein. This is nearly of the same magnitude as found by Leitch & Godden. Our calculations have partly been based on Swedish material published by Dahl (1953) and partly on analytical data of Sahyun (1948) and the American Meat Institute Foundation (1960) where Scandinavian information is lacking. The authors have also estimated that at present under Norwegian conditions around 40 per cent of the edible offal protein is actually utilised as ingredients in food products for human consumers. This equals about 1 per cent of the carcass weight.

Estimation of edible protein content in carcasses of various types

The literature dealing with the chemical composition of carcasses is scarce compared to the huge amount of data available on measurements and also on physical composition. The classical old data of Missouri steers have been thoroughly examined by Watson (1943) who has given the results in percentage of live weight. The edible protein content was found to be relatively constant (8.5 per cent of live weight, as mentioned earlier). In trying to assess the edible protein in carcasses, from more recent experiments, we came to the conclusion that it is more reliable to express the edible protein in carcass as a percentage of the carcass weight. Such data have been compiled from the literature and from unpublished material from our institute, and the estimated average figures arrived at are given in Table 1. These data will, in the following pages, be used in estimating the efficiency of protein production in those cases where such analyses have not been performed. In fact, the majority of reports do not give information on the protein content of carcasses. Edible protein in byproducts or offals is not included in the figures in Table 1. Correction for edible offal can, of course, be made if desired. According to the analysis, made above, from 1.0 to 2.5 percentage units may be added in accordance with the conditions in each case. There is surprisingly only a small variation in the edible protein content in the different categories of carcasses. This might be explained by the relatively high percentage of bone when the carcass is characterised as thin. On the other hand fat carcasses, with a smaller bone proportion, contain less muscle and protein in the edible part. On the whole, bull carcasses rank highest in protein, and fat carcasses of older lambs, sheep and cows are the lowest.

EFFICIENCY OF PROTEIN PRODUCTION

This has been considered in the following stall-fed systems:
  (i) dairy herd (milking cows and goats),
  (ii) veal production (including bobby veal),
  (iii) dairy beef production (bulls, steers, etc.),
  (iv) lamb production (based on milk replacers and concentrates).

Dairy herd

Intensive feeding of the dairy herd, without turning the cows on to pasture, has been used more and more during the last two decades. The system may be practised with or without an equivalent number of replacement heifers in the herd. Including the protein in edible cow carcass (quarter cow per year) and the carcasses of the extra calves killed at a very young age, the annual protein production of a cow yielding 6000 kg milk per annum is set at 220 kg, of which only 10 kg originates from meat. The total efficiency of protein production will amount to 220/560 = 39 per cent according to Norwegian feeding standards, and for each 100 feed units (FU) total feed 4.6 kg protein are produced in milk and edible meat. If milk production alone is considered, without raising heifers, the efficiency would reach 43 per cent.

Table 1. *Estimated protein contents of carcasses (protein in edible carcass as a percentage of carcass weight)*

New-born calves	13
Intensively fed calves	15
1–2 years old bulls	14–15
Steers (fat)	12–13
Steers (medium)	13–14
Steers (thin)	14–15
Cows	12–14
Lambs (young)	13
Lambs (older, fat)	11–12
Lambs (older, medium)	12–13
Mutton	11–12

The literature referred to for the calculation of the estimated figures in Table 1:

Calves, veal: Haigh, Moulton & Trowbridge (1920); Dahl (1953); Mandrup-Jensen & Kousgaard (1969); Danilov (1969); Homb (unpublished).
Beef: Chatfield & Adams (1940); Hopper (1944); Hankins (1946); Sahyun (1948); Dahl (1953); Homb & Offergaard (1957); Niinivaara & Antila (1968); Mandrup-Jensen & Kousgaard (1969); Haugland (unpublished).
Lamb and mutton: Chatfield & Adams (1940); Hankins (1947); Sahyun (1948); Wallace (1955); Homb & Offergaard (1957); Nedkvitne et al. (1965); Niinivaara & Antila (1968); Danilov (1969).

In order to find out how much milk cows were able to yield under the most favourable nutritional and environmental conditions a demonstration was performed on three farms in Denmark during the years 1947–55 (Nielsen, 1961). For groups of 10–12 cows the average yearly milk yield amounted in all cases to 10 000–11 000 kg. Though some of the groups had access to pasture for a shorter period, others were stall-fed throughout the year. Breirem (1955) has pointed out the high efficiency of energy utilisation in these demonstrations, finding a rectilinear relationship between feed units consumed and milk production. The efficiency of protein production was also high, about 45 per cent, as estimated by the authors, and per 100 FU total feed 4.2 kg milk protein were produced.

In a three-year study in Denmark stall feeding has been compared with the traditional indoor feeding in winter combined with grazing during summer (Klausen et al. 1968). Three groups, each consisting of twelve cows, were fed indoors throughout the year. Annual milk yield was 5000–6000 kg. Based on the reported data the efficiency of protein production is estimated to be 40–44 per cent. No inputs for raising heifers were included in this or the above-mentioned Danish report. About 4.0 kg milk protein were produced per 100 FU.

Review of the literature on this aspect reveals total protein efficiency figures of 30 to 47 per cent (Leitch & Godden, 1941; Byerly, 1959, 1966, 1967, 1968; Holmes, 1970; Huber, 1971). The highest efficiency is obtained by theoretical considerations of very high-yielding cows (Byerly, 1968). There is a remarkably good agreement between these American figures and the Danish observations mentioned above. Holmes (1970) found 22 per cent efficiency when basing his estimations on the total protein content of the ration of the whole dairy herd under British conditions. When recalculated by the authors on the basis of digestible protein it works out to 30 per cent. In close agreement with this, Byerly (1966) concludes that at current levels of feeding the protein efficiency of American cows may be about 30 per cent.

The high protein efficiency figures that may be obtained in milk production, under favourable conditions, are not typical for cattle but are related to milk production *per se* (Byerly, 1967). The goat and sheep also have about the same efficiency of protein utilisation in lactation. Species differences in efficiency count relatively little in lactation as well as in liveweight gain (Mayer, 1948–9, quoted by Byerly, 1967). The considerable difference in protein efficiency of body weight gain and milk production is an interesting question, and is primarily connected with the production capacity per unit of time. A 450 kg cow is able to produce 1.05 kg milk protein per day (30 kg milk), while a bull can at the most produce one-fifth of this amount in its body of which only one-half or a little more is edible by human beings. The udder has apparently by selection become a highly specialised organ with an enormous capacity as a protein builder as well as a carbohydrate and fat builder in co-operation with the digestive and circulatory mechanisms of the cow. A high potential for feed intake in lactating cows is inseparably connected with high production (Preston, 1968). As to the energy cost of forming protein, milk production again is superior to protein deposition in the body itself (Breirem, 1946; Blaxter, 1961; Flatt et al. 1972).

Milk protein is of high quality and this reflects the specialised nature of the mammary function. As feed for the young calf milk protein in one way or another is indispensable. During the 1960s great stress was laid on developing milk replacers with protein sources other than milk. Soyabean protein is one promising source, another is finely ground fat-extracted fish meal. The latter, according to American and Norwegian experiments, is not able to compete with the milk protein during the very first part of the calf's life. After three weeks, however, the calves seem to have obtained a digestive system capable of utilising fish protein at a rate allowing normal growth (Matre, 1971).

Low protein supply in the ration will automatically lead to a higher estimated efficiency of protein utilisation, in any case for a certain period of time. This might, however, be followed by a lower degree of energy utilisation. Based on Møllgaard's and Lund's experiments, Reid (1962, quoted by Crasemann & Breirem, 1969) has found a lower percentage of the metabolisable energy being converted to net energy under such circumstances. For growing swine a parallel to this is not found. Further, as the greater part of the nitrogen supply in the feed is covered by nitrogen sources, other than protein, extremely high figures for protein utilisation may be obtained (on the assumption that only protein in the ration is considered). In Virtanen's (1966) experiments some cows could produce 160 kg milk protein per year on a ration of urea and ammonium citrate as the only nitrogen sources. The possibility of obtaining higher efficiency figures for protein production by the use of urea-feeding will be dealt with later.

Lower milk yields, overfeeding protein and using liberal amounts of whole milk as well as skim milk (or other milk products) in calf rearing are reasons why protein utilisation under practical conditions is lower than the theoretical calculations suggest or what is obtained under favourable conditions.

It may be reasonable to suggest that for most practical purposes the efficiency of protein utilisation in the dairy enterprise is 30–40 per cent and that 3.5–4.0 kg of milk protein can be produced by 100 FU. In these figures due allowance has obviously been made for the milk protein which has to be fed back to the calves (as whole milk, skim milk, etc.). As the amount of this protein in the diet of calves is steadily declining and as the average milk yield per cow is increasing from year to year, it is supposed that the highest figures mentioned may be taken as a target for the future, as well as the best possible obtainable today.

Several authors have pointed to the accepted practice that the cow is able to cover its maintenance needs by forage, and this is even considered a natural way in feed planning (Breirem, 1955; Reid, 1969). Partial efficiency figures, as explained earlier, are relevant in many cases. Partial protein efficiency may be as high as 60–70 per cent (50–60 g digestible protein in the feed per kg of milk). Moore, Putnam & Bayley (1967) refer to a Wisconsin study where 96 per cent of protein in the cereal and oilseed rations was returned as milk protein. This high figure was obtained because the protein in forages and byproducts was not included in the input.

The milking goat herd used to be connected with pasture, kidding taking place in spring. Modern intensive goat milk production in Norway has but little to do with grazing. The peak of production is from January to midsummer. Ad-libitum feeding of hay or silage usually covers the maintenance requirement and concentrates have to be supplied according to milk yield of each individual goat (Opstvedt, 1967). Annual milk yields of 600 kg for a 50 kg goat are usual and this leads to about the same efficiency of protein production as in dairy cows. This is in agreement with the theoretical calculations by Byerly (1967). Practically all milk from goats is consumed by human beings, either as liquid milk or as cheese; for instance, in Norway where goat's milk obtains a higher price than cow's milk because goat's milk is used for a special type of cheese, except for colostrum the kids are often raised on milk substitutes and skimmed cow's milk.

Veal production

The newborn animal seems to possess great ability to utilise protein as well as energy, from colostrum and from milk of its own species. This applies apparently to all mammals, and it has been clearly demonstrated by Wöhlbier (1928) in swine and Blaxter (1950) in calves. It has been shown that during the late prenatal stage energy is utilised to a limited degree, and one hypothesis is that parturition represents a change to a compensatory growth, with small costs of feed per unit of gain (Breirem & Homb, 1972). The very high percentage of nitrogen retention obtained during the first few days of the calf's life seems gradually to decrease even when milk feeding is continued. It may be mentioned that calves, in the pre-ruminant stage, utilise metabolisable energy for growth with an efficiency of 66–69 per cent (van Es et al. 1969) and are thus very similar to monogastric animals. According to Norwegian practice milk-fed veal calves are allowed to drink milk (whole milk, milk substitute) two or three times a day in amounts close to ad-libitum. As an average of ten experimental groups, 32 per cent of the digestible protein in the milk was found in the gain of edible carcass protein, with very little variation between experiments and groups. If, however, the byproducts (offal) actually eaten are also included the percentage utilisation of the protein would be 35 and if all edible byproducts are considered this figure would rise to 39. These estimations have been made on the basis of gain in the edible carcass from the very young calf stage to slaughter, which in this case is at about ten weeks of age.

Danish practice is to use a moderate amount of concentrates (about 50 kg per animal) in addition to whole milk or skim milk substitutes, and the feeding period is extended to about 14–17 weeks. Consequently, the estimated efficiency figures are somewhat lower, namely 25–30 per cent (Brolund Larsen, 1958; Lykkeaa & Sørensen, 1966; Sørensen & Lykkeaa, 1967). This is in accordance with the Norwegian results (Homb, 1960; Matre, 1971), and the explanation may lie in a lower degree of fat digestion and, therefore, also a lower energy intake.

Milk-fed veal has declined in importance in countries with a shortage and high prices for calves (Lykkeaa & Sørensen, 1966; Homb, 1970). One reason might also be a decreased demand for this luxury type of meat. If, however, all available calves (nearly three-quarters of the total) are fed 700 kg whole milk or skim milk (or milk replacer based on skim milk) to a carcass weight of 60 kg, intensive milk production giving 40 per cent efficiency of protein utilisation would give only about 38 per cent. Because of the relatively high efficiency of milk-fed veal calves the decline is thus very moderate.

Bobby veal (older calves)

In the Scandinavian countries a special production system developed rapidly during the 1950s. Preferred carcass weights in Norway are 70–90 kg, and in Sweden and Denmark 100–140 kg. To obtain cheaper feeding the amount of milk included in the diet is smaller than in the milk-fed veal. Concentrates, and in Denmark roots, make up a great part of the ration as a rule, besides some skim milk. The Norwegian type of production system, with 4–5 months old calves receiving 8 l per day skim milk, concentrates *ad libitum*, plus a very small amount of hay, has shown a maximum of 26 per cent protein utilisation (Opstvedt, 1964; Haugland & Homb, 1966) and is estimated to be 20 per cent when the milk feeding has been discontinued at an early age or when the calves have been fed for a longer period of time.

An analysis of comprehensive data from the Danish progeny tests (Nielsen, Nielsen & Dissing Andersen, 1969) has resulted in an estimated protein efficiency of 18 per cent. The average carcass weight was 136 kg and the protein supply was liberal. Higher figures would have been found for protein efficiency if the protein level had been lower. An interesting experiment (Sørensen & Lykkeaa, 1968) showed clearly the protein efficiency declining from 23 to 14 per cent as the protein level was raised from 117 to 187 g digestible protein per FU. Daily gain and feed conversion was almost equal in all treatments that were included. 100 FU in the ration produced 2.7 kg protein in the edible carcass. It may, however, be pointed out that the calves in this study were of higher body weight. At 60–100 kg live weight Preston et al. (1965) found higher nitrogen retention on an all-concentrate diet containing 19.4–21.7 per cent crude protein than on one containing 14.8–16.8 per cent. The protein efficiency ratio, however, in this case also diminished as the protein level in the feed increased.

It may thus be seen that in this type of veal production 20–25 per cent efficiency of protein utilisation is obtained depending on the age of calves and whether skim milk is used or not. Even when this type of production is considered as a system by itself it may be combined with a milk production unit. The combination in this case, however, will also show a lower efficiency than milk production alone. The calculated decline is from 40 per cent for milk production to 37 per cent for the combination with three-quarters of the calves born producing 100 kg carcass each. It is presumed that the milk consumption per calf is 700 l. The protein efficiency of these calves was assumed to be 22 per cent.

Dairy beef production

Traditionally, dairy beef has made up the bulk of beef in the Scandinavian countries. Whether the cattle are defined as dairy or dual-purpose breeds is only a question of expression; the main product is milk. The calf cost is so much lower than in the purebred beef breed system that this more than counteracts the reduced tenderness of the meat (Wellington, 1971). In Great Britian and even in the USA an increased interest in dairy beef is developing (Hallman, 1971). In 1969 it was found that 12 per cent of the feeder cattle in USA were of dairy breeding. When protein deposition is discussed, Friesian steers undoubtedly rank higher than the beef breeds (Preston et al. 1963; Cole et al. 1963; Preston, 1968; Garrett, 1971). It is more difficult to evaluate the significance of increased juiciness and marbling of meat obtained from beef breeds (Wellington, 1971). The low efficiency of pure beef production has been established by several workers (e.g. Byerly, 1967). Furthermore, this production method is not of any interest as a stall-feeding system, and will, therefore, not be discussed here.

Of great interest in intensive dairy beef production is the question of bulls versus steers. Greater protein production in bulls may have to be judged against better marbling and grading (Wellington, 1971), juiciness, taste, etc. (Homb, 1958; Field, Schoonover & Nelms, 1964) found in steers in several trials. The preference of consumers for any of these also depends on tradition. In spite of the lack of marbling, the Norwegian market favours the meaty (and protein-rich) bull carcasses, while the American tradition is for the opposite. American consumers appreciate fully the steak of a first class steer (or heifer) with such fat covering that it does not dry out to any noticeable extent, while most people in the Scandinavian countries have no real opportunity to buy the luxury type of beef available in the UK and the USA and therefore do not demand it. The advantage of bulls in the efficiency of protein utilisation lies partly in the higher daily muscle development and partly in the more efficient feed conversion. This has been shown in many experiments where an intensive feeding system was used (Homb, 1958; Nichols et al. 1964; Berg, 1969; Turton, 1969; Kay, 1969). Table 1 shows the differences in the edible protein content of carcasses of bulls and steers.

Of other factors affecting the protein efficiency in intensive dairy beef production, the content of protein in the ration should be stressed. Homb (unpublished) in three trials comprising 72 bulls of the NRF breed fed rations with varying proportions of hay and concentrates, found edible protein in the carcasses to amount to 13–15 per cent of the digestible protein in the ration. The efficiency figures were almost unaffected by the ration components, possibly due to an unnecessarily high protein level in the concentrates. The same can be said of Haugland's experiments (also unpublished) which showed an efficiency of protein utilisation of 15–16 per cent. The results of these experiments are based on comprehensive data for bulls which received grass silage *ad libitum* besides a fixed ration of concentrates. All half-carcasses were analysed for protein and fat. In these Norwegian trials 100 FU produced 1.7–2.0 kg edible carcass protein. Other Norwegian experiments with actively growing bulls have shown protein efficiency figures of 18–20 per cent when the protein level was lower (Saue et al. 1963).

A very useful basis for evaluating protein efficiency at varying protein levels in the rations has been obtained by analysing data from Danish experiments (Kirsgaard et al. 1969). As an average for 20–24 bulls per treatment, protein efficiency has been estimated to vary from 12 to 20 per cent when the content of digestible protein declined from 133 to 81 g per FU for the interval from 193 to 507 kg liveweight. Almost all the protein originated from concentrates. This problem has also been analysed in a Rowett experiment with Friesian bulls by Kay & MacDearmid (1969) who came to the conclusion that no improvements were achieved by increasing the crude protein content of the dry matter above 14.5 per cent for animals up to 250 kg liveweight and above 12 per cent for heavier bulls. The data given in this interesting paper have been used in the estimation of the efficiency of protein utilisation which was found to be 18 per cent. This is in reasonable agreement with the Danish results. A number of experiments with intensively fed young bulls in Europe have been analysed, resulting in 15–20 per cent efficiency of protein utilisation (Obraćevic, 1962; Norrman, 1969).

Intensive beef production with all-concentrate rations has been developed in several countries during the 1960s (Kay, 1969). Of special interest from different angles is barley beef production with steers in the UK, which has been thoroughly analysed by the Rowett group. Allen (1968) has, in a review, given the targets for this type of production based on Friesian steers 10 months old as 400 kg liveweight and 230 kg carcass weight. This gives an estimated efficiency of protein utilisation of 12–13 per cent, provided 5.5 kg concentrates (11 per cent digestible protein) per kg live weight gain are consumed. According to Allen, one-third of the members of the Beef Recording Association have obtained results equal to this target. This agrees fairly well with Kay (1969) who states a feed conversion of 5.2 as being 'not unusual, although much poorer conversions than this are frequently found in practice'. From a physiological point of view an all-concentrate ration is hardly an ideal situation for the ruminant. Factors such as processing methods for the barley (coarsely or finely ground, pelleted, steam-treated), feed additives, buffers, minerals, fish meal, etc., have been examined and found to be of importance in order to reach the best possible physiological conditions enabling effective production and a high degree of protein utilisation (Kay, 1969). Though the problems arising are greater when maize is fed, barley beef production has not been free from pathological conditions, such as rumenitis, bloat and liver abscesses, and an abnormal mortality rate has been reported (Allen, 1968). The high incidence of liver abscesses has to a certain degree been counteracted by adding antibiotics (Kay, 1969).

No doubt, the animals seem to desire some form of roughage. When they have access to straw (bedding) this will satisfy the strongest desire. Rumenitis can be counteracted by using ground straw as an ingredient in the feed mixture. Lamming, Swan & Clarke (1966) and Lamming & Swan (1967) found that diets diluted with up to 30 per cent ground barley straw in a pelleted ration still contained sufficient energy and nitrogen to sustain efficient growth rates and allow satisfactory carcass development. This was not the case, however, with 50 per cent dilution. Our estimate of the protein utilisation by these steers, from about 300 to 440 kg liveweight, comes to 12–13 per cent, being almost constant for straw percentages from 0 to 30. Inclusion of 40–50 per cent straw gave an estimated efficiency of protein utilisation of 10–11 per cent.

In experiments published by Kay, Massie & MacDearmid (1971) chopped dried grass, given with concentrates, had a positive effect on feed consumption, but led to a lower growth rate. Daily carcass gain dropped gradually from 625 g to 430 g as dried grass increased from 0 to 100 per cent of the total ration. The dry matter of the chopped grass was digested to the extent of 69.5 per cent, while that of the pelleted concentrate mixture was 78.2 per cent. The group on an all-concentrate ration showed an estimated protein efficiency of 13–14 per cent, but this dropped markedly with the increasing amount of dried grass.

Ruane & Caffrey (1967) gave their animals 1 lb of hay daily to avoid digestive disturbances. The barley group converted about 14 per cent of the digestible feed protein into edible carcass protein on an ordinary protein level diet, but only 11–12 per cent when the protein content was raised to 15 per cent from 24 weeks of age.

In contrast to basing intensive beef production on concentrates, Lonsdale, Poutiainen & Tayler (1971) and Poutiainen, Lonsdale & Outen (1971) at Hurley have tried artificially dried grass as the sole feed for young Hereford × Friesian steers. The experiments started at about 135 kg live weight and lasted 70 days. A thorough examination of the feed and the animals at slaughter enables us to calculate the efficiency figures more accurately. With chopped grass the protein efficiency was 13 per cent and with coarsely milled grass 16 per cent. Inclusion of 50 per cent rolled barley in the wafers led in both cases to 18–19 per cent efficiency of protein utilisation, though the digestibility and feed intake were about the same whether barley was included or not. The reason for the better protein utilisation in the latter case is probably the lower percentage of protein in the wafers containing barley. Chopped grass (wafered) did not produce as much carcass gain as chopped grass plus barley. The high figures obtained for protein utilisation in these experiments may partly be explained by the fact that the trial was limited to a period when the animals in general had a maximum nitrogen retention. These data represent a valuable gain to our knowledge of the possibility of grass as the sole or main feed in stall-fed young ruminants. It may be pointed out that the grass was of excellent quality, the dry matter digestibility being 73–75 per cent.

Environmental factors may also influence the efficiency of protein utilisation in stall-fed growing ruminants. Saue et al. (1963) compared two different types of housing for actively growing young bulls on equalised feed intake. Tied-up animals in a conventionally insulated Norwegian stall grew faster than loosely housed animals. Estimated efficiency figures for protein were 3 percentage units higher in the former (19 versus 16). The experiments were carried out during the whole winter in a cold climate. Other results along the same lines have been discussed by Saue et al. (1963).

A stall-feeding system for beef-producing dairy heifers is rare. Information collected by the authors from the available literature tends to show that the protein utilisation in such heifers is lower than in bulls and probably lower than in steers too.

From the discussion on beef production it may be suggested at the end that 12–20 per cent of the digestible protein in the feed will be converted into edible carcass protein. Best utilisation is obtained in rapidly growing young bulls on a high feed intake with a moderate protein level, while a lower degree of utilisation occurs in steers and when the protein supply is liberal. Forage alone, even high-quality dried and processed grass, does not seem to give maximum utilisation figures. One of the reasons for this is the high protein content of the best quality grass. 100 FU in the ration has produced only 1.5–2.0 kg edible carcass protein.

It may be mentioned that conversion figures would be somewhat higher with this stall-fed system if protein in animal byproducts is included in the output. Further, the inclusion of urea in the ration can also improve the efficiency of protein utilisation.

Lamb production (artificial lamb rearing)

It has been clearly established that the number of lambs produced per ewe annually is a very important factor affecting the efficiency of lamb production (Wallace, 1955; Blaxter, 1968; Owen, 1969; Spedding, 1969). Though traditional sheep husbandry is closely connected with pasture, experiments in progress in the physiology of reproduction make us think how to manage lambs which cannot be taken care of by a ewe with only two teats. More and more lambs have to be looked after by means other than the nursing method. Looking into the future, it may be an ecosystem by itself to raise lambs by artificial means. One method is to use milk replacers for some time and then change to a clean pasture (Spedding, 1969; Treacher & Penning, 1970). Another system is indoor feeding throughout to slaughter (Large, 1965; Lawlor & Crowley, 1968). Regardless of which system is chosen, in artificial rearing, an adequate colostrum supply is essential (Owen, 1969). In the absence of colostrum from ewes, colostrum of cows although less effective has proved to be a valuable substitute (Nedkvitne, 1970; Nedkvitne & Haugland, 1970). A further indoor feeding system starts with a suitable milk replacer given either *ad libitum* from a feeder, or four times and, later, twice a day. After a certain amount of milk replacer has been fed, a dry ration (often pelleted) of concentrates is supplied. The protein of fish meal seems to be of greater value than that of soyabean in the feeding of lambs as well, and barley is more palatable than oats (Davies & Owen, 1967; Davies, 1968). Other factors in this type of lamb feeding have been thoroughly discussed by Owen (1969). The data for estimating the efficiency of protein production are more scanty than for intensive beef production. A feed conversion ratio of 3.21 for the whole fattening period, when using a pelleted 85 per cent barley ration, indicates, according to Owen et al. (1967), the commercial feasibility of intensive lamb production. Provided the content of digestible protein in dry feed is 14 per cent, the efficiency of protein utilisation is estimated to be 16–18 per cent. The protein content of the ration could possibly be reduced without depressing the growth rate, but one should remember that wool growth also requires protein in sheep, in contrast to growing cattle (Spedding, 1970). During the first few weeks, when liquid milk replacers make up the main feed, it is assumed that 25–30 per cent of the digestible feed protein is retained in the edible carcass, and for the whole period from two days of age to about 35 kg live weight the efficiency figure may be about 20 per cent.

THE EFFECT OF NON-PROTEIN NITROGEN (NPN)

In the experiments referred to, as a basis for estimating the efficiency of protein production the ration protein has consisted of only minor quantities of NPN, and in all these cases the NPN has been from natural sources. There is little doubt that the use of urea and other NPN sources is of increasing importance as a part of the world's feed budget. In future one may expect that a certain part of the nitrogen supply in the diet of ruminants will consist of NPN. The question naturally arises as to how much NPN can be used in an intensive stall-feeding system without interfering too much with the efficiency in one way or another.

During the last two decades a number of review articles on NPN and urea have appeared, for instance by Loosli & McDonald (1968). Ørskov (1970 a, b) has recently presented his viewpoint on the question raised above. His results provide a nice illustration of the effect of energy level on the utilisation of NPN. Ørskov also stresses the difference between the dietary protein:energy ratio required by the host animal in full production and that required by the rumen micro-organisms for optimum growth. He refers to several studies giving evidence to suggest that microbial nitrogen synthesis is limited by the thermodynamic factors which affect anaerobiosis.

It goes without saying that modern methods of stall-feeding ruminants are ultimately connected with high intensity. Low-cost pasture is one of the main factors behind the profitability of a moderate feed level, which may exist under certain conditions. In stall-feeding systems labour is often involved to such a degree that high intensity may represent the only alternative. Though very interesting from a theoretical and fundamental point of view, Virtanen's (1966) studies with NPN as the sole nitrogen supply to milk cows have hardly any bearing on the situation in developed countries today. Apart from the effect on peak lactation the daily milk yield under such a feeding regime will probably not exceed 10 kg (Flatt et al. 1972; Ørskov, 1970a), and it seems likely that difficulties may also arise with fertility. Urea as the sole nitrogen source, though leading to extremely high efficiency figures, will therefore not be discussed here.

To have a part of the nitrogen requirement covered by urea may, however, be useful. For young cattle Valeur (1969) in his review recommends 30–33 per cent NPN as the maximum. According to Loosli & McDonald (1968) 1 per cent of urea in the total ration should not be exceeded. In practical rations this will equal about 20 per cent of the nitrogen requirement. Even at this level there may be some limitations. Cows with daily milk yields higher than 20 kg do not seem to respond to urea under Danish feeding conditions (Møller, 1969), nor probably in the USA (Huber, 1971). Another point mentioned by Loosli & McDonald is the poor palatability of concentrate mixtures containing urea. This factor should not be overlooked in intensive milk and beef production. The content of NPN from natural feedstuffs also seems to play a role in this connection, as claimed by Møller (1969). Under Norwegian conditions heavy nitrogen fertilisation of the grass is thought to be more profitable than to feed urea (Breirem, 1970).

This discussion leads to the conclusion that the efficiency figures for protein utilisation mentioned earlier may be increased by 0–20 per cent in accordance with the conditions in each case, if urea is included in the feeding regime without interfering with the intensity.

This topic should not be closed without making a reference to the extensive production systems that may be developed on the basis of the results of a large volume of research work on the utilisation of urea, when low grade forage is used. Though much doubt is raised on the economic aspect of the question (Loosli & McDonald, 1968), it is reasonable to suppose that urea feeding might become important in the future planning of animal husbandry in developing countries.

PRODUCTION OF PROTEIN PER UNIT OF AREA

Both the farmer and the community may be more interested in the production of milk or meat per unit area than production in relation to protein in the feed. Blaxter (1968) and Holmes (1970) have both stressed this point.

Under certain conditions a herbage may give a high yield of dry matter as well as protein, but the efficiency of protein production in animals may not necessarily be of an impressive magnitude.

If a certain area of a field is used mostly for grain and only a minor part for herbage, the production of protein in milk or meat will make up a greater part of digestible feed protein than when most of the field is covered with grass. An example will illustrate this. It is supposed that 1 ha land gives 3500 kg barley or 10000 kg dried grass (high quality). This level of crop yield may be possible under semi-humid climatic conditions. The grain as well as the grass are used in intensive raising of young bulls. Norwegian experiments indicate that the feed conversion ratio is about constant within a wide range of proportion, grass:concentrates (Homb, 1970), and according to results obtained at Hurley artificially dried grass possesses a surprisingly high productivity when fed to growing cattle, even without the grain supplement (Lonsdale et al. 1971). Further assumptions are that barley contains 7.5 per cent digestible protein and dried grass 10 per cent, 2000 FU are supposed to produce a 240 kg bull carcass with 14.5 per cent edible protein (1 FU = 1 kg of barley = 1.5 kg of dried grass). Table 2 summarises the results of the effect of four different proportions of barley and dried grass on the production of bull carcasses per unit of land.

Table 2. *Efficiency of production of bull carcasses using different proportions of barley and grass*

% of FU from dried grass	% of dried grass in the dry ration	Efficiency of protein utilisation (%)	Area needed per 240 kg bull carcass (ha)	Carcass weight produced (kg/ha)	Edible carcass protein (kg/ha)
15	21	20	0.53	450	65
30	39	18	0.49	490	71
45	55	16	0.45	530	77
60	69	15	0.41	580	84

Even when the protein efficiency figures are decreasing as more and more grass is included in the ration, the protein production per ha is increasing with this type of agricultural production. The same could be shown in the case of the milk production system, which is estimated to give 210, 230 and 250 kg milk protein per ha when respectively 30, 45 and 60 per cent of FU originate from grass, with 54, 48 and 44 per cent protein utilisation. This estimate is based on milking cows yielding 6000 kg milk per year, with a feed expenditure of 4000 FU, no consideration being made for the replacement of the herd. These values should not, therefore, be used as being representative of practical farming.

According to Holmes (1970) crop production is superior to any form of animal production, in terms of output per unit area of land. The difference between the output of protein is considerable. As examples from his estimates may be given 115 kg ha^{-1} protein from dairy cows (practical conditions in UK), and 350 kg ha^{-1} wheat protein. It is worth mentioning that the production of lysine, according to these figures, is about equal in these two alternative systems for land use, while the wheat crop gives three times as much S-bearing amino acids as the dairy herd. From the figures in Table 2 it appears that beef production, even when practised in an intensive system, is inferior to the dairy herd as far as efficiency of protein production per ha is concerned. Of the animal production systems discussed here, only the dairy herd shows efficiency figures that can be compared with those observed in vegetable or grain crops.

The results shown in Table 2 are subject to variation depending on factors like soil fertility, climate and irrigation. The example given serves solely as an illustration of how to evaluate the productivity by replacing grain with grass. The calculations show that one should not rely only on the efficiency obtained in animal production. The animal represents only one step in the production system as a whole which starts with plant growth and is brought to completion when the animals are milked or slaughtered.

GENERAL CONSIDERATIONS

The evolution of ruminants from monogastric animals is a result of necessity. The fore-stomach is well suited to digest and utilise fibrous plant material. However, the ruminant digestion cannot compete with digestion in monogastric animals in the case of concentrates. This is a limitation with stall-fed ruminants which according to earlier arguments, as a rule, have to be intensively fed. Except for milk-fed veal, which never can be a system of great interest and which represents the non-ruminant situation, meat production in cattle is inferior to poultry and pigs in regard to the efficiency of utilisation of both energy and protein. Although dairy beef is far superior to the one cow–one calf system, the 15–20 per cent protein efficiency obtained in young bulls may be considered a waste of protein when the bulk of the ration consists of grain. This would also be true when using offspring of dairy cows and beef sires, even though these as a rule give carcasses with better conformation and in some cases somewhat better growth rates (Lindhé & Hemingsson, 1968). The situation is somewhat better if, for example, one-half of the feed requirement is covered by forages which do not always compete with grain in the farm area. Changing from grain to grass and legumes may also be taken as an improvement of the soil fertility. However, when the forage part exceeds a certain percentage the total efficiency is decreased.

The broiler flock has its advantages in rapid growth with a high percentage of protein. It ranges higher in the efficiency of protein utilisation than other meat-producing animals. Egg production has been found to have nearly the same high efficiency as the broiler flock (Holmes, 1970). Though pork has a narrower muscle:fat ratio, protein production in the pig is, surprisingly, more efficient than in ruminants. This may partly be explained by the high feed intake (up to three times the maintenance requirement, while in young cattle and lambs it is about twice). A 50 kg pig may deposit 60 g edible carcass protein per day on a ration of 2 FU and 225 g digestible protein, while a 400 kg bull may have a retention of 90 g edible carcass protein per day on a ration of 6 FU and 500–600 g digestible protein. The modern pig is a more meaty animal than its ancestors but still pig carcasses are relatively rich in fat (15 per cent protein and 25–30 per cent fat in edible carcass). Even if it is stated that the modern pig is a fair converter of feed protein to edible carcass protein, the surplus it gives by depositing fat should be born in mind. Compared to beef it is possible to produce pork at lower cost per kg carcass weight, and the market price is, therefore, lower. This is an important point in some countries, like Norway, where the sale of pork shows an increasing trend while the beef sale per head is constant. The opposite has been the case in the USA where the consumption of beef (and poultry) has increased during the post-war period, while pork sales have declined (Blaxter, 1970).

As mentioned earlier lactation is the most efficient way of forming animal protein. This may be attributed to the specialised structural and functional characteristics of the mammary gland. From the viewpoint of protein production, for the majority of consumers in the world, milk production should, therefore, be given a high priority. Countries fairly high in national income, though not at the top, like Finland and Norway, have predominantly a lacto-vegetarian diet. Each Norwegian consumes 24.2 g milk protein (milk, cheese, etc.) besides 14.5 g meat protein per day (Breirem, 1971). This is the result of the Government policy which gives priority to milk production. One g protein is decidedly cheaper in milk than in any kind of meat. Regions with the highest living standard can afford to have a much higher consumption per inhabitant.

High income groups have the highest demand for meat (Blaxter, 1970). It is primarily a question of muscle (Blaxter, 1968) which seems to appeal to the taste of most consumers, and this taste can only be satisfied by the wealthier class of people. A questionnaire in France referred to by Blaxter (1970) verified this. More than half the people had a desire to buy more meat if their incomes had been higher.

A very important advantage of milk production has already been mentioned. This refers to the possibility of using forages to cover the maintenance requirement. An intensive stall-feeding system may be based on one-half of the feed units from forages and the other half from concentrates. This point is in favour of an intensive use of the land, as mentioned above.

Stall-feeding ruminants might be discussed as an alternative to any kind of grazing systems. For the dairy herd this question has been thoroughly analysed by several authors (Huffman, 1959; Homb, Saue & Breirem, 1960; Runcie, 1960; Mo & Waehre, 1968). The most striking advantage of stall-feeding the dairy herd throughout the year is better utilisation of grassland and a better control of feeding and management (Shaudys, 1961). An interesting finding is that, by adding formic acid to the grass at cutting, one harvesting per day led to at least as high a feed intake as when the cows received fresh grass twice a day (Mo & Waehre, 1968).

CONCLUDING REMARKS

Animals usually do not have a good reputation as converters of protein; on average the efficiency is low. Consumption of plant protein by human beings is recommended as being the most effective, considering the world's protein supply as a whole. As may be observed from Table 3, the milking herd is by far the best converter of plant protein to a high-quality protein, as 30–40 per cent may be found in the milk. Replacing 10–20 per cent of the feed protein by urea tends to increase the efficiency of protein utilisation. The goat herd possesses about the same efficiency as cows but requires more labour to produce an equal amount of milk. Milk production per unit area is also superior to other ruminant products. In times of emergency, e.g. war, when resources need to be used judiciously, the dairy herd should have a high priority in human nutrition. With the inclusion of milk-fed veal or raising of dairy bulls in the system, the efficiency of protein conversion decreases. Though milk-fed veal shows a high protein efficiency, it should be remembered that the true efficiency, calculated on the basis of the plant protein supply to the dairy cows, is low.

Intensive dairy beef production (bulls or steers) converts 15–20 per cent of the digestible protein in the feed into edible carcass protein. Though this is better than the one cow–one calf system, fattening the young ruminant is inferior to milk production. This can be explained by the lower rate of

Table 3. *Estimated efficiency of protein utilisation by different ainmal production systems.* (*Protein in edible products as percentage of digestible protein in the feed and kg edible protein per 100 FU*)

Stall-feeding system	Protein efficiency (%)	kg edible protein per 100 FU
Dairy herd (best conditions) including raising replacement heifers	40–45[1]	4.0–5.0
Dairy herd (best conditions) including raising replacement heifers + milk-fed veal or older calves	37–43[1]	—
Dairy herd (most practical conditions)	30–40[1]	3.5–4.5
Veal production (milk-fed young calves)	25–35[2]	—
Veal production (bobby veal, older calves)	20–25[2]	2.5–3.0
Young dairy bulls	15–20[1,2]	1.5–2.0
Young dairy steers	12–18[1,2]	1.5–1.8
Artificial lamb rearing	20[2]	—

[1] By including a small percentage of urea in the ration, the efficiency figures may be increased by 0–20 % in accordance with the conditions in each case.

[2] By including edible protein in animal byproducts commonly consumed, the efficiency figures can be increased by 1–2 units. If all edible protein in byproducts were consumed, the efficiency figures could be raised by 2–4 units.

protein deposition per unit of time in the carcass than in the formation of milk protein in the highly specialised mammary gland. Compared to broiler and pork production the young growing stall-fed ruminant represents a costly production system, both physiologically and economically, thus resulting in a higher price per kg carcass weight. Higher living standards demand meat, i.e. muscle, as a part of the diet not so much because of protein but for the meat flavour and the texture. For nutrition, in many countries milk and milk products are more important than meat, especially where the lacto-vegetarian type of diet is predominant.

REFERENCES

ALLEN, D. M. (1968). Use of concentrates. In *The Development of the Production of Beef and Veal*, pp. 63–70. Dubrovnik, 16–20 Sept. 1968. OECD.

AMERICAN MEAT INSTITUTE FOUNDATION (1960). *The Science of Meat and Meat Products*. W. H. Freeman & Co., San Francisco.

BAILEY, M. E., & ZOBRISKY, S. E. (1968). Changes in proteins during growth and development of animals. Publication 1598. National Academy of Sciences, Washington, DC, pp. 87–125.

BERG, R. T. (1969). Growth and carcass composition of heifers, steers and bulls. *University of Alberta Feeders' Day*, Report 48.

BLAXTER, K. L. (1950). The protein and energy nutrition of the young calf. *Agricultural Progress* **25**, 85–93.
BLAXTER, K. L. (1961). Efficiency of feed conversion by different classes of livestock in relation to food production. *Federation Proceedings*, **20**. *Proceedings of the Fifth International Congress of Nutrition*, Part III. Supplement No. 7.
BLAXTER, K. L. (1968). Relative efficiencies of farm animals in using crops and by-products in production of foods. *Proceedings of the Second World Conference of Animal Production*. University of Maryland, 31–40.
BLAXTER, K. L. (1970). Domesticated ruminants as sources of human food. *Proceedings of the Nutrition Society* **29**, 244–53.
BREIREM, K. (1946). Aktuelle problemer i vårt husdyrhold. *Landbruket i Norge*, 76–86. Reprint no. 46, Institute of Animal Nutrition, Agricultural College of Norway.
BREIREM, K. (1955). Home-grown feeds in the feeding of cattle. *FAO Agricultural Development Paper*, No. 51, 67–99, 137–42.
BREIREM, K. (1970). Lectures on *Fóring av melkekyr*. 1. *Næringsbehovet og fórmidler*. Agricultural College of Norway.
BREIREM, K. (1971). Proteinproduksjonen i Norden i relasjon til verdens protein-forsyning. *Nordiske Jordbrugsforskeres Forening*. Congress at Uppsala, June 1971.
BREIREM, K., & HOMB, T. (1972). Energy requirements for growth. In *Handbuch der Tierernährung*, volume II, pp. 547–84. Paul Parey, Hamburg.
BROLUND LARSEN, J. (1958), Animalsk fedt til malkekøer og fedekalve. *Forsøgslaboratoriet*. Bulletin No. 303.
BYERLY, T. C. (1959). The measurement of feed efficiency. Talk at the *International Animal Feed Symposium, Washington, DC*, 5 May 1959.
BYERLY, T. C. (1966). The role of livestock in food production. *Journal of Animal Science*, **25**, 552–66.
BYERLY, T. C. (1967). Efficiency of feed conversion. *Science*, **157**, 890–5.
BYERLY, T. C. (1968). Animal husbandry as a profession. *Proceedings of the Second World Conference of Animal Production*, University of Maryland, 220–8.
CALLOW, E. H. (1945). The food value, quality and grading of meat with special reference to beef. Reprinted from *British Society of Animal Production* 41–56.
CALLOW, E. H. (1947). Comparative studies of meat. 1. The chemical composition of fatty and muscular tissue in relation to growth and fattening. *Journal of Agricultural Science*, **37**, 113–31.
CHATFIELD, C., & ADAMS, G. (1940). Proximate composition of American food materials. *US Department of Agriculture Circular*, No. 549.
COLE, J. W., RAMSEY, C. B., HOBBS, C. S., & TEMPLE, R. S. (1963). Effects of types of breed of British Zebu and dairy cattle on production, palatability and composition. I. Rate of gain, feed efficiency and factors affecting market value. *Journal of Animal Science* **22**, 702–7.
COOPER, M. McG. (1955). Beef production. *Animal Production*, **14**, 1–7.
CRASEMANN, E., & BREIREM, K. (1969). Nahrungsbedarf und Nährwert. In *Handbuch der Tierernährung*, volume I, pp. 580–93. Paul Parey, Hamburg.
DAHL, O. (1953). Köttets kemi. In *Boken om kött*, 144–210. Bengt Forsbergs Förlag, Malmö.
DANILOV, M. M. (1969). *Handbook of Food Products. Meat and Meat Products*. Moscow, 1964. Israel Program for Scientific Translations, Jerusalem.
DAVIES, P. J. (1968). The effect of cereal and protein source on the energy intake and the nitrogen balance of fattening lambs given all-concentrate diets. *Animal Production*, **10**, 311–17.

DAVIS, D. A. R., & OWEN, J. B. (1967). The intensive rearing of lambs. 1. Some factors affecting performance in the liquid feeding period. *Animal Production,* **9**, 501–8.
FIELD, R. A., SCHOONOVER, C. O., & NELMS, G. E. (1964). Performance data, carcass yield, and consumer acceptance of retail cuts from steers and bulls. *University of Wyoming Bulletin,* 417.
FLATT, W. P., MOE, P. W., MOORE, L. A., BREIREM, K., & EKERN, A. (1972). Energy requirements of cows for lactation. In *Handbuch der Tierernährung,* volume II, pp. 341–92. Paul Parey, Hamburg.
GARRETT, W. N. (1971). Energetic efficiency of beef and dairy steers. *Journal of Animal Science* **32**, 451–6.
HAIGH, L. D., MOULTON, C. R., & TROWBRIDGE, P. F. (1920). Composition of the bovine at birth. *Missouri Agricultural Experimental Station Research Bulletin,* 38.
HALLMAN, L. C., JR (1971). Raising dairy calves for beef purposes. *Journal of Animal Science,* **32**, 442–5.
HAMMOND, J. (1936). The problem of quality in relation to meat production. *Scottish Journal of Agriculture,* **19**, Reprint.
HANKINS, O. G. (1946). Estimation of the composition of beef carcasses and cuts. *US Department of Agriculture Technical Bulletin,* No. 926.
HANKINS, O. G. (1947). Estimation of the composition of lamb carcasses and cuts. *US Department of Agriculture Technical Bulletin,* No. 944.
HAUGLAND, E., & HOMB, T. (1966). Fóringsforsøk med mellomkalv. Bulletin No. 132, Institute of Animal Nutrition, Agricultural College of Norway.
HOLMES, W. (1970). Animals for food. *Proceedings of the Nutrition Society,* **29**, 237–44.
HOMB, E., & OFFERGAARD, E. (1957). Innhold av endel næringsstoffer i kjøtt og kjøttvarer. *Forskning og Forsøk i Landbruket* **8**, 61–75.
HOMB, T. (1958). Sammenligning av okser og kastrater i kjøttproduksjonen. Bulletin No. 87, Institute of Animal Nutrition, Agricultural College of Norway.
HOMB, T. (1960). Forsøk over gjøkalvfóring. Bulletin No. 99, Institute of Animal Nutrition, Agricultural College of Norway.
HOMB, T. (1970). *Produksjon av storfekjøtt. Kvalitet og fóring.* Forlag Buskap og Avdrått, Gjøvik.
HOMB, T., SAUE, O., & BREIREM, K. (1960). Feeding dairy cattle during summer time. Vortrag des internationalen Ferienkurses für junge Tierzuchtwissenschaftler vom Sept. 14–28, 1960. Mariensee. *Schriftenreihe des Max-Planck-Instituts für Tierzucht und Tierernährung.* Sonderband VII, 1–37.
HOPPER, T. H. (1944). Methods of estimating the physical and chemical composition of cattle. *Journal of Agricultural Science,* **68**, 239–68.
HUBER, J. T. (1971). Personal communication.
HUFFMAN, C. F. (1959). Summer feeding of dairy cattle. A review. *Journal of Dairy Science,* **42**, 1495–551.
KAY, M. (1969). Intensive beef production. *World Review of Animal Production,* **5**, 64–70.
KAY, M., & MACDEARMID, A. (1969). The effects of diets containing different levels of crude protein on food intake and growth rate of Friesian bulls. *Proceedings of the Symposium of the Meat Research Institute,* April 1969, 63–8.
KAY, M., MASSIE, R., & MACDEARMID, A. (1971). Intensive beef production. 12. Replacement of concentrates with chopped grass. *Animal Production,* **13**, 101–6.
KIELANOWSKI, J. (1962). Objektive Voraussetzungen einer langfristigen Planung der tierischen Produktion. *Archiv für Tierernährung* **12**, 17–26.
KIELANOWSKI, J. (1971). Protein requirements of growing animals. In *Handbuch der Tierernährung,* volume II. Paul Parey, Hamburg. (In Press.)

KIRSGAARD, E., BROLUND LARSEN, J., KLAUSEN, S., & AGERGAARD, E. (1969). Forskellige proteinmængder. *Forsøgslaboratoriet. Årbog*, 403–7.

KLAUSEN, S., AGERGAARD, E., KIRSGAARD, E. & BROLUND LARSEN, J. (1968). Besætningsforsøg med køer på stald hele året og med forskellige foderrationer. *Forsøgslaboratoriet. Årbog*, 504–8.

LAMMING, G. E., & SWAN, H. (1967). The winter feeding of yarded beef cattle. In *Progress in Livestock Nutrition* 1962–67, 7–17. US Feed Grains Council.

LAMMING, G. E., SWAN, H., & CLARKE, R. T. (1966). Studies on the nutrition of ruminants. 1. Substitution of maize by milled barley straw in a beef fattening diet and its effect on performance and carcass quality. *Animal Production*, 8, 303–11.

LARGE, R. V. (1965). The artificial rearing of lambs. *Journal of Agriculural Science*, 65, 101–8.

LAWLOR, M. J., & CROWLEY, J. P. (1968). The indoor fattening of early weaned lambs. *EAAP Symposium on Animal Management*.

LEITCH, I., & GODDEN, W. (1941). Imperial Bureau of Animal Nutrition. *Technical Communication*, No. 14. Rowett.

LINDHÉ, B., & HEMINGSSON, T. (1968). Crossbreeding for beef with Swedish Red and White Cattle. Part II. Growth and efficiency under standardized conditions together with detailed carcass evaluation. *Lantbrukshögskolans Annaler*, 34, 517–50.

LONSDALE, C. R., POUTIAINEN, E. K., & TAYLOR, J. C. (1971). The growth of young cattle fed on dried grass alone and with barley. 1. Feed intake, digestibility and body gains. *Animal Production* 13, 461–71.

LOOSLI, J. K., & MCDONALD, F. W. (1968). Nonprotein nitrogen in the nutrition of ruminants. Food and Agriculture Organisation, Rome.

LYKKEAA, J., & SØRENSEN, M. (1966). Sødmælkskalve. 1. Antibioticaforsøg. *Forsøgslaboratoriet. Årbog*, 355–7.

MANDRUP-JENSEN, J., & KOUSGAARD, K. (1969). Slagte- og kødkvalitetsundersøgelser. *Forsøgslaboratoriet*. Bulletin No. 372, 37–72.

MATRE, T. (1971). Manuscript to be published.

MO, M., & WÆHRE, O. (1968). Grønnfóring som alternativ til beiting i melkeproduksjonen. *Husdyrforsøksmøtet, NLH*, 5–6 Dec. 1968, 24–31.

MOORE, L. A., PUTNAM, P. A., & BAYLEY, N. D. (1967). Ruminant livestock. Their role in the world protein deficit. Reprinted from *Agricultural Science Review*, 5, No. 2.

MOULTON, C. R. (1923). Age and chemical development of mammals. *Journal of Biological Chemistry* 57, 79–97.

MØLLER, P. D. (1969). Urinstoffets anvendelse i kvægets fodring. 1. Urinstof til mælkeproduktion. *Forsøgslaboratoriet. Årbog*, 326–35.

NEDKVITNE, J. J. (1970). Mjølkeerstatning til lam. Reprint No. 353, Institute of Animal Nutrition, Agricultural College of Norway.

NEDKVITNE, J. J., HAUGLAND, E., BUCH-HANSEN, T., & DYNGELAND, J. (1965). Granskingar over haustfeiting av lam i Rogaland 1964. *Årsmelding for Rogaland Fellessalg* 1964, 55–68.

NEDKVITNE, J. J., & HAUGLAND, E. (1970). Kunstig oppal av lam. *Nordisk Jordbruksforskning* 52, 70–1.

NICHOLS, J. R., ZIEGLER, J. H., WHITE, J. M., KESLER, E. M., & WATKINS, J. L. (1964). Production and carcass characteristics of Holstein-Friesian bulls and steers slaughtered at 800 or 1000 pounds. *Journal of Dairy Science*, 47, 179–85.

NIELSEN, E. (1961). Oversigt over kvægforsøg 1933–1959. *Forsøgslaboratoriet*. Bulletin No. 328.

NIELSEN, E., NIELSEN, A., & DISSING ANDERSEN, AA. (1969). Afkomsprøver for kødproduktion. II. *Forsøgslaboratoriet.* Bulletin No. 372.
NIINIVAARA, F. P. & ANTILA, P. (1968). Köttets näringsvärde. *NJF's köttsymposium i Stockholm*, Sept. 16–17, 1968, 31–45.
NORDRUM, E. (1971). Personal communication.
NORRMAN, E. (1969). Uppfödning av kalvar och ungnöt. In *Nötkreatur*, 291–412. LT's Förlag, Stockholm.
OBRAĆEVIC, C. (1962). Food efficiency in fattening young bulls – Experiments in Yugoslavia. Reprinted from *Journal of Scientific Agricultural Research*, **15**, 117–26.
OPSTVEDT, J. (1964). Fortsatte forsøk med syrna skumma mjølk til kalver. Bulletin No. 123, Institute of Animal Nutrition, Agricultural College of Norway.
OPSTVEDT, J. (1967). Fóringsforsøk med geiter. Bulletin No. 134, Institute of Animal Nutrition, Agricultural College of Norway.
ØRSKOV, E. R. (1970a). Nitrogen utilisation by the young ruminant. *Proceedings of the Fourth Nutrition Conference*, Ed. Swan & Lewis, 20–35. J. & A. Churchill, London.
ØRSKOV, E. R. (1970b). Forskellige metoder til forbedring af proteinudnyttelsen hos drøvtyggere. *Ugeskrift for Agronomer*, **115**, 264–70.
OWEN, J. B. (1969). Sheep Production – View of the Future. Seale-Hayne Agricultural College, Newton Abbot, Devon.
OWEN, J., DAVIES, D. A. R., MILLER, E. L., & RIDGMAN, W. J. (1967), The intensive rearing of lambs. 2. Voluntary food intake and performance on diets of varying oat husk and beef tallow content. *Animal Production*, **9**, 509–20.
POUTIAINEN, E. K., LONSDALE, C. R., & OUTEN, C. E. (1971). The growth of young cattle fed on dried grass alone and with barley. 2. Effects on digestion. *Animal Production*, **13**, 473–84.
PRESTON, R. L. (1968). What is needed to break through the efficiency barrier in beef cattle? *Feedstuffs*, **40**, No. 13, 26.
PRESTON, T. R., BOWERS, H. B., MACLEOD, N. A., & PHILIP, E. B. (1965). Intensive beef production. 6. A note on the nutritive value of high moisture barley stored anaerobically. *Animal Production*, **7**, 385–7.
PRESTON, T. R., WHITELAW, F. G., AITKEN, J. N., MACDEARMID, A., & CHARLESON, E. B. (1963). Intensive beef production. 1. Performance of cattle given complete ground diets. *Animal Production*, **5**, 47–51.
REID, J. T. (1969). The future role of ruminants in animal production. In: Phillipson, A. T. (Ed.), *Physiology of Digestion and Metabolism in the Ruminant*, 1–22. Newcastle upon Tyne: Oriel Press.
REID, J. T. (1970). Will meat, milk and egg production be possible in the future? *Proceedings of the 1970 Cornell Nutrition Conference*, 50–63
RUANE, J. B., & CAFFREY, P. (1967). Intensive beef production in Ireland. In *Progress in Livestock Production*, 19–31. US Feed Grains Council.
RUNCIE, K. V. (1960). Utilization of grass by strip-grazing and zero-grazing with dairy cows. *Proceedings of the Eighth International Grassland Congress*, 644–8.
SAHYUN, M. (1948). *Protein and Amino Acids in Nutrition.* Appendix. Table 2. Reinhold Publishing Corporation, New York.
SAUE, O., TOLLERSRUD, S., CHRISTENSEN, H., & HOMB, T. (1963). Forsøk med fóringsokser og kviger i båsfjøs og uisolert bingefjøs. Bulletin No. 110, Institute of Animal Nutrition, Agricultural College of Norway.
SHAUDYS, E. T. (1961). Labour, equipment and costs of using rotational grazing and green chop pasture systems in Ohio. *Ohio Agricultural Experiment Station Research Bulletin* 878.

SØRENSEN, M., & LYKKEAA, J. (1967). A. Sødmælkskalve. Tilskud af korn. *Forsøgslaboratoriet. Årbog*, 491–4.
SØRENSEN, M., & LYKKEAA, J. (1968). Skummetmælkskalve. 1. Forskelligt proteinindhold i foderet. *Forsøgslaboratoriet. Årbog*, 553–75.
SPEDDING, C. R. W. (1969). The agricultural ecology of grassland. *Agricultural Progress*, 44, 7–23.
SPEDDING, C. R. W. (1970). *Sheep Production and Grazing Management.* Baillière, Tindall & Cox, London.
TREACHER, T. T., & PENNING, P. D. (1970). Die Entwicklung eines automatischen Systems zur künstlichen Aufzucht von Lämmern. *EAAP study meeting, Budapest.* Reprint.
TURTON, J. D. (1969). The effect of castration on meat production from cattle, sheep and pigs. *Proceedings of the Symposium of the Meat Research Institute*, April 1969, 1–50.
VALEUR, C. M. (1969). Urinstof til opdræt og kødproduktion. *Forsøgslaboratoriet. Årbog*, 336–9.
VAN ES, A. J. H., NIJKAMP, H. J., VAN WEERDEN, E. J., & VAN HELLEMOND, K. K. (1969). Energy, carbon and nitrogen balance experiments with veal calves. *Energy Metabolism of Farm Animals.* Ed. K. L. Blaxter, J. Kielanowski & G. Thorbek. Oriel Press Ltd, Newcastle upon Tyne.
VIRTANEN, A. J. (1966). Milk production of cows on protein-free feed. *Science* 153, 1603–14.
WALLACE, L. R. (1955). Factors influencing the efficiency of feed conversion by sheep. *Proceedings of the Nutrition Society*, 14, 7–13.
WATSON, D. M. S. (1943). Beef cattle in peace and war. *Empire Journal of Experimental Agriculture*, 11, 191–228.
WELLINGTON, G. H. (1971). Dairy beef. *Journal of Animal Science* 32, 424–30.
WISMER-PEDERSEN, J. (1968). Kødets organoleptiske egenskaber. *NJF's Køttsymposium i Stockholm*, 16–17 Sept. 1968, pp. 25–30.
WÖHLBIER, W. (1928). Stoffwechselversuche zum Eiweissansatz bei saugenden Ferkeln. *Biochemische Zeitschrift* 202, 29–69.

16

ECOLOGICAL FACTORS AFFECTING AMOUNTS OF PROTEIN HARVESTED FROM AQUATIC ECOSYSTEMS

By H. A. REGIER*

Fisheries Resources Division, Department of Fisheries,
Food and Agriculture Organisation, Rome

INTRODUCTION

Fish 'producers' divide rather naturally into two groups: aquaculturists and those who harvest organisms in the wild. Aquaculturists or fish culturists use approaches similar to those applied to the husbandry of domesticated mammals and birds. A review of progress in efficiency and other aspects may be found in the 1966 FAO symposium proceedings edited by Pillay (1967, 1968) and the work by Bardach & Ryther (1968). Strong efforts are being directed in many parts of the world to improve the methods and extend their application in close physical, social and economic association with agriculture. These are reviewed in the following section.

Fishermen that angle, trap, net, gather, or poison animals to obtain a harvest, of course resemble the hunters and gatherers of terrestrial systems. It should not be lightly inferred that such fisheries systems are necessarily inefficient. The aquatic world is alien to us as terrestrial mammals with its three-dimensional space (cf. effectively two dimensions for most terrestrial mammals and agriculturists) that is filled with a high density and relatively opaque liquid. Except for very shallow areas it is very largely beyond the sort of control that we can exert on land by constructing buildings and fences, burning and cultivating, controlling competitors and predators, all under continuous visual monitoring. For much of the oceans it may never be feasible to proceed beyond a 'hunting economy'. Perhaps for this reason the approaches to studying and managing fish and fisheries in the wild until now owe little to theories of terrestrial ecology and to practices on land. It may be that an energetics–trophodynamics context will provide the framework in which these can be brought together, but more on this later.

* Present address: Ramsay Wright Zoological Laboratories, University of Toronto, Toronto, Ontario, Canada.

THE PHYSIOLOGY AND ECOLOGY OF FISH AND SHELLFISH FROM A CULTURE VIEWPOINT

Aquatic organisms live in a medium of about the same density as their own and so require less ponderous skeletal structures for support and locomotion than is needed by birds and terrestrial mammals. Being poikilothermic, they need not spend part of their caloric intake in maintaining a constant body temperature (Brett, 1970). In addition they may be placed in a medium with an osmotic pressure similar to that of their own body fluids in order to minimise expenditure of energy for osmoregulation. Perhaps because of these factors, conversion efficiencies of about 1.0 to 1.2 (kg dry balanced diet per kg wet fish flesh produced) can be achieved without much difficulty. Hastings (1969) has reviewed many aspects of fish nutrition and Table 1 is based on his review. Stansby (1962) has summarised data on the composition of different fish.

Table 1. *Efficiency of various diet ingredients fed to chinook salmon*, Oncorhynchus tschawytscha (*data from W. E. Shanks, quoted by Hastings, 1969*)

Diet ingredient	Weight gain (g)	Protein fed (g)	Weight gain / Protein fed	Percent protein utilisation
Casein	63.2	26.5	2.38	31.4
Liver	114.6	37.7	3.04	39.4
Cottonseed meal	13.7	15.4	0.91	10.0
Fish meal	96.2	36.3	2.65	35.4
Brewer's yeast	21.1	25.0	0.85	10.0

Whether for physiological reasons given above, or others not specified, the cost of producing fish usually compares favourably with other forms of flesh production. The Government of Hungary provided the following data for the relative costs of production (forints per kg) during 1968: beef, 22.92; poultry, 20.07; pork, 15.71; and pond-reared fish, 15.41.

Sessile organisms such as molluscs spend less energy than others on locomotion and have food brought to them by water currents, but this may be partly compensated for by the need to expend energy to filter minute organisms from large volumes of water. In any case, little is known about the conversion efficiency of sessile organisms. The problem has received little practical stimulus since such organisms generally feed on phytoplankton growing wild in areas where little control can be exercised on the phytoplankton and where it is essentially a free resource.

Genetic selection has contributed to greater efficiency of growth and production, notably with the common carp, *Cyprinus carpio*, and the rainbow trout, *Salmo gairdneri*. Efforts in this direction are at a very preliminary state compared to those in other areas of animal husbandry.

SOME PRACTICAL ASPECTS OF AQUACULTURE SYSTEMS

The efficiency of aquaculture depends partly on the degree of human control being exercised. Major factors that may be manipulated include water level, plant nutrient inputs, temperature, production of fish food organisms, control of competitors, prevention of epizootics and predation, and so forth. In practice much of the management is based on extensive experience rather than on intensive scientific comprehension.

The trend in aquaculture is toward an association of compatible and complementary species. For example, in Indian fish culture four species are often used: the catla, *Catla catla*, feeding at the surface; the rohu, *Labeo rohita*, feeding pelagically at mid depths; the mrigal, *Cirrhina mrigala*, feeding at the bottom; and the calbasu, *Labeo calbasu*, feeding mainly on molluscs. In traditional Chinese fish culture a somewhat similar combination consists of: the silver carp, *Hypophthalmichthys molitrix*, which feeds on plankton; the big head, *Aristichthys nobilis*, which consumes only macroplankton; the grass carp, *Ctenopharyngodon idella*, which feeds on coarse vegetable matter; the black carp, *Mylopharyngodon piceus*, which feeds on molluscs; the mud carp, *Cirrhina molitorella*, which consumes worms and organic muds; and the common carp, *Cyprinus carpio*, which is omnivorous and acts as a general scavenger. Even in monocultures, different age or size groups when raised together utilise different types of food organisms and resemble mixed cultures in this respect.

In semi-controlled conditions often the greatest expense in aquaculture relates to purchasing and maintaining pond area or volume, inflow of adequate water quantities, or various fences or floats. Thus 'efficiency' is often given as kg per hectare when related to area, and as kg per cubic metre per hour when related to water flow; both are expressed on a yearly basis. In some Japanese running-water ponds about 70 kg m^{-3} h^{-1} may be produced. In raceway culture of rainbow trout yields of 30000 kg ha^{-1}/ 150 m^3 h^{-1} have been reported; reduced proportionately this amounts to 200 kg ha^{-1}/m^3 h^{-1}. In experimental circulating water systems it has been possible to raise five common carp to about 1 kg weight each in one 40 l tank, i.e. a fish mass to water mass ratio of about 1 to 8. In themselves these data are not very informative on the question of economic or ecological efficiency, but do point to the fact that fish species can be found that are well-adapted to intensive culture where space or water costs are high.

Similar findings apply to raft cultures of molluscs, particularly mussels, where costs depend heavily on construction and maintenance of rafts. If one can conceive of hectares as being a convenient measure of the size of rafts, the annual yield of *Mytilus edulis* has been reported as about 300000 kg ha^{-1} edible flesh from some Spanish estuaries (Pinchot, 1970). There appear to be many fertile estuaries in the world where nearly 'worthless' phytoplankton could be harvested in this manner.

AN OVERVIEW OF FISHERIES SCIENCE RELATING TO 'WILD' ECOSYSTEMS

We are concerned in this section with wild aquatic ecosystems. In his survey of the Indian Ocean, Cushing (1971) identified five major types: upwelling areas and associated offshore divergences; offshore oceanic areas; the coral seas; the mangrove swamps; and the coastal areas outside upwelling areas. To extend the coverage to all aquatic ecosystems we might add another five categories; estuaries of the temperate zones; rivers with their flood plains; reservoirs; lakes and swamps; and ponds under aquaculture.

Margalef (1968) has inferred that 'ecosystems reflect the physical environment in which they have developed, and (the models and methods of) ecologists reflect the properties of the ecosystem, in which they have grown and matured'. Three dominant approaches to fisheries research and development during the past two decades can be visualised in Margalef's context: (i) the Lowestoft way has been successful with benthic stocks of the North Sea and Grand Banks (see, e.g. Beverton & Holt, 1957); (ii) Ricker (1958), Fry (1949), and their North American colleagues began their work with smaller freshwater stocks; (iii) Schaefer (1970) chose a somewhat different set of variables when he studied Pacific pelagic species.

With respect to the other three types of marine ecosystems, scientific work in fisheries of estuaries and upwelling areas is just getting well under way and the question of appropriate variables and models is still unsettled in each. Mangrove swamps and coral seas are still largely unstudied from a fisheries viewpoint. Fisheries workers on temperate freshwaters have applied models and methods of Ricker and Fry widely. Other approaches that have so far attracted limited interest include use of population dynamics indices (Abrosov, 1969), abiotic limnological indices stemming from Rawson (Ryder, 1970), and indices of community structure (Swingle & Swingle, 1967).

Except for the last two, all the approaches referred to above, and in general their myriads of adaptations, are aimed at maximising the efficiency of a harvest regime on an individual stock under the implicit assumption that the community, of which the stock is a component, reacts to the fishery in a way that can be effectively, if implicitly, modelled as a function of the vital variables (parameters) of the stock under consideration. I labour this point to emphasise that a concept of ecological or biological efficiency as in a trophodynamic context has played essentially no role in modelling and managing fish and fisheries in wild ecosystems by these methods to date. Swingle's approach, aimed at the community level, has overtones of the ecological efficiency of a fish prey to fish predator part of a system. Rawson's method incorporates implicitly a measure of the efficiency of the jump from plant nutrient materials to landed catch of fish, with no intermediate stratification attempted.

It will be some time before the interrelationships between the various models in use now and in the past, and any proposed more comprehensive models, can be clarified to permit a transformation of the inferences within the limited contexts into the broader one, in order to provide, say, estimates of the efficiency of protein production. The most immediate practical application of such a broad model would relate to the need for approximate estimates of the fishery potential of areas now only lightly exploited. From such estimates, developers can infer the appropriate scale of socio-economic investment to ensure an efficient step-wise approach to well-balanced development and management.

Quite a number of workers have estimated the fisheries potential of the world's oceans within a trophodynamic context; Ryther (1969) has briefly reviewed some of these attempts. Cushing (1971) estimated tertiary production of the Indian Ocean by regions to provide a measure of the approximate geographical distribution of fish production; this attempt and others are examined in the following section.

At FAO a number of us have recently proposed a broad model that includes elements of the Rawson approach and some concepts employed by Margalef (Regier & Henderson, 1973). This model may shed light on some of the difficulties with the trophodynamic model, as will be indicated below.

Lest the preceding paragraphs imply that fisheries workers are waiting for a more comprehensive context to be clarified before estimates of potential are provided to the developers we may quote Alverson, Longhurst & Gulland (1970) who 'believe that the question of potential fish production can be best answered through a more pragmatic approach (than that of Ryther, 1969) based on knowledge of present commercial fish stocks and on extrapolation from exploratory surveys and other direct evidence'. And in fact they have contributed to a fairly comprehensive co-operative attempt of this kind under FAO auspices (Gulland, 1971).

TROPHODYNAMIC METHODS APPLIED IN FISHERIES

Largely because of its implications for human population policies, the question of how much protein the sea can provide has stirred interest and controversy among scientists and laymen. Each new estimate is defended or attacked by one side or the other of population protagonists, and presumably all scientists publishing on the problem are fully cognisant with this political process to which they periodically contribute some fuel.

One of the more interesting recent events in the scientific part of this dialogue was the strong response of Alverson, Longhurst & Gulland (1970) to a paper by Ryther (1969). Very briefly, Ryther used the most recent summary of primary productivity data for the world's oceans, some of his own inferences on the average number of trophic steps, and the average ecological efficiency between photosynthesis and fish harvesting for different types of marine ecosystems. He concluded that the potential marine fish harvest, utilising all *conventional* resources, was unlikely to be appreciably greater than 100×10^9 kg (see Table 2), or about twice the amount, 60.5×10^9 kg, actually harvested in 1967. He then attempted to check various of his regional estimates by comparing them with actual catches and fishery workers' inferences on the relative level of current exploitation based on conventional models and indices. He felt that the latter in general bore out his estimates based on trophodynamic concepts and data.

His critics – Alverson, Longhurst & Gulland (1970) – made three major points. (1) Ryther had used three as an average number of trophic steps from photosynthesis to fish harvested for coastal areas and five as the number for oceanic areas, both of which his critics felt were too large by perhaps one unit, and they referred to data consistent with their view. It is not clear what their position was on Ryther's use of 1.5 trophic levels in upwelling areas. It may be noted that Cushing (1971) states that there are four main trophic levels in the surface waters of the deep ocean, and also in the 'coastal upwelling system'. (2) They warned that the data and model were at this stage too imprecise for usefully accurate assessment of potentials. (3) They detected a series of inaccuracies in the data that Ryther had accepted for purposes of his checks and they suggested that these inaccuracies had contributed to his 'pessimistic' conclusions.

In referring to the preliminary status of the trophodynamic model and existing estimates, Alverson, Longhurst & Gulland (1970) mentioned that 'Ryther's estimates could easily be in error by a factor of 1 to 2 orders of magnitude'. To date, relatively little attention has been directed to the statistical aspects of such estimates. In the absence of such studies both optimists and pessimists may, on occasion, throw up their hands with statements of possible errors in terms of orders of magnitude.

Table 2. *Use of a trophodynamic model to estimate fish production by three ocean provinces defined according to level of primary organic production (after Ryther, 1969)*

Variable	Open ocean	Coastal zone[1]	Upwelling areas	Total
Percentage of ocean	90.0	9.9	0.1	100
Area ($km^2 \times 10^6$)	326	36	0.36	362
Mean primary production ($g\ m^{-2}$ per annum)	50	100	300	—
Total primary production ($kg \times 10^9$ per annum)	16.3	3.6	0.1	20.1
Trophic levels	5	3	1.5	—
Mean efficiency, %	10	15	20	—
Fish production ($kg \times 10^9$ fresh weight)	1.6	120	120	242

[1] Includes offshore, non-upwelling areas of high productivity.

Ryther's three critics suggest that the potential of conventional resources 'will amount to well over 100 million metric tons'. In addition: 'To this we must add the crustaceans, the molluscs (including cephalopods), and a whole range of small fishes whose potential as food appears larger by perhaps an order of magnitude than that of better-known fish.' Subsequently they state: 'In the end, Ryther may be right but for the wrong reason. If the world catch of sea fish levels off in the next decade, this is likely to be due to the collapse of major fisheries because of climatic cycles, to overfishing, to oceanic pollution, to a failure to resolve problems of international jurisdiction, or to a combination of these factors rather than to inadequacy of unexploited resources.' That last statement, defensible in a conventional fisheries theoretical context, seems to defy interpretation within a trophodynamic context. If we are at this point in time only removing a fraction that is about one-half of one order of magnitude of the full potential, why should improper management of the particular stocks now intensively fished seriously impede expansion of the exploitation to the (postulated) remaining 95 per cent of unutilised or grossly underutilised stocks?

Attention has often been drawn to the fact that in their ontogeny fish pass through a number of trophic levels. I have not yet encountered a trophodynamic model of fish production that takes this observation fully into account though Petipa, Pavlova & Mironov (1970) have attempted to do so for pelagic invertebrates. Much of the biomass of piscivorous fish, at present in trophic level X, was developed some time ago when the fish were in trophic levels $X-1$ or $X-2$. Alverson, Longhurst & Gulland (1970) imply that for some stocks the maximum biomass of separate cohorts is achieved before they are recruited into a size class where they are taken by the fishery. The 'ecological efficiency' of such groups is less than zero, since the biomass of such a component is a decreasing function of time, on average.

I have chosen to review Ryther's paper, and a criticism of it, partly because *bona fide* public controversies in fisheries are sufficiently rare that one should not lightly ignore them. In addition, these authors do rather clearly identify some of the present difficulties with the trophodynamic approach to fisheries problems. With the foregoing in mind, we may review Cushing's (1971) related work with the Indian Ocean.

Ryther (1969) identified three major oceanic provinces according to relative primary productivity and then examined trophic characteristics of each before estimating potential fish harvests. Cushing began instead by identifying five kinds of ecosystems, already mentioned above, and subsequently related primary and secondary production levels to these. He then extrapolated to estimates of tertiary production (Figure 1) without reference to special trophic level characteristics of the different kinds of ecosystems. As indicated earlier, he postulated or inferred in the first paragraphs of his work the existence of four trophic levels in both offshore oceanic systems and in 'coastal upwelling systems', and did not specify levels for coral seas and mangrove swamps.

Cushing noted evidence 'that the transfer coefficient from primary to secondary production decreases from about 15 per cent in areas of low primary production to about 5 per cent in areas of high primary production' but did not use these data to estimate tertiary production. His estimates of efficiency seem to be inconsistent with Ryther's (1969) suggestion that 'when food availability is low, the added costs of basal metabolism and external work relative to assimilation may have a pronounced effect on growth efficiency', and are clearly at odds with Ryther's efficiency estimates shown in Table 2. Consistent with Ryther's suggestion is Henderson's (1971b) tentative inference that more fertile lakes show greater transfer efficiencies between phytoplankton production and fish production than do less fertile waters. Henderson's data indicated that fish yields

PROTEIN FROM AQUATIC ECOSYSTEMS 271

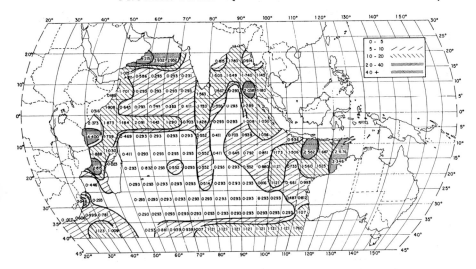

Figure 1a. The distribution of tertiary production, in million tons wet weight per 5° square, estimated as average of 1 % of primary production and 10 % of secondary production during the northeast monsoon, 16 October–15 April (from Cushing, 1971).

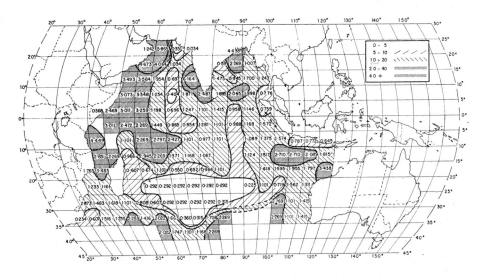

Figure 1b. The distribution of tertiary production, in million tons wet weight per 5° square, estimated as average of 1 % of primary production and 10 % of secondary production during the southwest monsoon, 16 April–15 October (from Cushing, 1971).

within a family of lakes may be roughly proportional to the square of primary production.

Cushing did not explicitly relate his estimates of tertiary production to the size of potential fish harvests, but used the estimates instead to identify large areas of the Indian Ocean that should be quite productive of fish and should be assigned high priority for survey and stock assessment, presumably within conventional fisheries contexts.

Henderson (1971 a) has employed a trophodynamic model in combination with the Rawson approach above to estimate the fish harvest potential of Kainji Lake, a new reservoir on the Niger River in Nigeria. He expanded from a relatively limited base of data on the ecosystem itself by interpolating estimates of relevant variables obtained by workers in other freshwaters. From a consideration of nutrient levels, phytoplankton production, number of trophic levels, temperature and transfer coefficients, each documented with Kainji Lake data and/or references from elsewhere, he obtained estimates of potential fish harvests that appear reasonable to fishery workers well-informed on these matters from a conventional fisheries' viewpoint.

This section may be terminated with the comments of Dickie (1970): 'In principle it is already possible to calculate the minimum food intake to support a specified amount of production at various levels in marine food chains. However, to estimate potential fisheries yield (with useful precision for management purposes) requires that this (reasoning) process be reversed and that the production or that part of it realised as yield be predicted from an observed food abundance. This latter process raises problems which are as yet unsolved...'

ON THE DIFFICULTIES WITH TROPHODYNAMIC METHODS

Any model based on trophic levels or a web of major trophic components that seeks to predict potential fish harvests should take into account changes in the community's trophic structure caused by the fishery (or other independent external stresses such as toxic pollution, enrichment or physical barring of migratory paths). It is well-known that the 'fishing-up process', as fishing intensifies in the early stages of exploitation, often results in a marked decrease in the abundance of larger species which are generally members of the highest trophic level. In ecosystems in which the particular species are pre-adapted to such a stress, the individual species can adapt physiologically to maintain high harvests at apparently greater ecological efficiency (Paloheimo & Dickie, 1970). Straightforward application of trophic level models would imply that a destruction of this highest level, as a result of stressing the stock to levels beyond which it is physiologically capable of adjusting, should lead to a marked increase in fish catches composed of species at lower trophic levels. Historical data on fisheries could be examined to test this hypothesis; a preliminary analysis (Regier, 1972) of a century of catch statistics on the fisheries of the Laurentian Great Lakes, where the highest trophic levels were destroyed, leads us to doubt that the above hypothesis is a useful one. The increased catch seems to be far below a figure of 5 or 10-fold, as inverses of possible ecological efficiencies of 20 per cent or 10 per cent respectively; none of the Great Lakes provided sustained catches greater by a factor of 1.5 after destruction of major large piscivores by the fishery and by sea lamprey, *Petromyzon marinus*, a parasitic invader. In four of the lakes (Ontario, Huron, Michigan, Superior) the commercial fisheries vanished almost completely because the small species were uneconomic, but even in Lake Erie where fishing remained quite intense on the small species the catches did not increase appreciably (Baldwin & Saalfeld, 1962, and recent supplements).

Part of the reason for the failure of the Great Lakes to respond as expected on the basis of a simplistic trophodynamic model may relate to re-adjustments of the structure of the food web in such a way that smaller fish species that earlier were at the X trophic level may 'graduate' or be forced into the $X+1$ level after the species that were at the $X+1$ level are effectively destroyed (Regier, 1972). Alternatively, or in addition, factors determining feeding efficiency may be so important in aquatic systems as to relegate trophic level factors to secondary status. Thus Kerr & Martin (1970) found that yield from a lake trout fishery was not diminished and may well have been increased by a lengthening of their food chain by one unit. The latter was accomplished by interposing an 'efficient' planktivorous coregonine, *Coregonus artedi*, between zooplankton and the normally piscivorous lake trout, *Salvelinus namaycush*.

Another aspect of the effect of fisheries exploitation on ecosystems is that chemical substances that were earlier plant nutrients and would again become such (following their excretion from living or decomposing fish) are obviously removed from the ecosystem in the fish harvested. In eutrophic ecosystems where the inflow rates of nutrients are not so low as to limit fish production, perhaps as in some estuaries, rich inshore areas or near the centre of upwelling systems, the removal of nutrients in fish bodies should have little effect on subsequent harvests in that particular area. The case, however, in less fertile oligotrophic areas such as offshore oceanic areas, coral seas and large tropical oligotrophic or amictic lakes might be quite different. Here the incident light energy and temperature are clearly not limiting as far as primary production is concerned. Whatever the inflow or pumping mechanisms are by which nutrients are brought into the euphotic zone of such systems, unless they function more efficiently under fishery exploitation the nutrient levels will decrease as a function of harvest rates to the detriment of the fish production system.

This section may be read as an extended apologia for not being able to provide data on the *ecological* 'efficiency of protein production' in wild aquatic ecosystems. A germ of a model of *economic* efficiency of fisheries as a function of broad ecological variables has been proposed (Regier & Henderson, 1973) but has been neither clearly formulated nor tested. Intense interest has been directed to questions of the economic efficiency at the population level with fisheries directed towards and managed with respect to single stocks (Christy & Scott, 1966; Crutchfield & Pontecorvo, 1969).

In general, we agree with the views of Alverson, Longhurst & Gulland (1970), quoted earlier, that for the short-run future more accurate estimates of unutilised potential (as well as under- and overutilised potential) can be obtained by conventional fisheries methods than by means of trophodynamic models at their present state of refinement. Beyond that, we (Regier & Henderson, 1973; Regier & Cowell, 1972) have recently proposed a broad model, described briefly in the next section, that may be seen either as complementary to, or as an alternative to, a trophodynamics approach, and can accommodate conventional fisheries approaches within itself.

AN ALTERNATIVE ECOLOGICAL MODEL FOR FISHERIES APPLICATION

Ten years ago, when conventional fisheries approaches were well developed and the trophodynamic context was beginning to attract some interest in fisheries, I was assigned research duties on Lake Erie. (This was some seven decades after the first serious effects from pollution and excessive fishing intensity were noted and publicly reported, two decades after fishery ecologists began publicly announcing with alarm that the lake was transforming rapidly, and at least one decade before society took any significant remedial action.) After reviewing the extensive literature on the ecology of Lake Erie and the usual complement of fisheries models and methods, and conducting some preliminary research, it was noted that only small parts of the system could be handled conventionally (Regier, 1966). There was no effective way of dealing with the problem related to

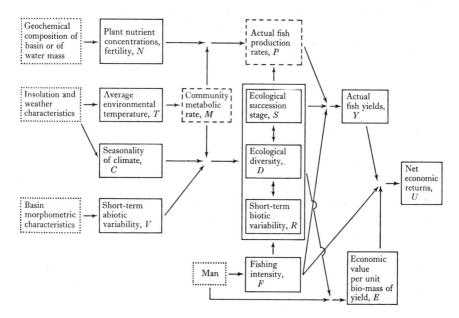

Figure 2. Schematic representation of the proposed general model.

the fact that the fish community in Lake Erie had been markedly unstable since about 1890 and that one after another of the most valued stocks had collapsed and some had become extinct (Regier, Applegate & Ryder, 1969). Somewhat oversimplified, conventional fisheries approaches presuppose the stability of stocks, or moderate deterministic cyclic variations associated with stock recruitment mechanisms, or both.

A decade of fretting about Lake Erie's problems, and about conceptual difficulties with many other heavily stressed aquatic ecosystems, led to the formulation of the general model in Figure 2, from Regier & Henderson (1973). Very briefly, the model incorporates concepts of succession, diversity, and stability as used by Margalef (1968, 1969) and others, together with those of Rawson's limnological approach, some additional considerations on the relationship of production rates of poikilotherms to temperature, and the effect of stress variability on community stability.

Some brief comments might serve to clarify the model. If for heuristic purposes we focus on stage of succession we note that for a fixed seasonal regime of major environmental stresses, the higher the temperature the nearer the average annual stage of succession will be towards the climax end of the succession sequence, and conversely. The opposite holds true in general for seasonality of climate, by which is meant a measure of the degree of between-season fluctuations in major abiotic factors, which may in turn be seen as a macro-measure of large-scale events of the same generic kind as measured by the short-term abiotic variability factor. The greater the fishing intensity, the more 'primitive' the successional or 'recessional' state. Fish production in the ecological sense is a function of the inflow rate of plant nutrients that contribute to phytoplankton production, average environmental temperature, and the biotic variables – successional state, diversity and short-term biotic variability. The further the fish community is driven into 'recession', the more the community becomes dominated by resilient, opportunistic species which are generally small, short-lived and fecund. These opportunists usually have a much lower economic value than larger species that dominate communities nearer a climax successional state. Thus the economic value of fisheries can be modelled to an approximate degree using very broad ecological variables.

It may be noted that present plans to test this model extensively begin with simple functions interrelating those boxes depicted in Figure 2 with continuous outlines; other boxes may be incorporated later. The boxes with dashed and dotted outlines were included for heuristic purposes. A trophodynamic model could be accommodated in part of the model – plant nutrients, community metabolic rate, actual fish production – by inserting additional boxes. To do so might increase the model's precision, but it is not now clear how great the improvement might be, if any. The model already implicitly incorporates various considerations usually included in a trophodynamic context.

The precision hoped for at the outset is to estimate, within a factor of two of the true values both in terms of biomass and economic returns, the potential annual harvests of aquatic ecosystems falling in the range of 0.1 kg ha^{-1} to 1000 kg ha^{-1}, a range of 10000-fold. Greater precision could presumably be attained by adding additional variables in the spirit of the model, perhaps trophodynamic considerations not already included implicitly in it, and conventional fisheries models at a population level.

This alternative model by Regier & Henderson (1973) has been presented here to highlight difficulties experienced with fisheries problems, and by way of contrast to identify some reasons why a simple trophodynamic model may not be usefully realistic.

CONCLUDING REMARKS

At present, fishery ecologists do not possess a general trophodynamic model that has been shown to be useful for *practical management* purposes though existing models may be useful at *early stages in fisheries development* (Cushing, 1971; Henderson, 1971b). Given adequate resources for research it would of course be possible to model many aquatic ecosystems to an acceptable degree of accuracy in a trophodynamic context; this is now being attempted within various projects under the International Biological Programme. It is not now clear whether the models and parameters thus produced will have important conceptual and practical advantages over the classical population dynamics approach of fisheries. I suspect that the energetics context will be as cumbersome to apply to events in communities transforming under the stress of fisheries – which, after all, is the fate of most communities subjected to fishing – as is the population dynamics context. Given a more comprehensive context, the latter two models may be accommodated as components within it, even though they appear too limited in themselves.

I have been able to say little about the efficiency of protein production in wild ecosystems as a function of trophodynamic events. Perhaps the question can only be expected to have operational meaning in systems under fairly intensive control by man.

Dr T. V. R. Pillay of the Fisheries Resources Division of FAO in Rome assisted me in a number of ways with this paper.

REFERENCES

ABROSOV, V. N. (1969). Determination of commercial turnover in natural bodies of water. *Problems of Ichthyology*, **9**, 482–9.
ALVERSON, D. L., LONGHURST, A. R., GULLAND, J. A. (1970). How much food from the sea? *Science*, **168**, 503–5.
BALDWIN, N. S., & SAALFELD, R. W. (1962). Commercial fish production in the Great Lakes 1867–1960. *Technical Report of the Great Lakes Fishery Commission*, No. 3.
BARDACH, J. E., & RYTHER, J. H. (1968). The status and potential of aquaculture. Springfield, Va.: Clearinghouse for Federal Scientific and Technical Information, Vols. 1 and 2.
BEVERTON, R. J. H., & HOLT, S. J. (1957). On the dynamics of exploited fish populations. *Fishery Investigations, London* (UK Ministry of Agriculture Fisheries & Food), **19**.
BRETT, J. R. (1970). Fish – the energy cost of living. In *Marine Aquaculture*; selected papers from the conference on marine aquaculture, pp. 37–52, Ed. W. J. McNeil. Corvallis: Oregon State University Press.
CHRISTY, F. T., JR, & SCOTT, A. (1965). *The Common Wealth in Ocean Fisheries*. Baltimore: Johns Hopkins Press.
CRUTCHFIELD, J. A., & PONTECORVO, G. (1969). *The Pacific Salmon Fisheries; a Study of Irrational Conservation*. Baltimore: Johns Hopkins Press.
CUSHING, D. H. (1971). Survey of resources in the Indian Ocean and Indonesian area. *FAO/UNDP/IOFC Indian Ocean Program* (IOFC/DEV/71/2).
DICKIE, L. M. (1970). Introduction to food abundance and availability in relation to production. In *Marine Food Chains*, pp. 319–24, Ed. J. H. Steele. Edinburgh: Oliver and Boyd.
FRY, F. E. J. (1949). Statistics of a lake trout fishery. *Biometrics*, **5**, 27–67.

GULLAND, J. A. (1971). *The Fish Resources of the Ocean*. London: Fishing News (Books) Ltd.

HASTINGS, W. H. (1969). Nutritional score. In *Fish in Research*, pp. 263–92. New York: Academic Press.

HENDERSON, H. F. (1971a). Estimation of potential catch of fish from Kainji Lake based on general ecology of the lake. (Unpubl. mimeo.)

HENDERSON, H. F. (1971b). Food-chain theory. *Journal of Tropical Hydrobiology and Fisheries*, 1, 69–83.

KERR, S. R., & MARTIN, N. V. (1970). Trophic-dynamics of lake trout production systems. In *Marine Food Chains*, pp. 365–76, Ed. J. H. Steele. Edinburgh: Oliver and Boyd.

MARGALEF, R. (1968). *Perspectives in Ecological Theory*. Chicago: University of Chicago Press.

MARGALEF, R. (1969). Diversity and stability: a practical proposal and a model of interdependence. In *Diversity and Stability in Ecological Systems. Brookhaven Symposia in Biology*, 22, 25–37.

PALOHEIMO, J. E., & DICKIE, L. M. (1970). Production and food supply. In *Marine Food Chains*, pp. 499–527, Ed. J. H. Steele. Edinburgh: Oliver and Boyd.

PETIPA, T. S., PAVLOVA, E. V., & MIRONOV, G. N. (1970). The food web, utilization and transport of energy by trophic levels in the planktonic communities. In *Marine Food Chains*, pp. 142–67, Ed. J. H. Steele. Edinburgh: Oliver and Boyd.

PILLAY, T. V. R. (1967). (Ed.) *Proceedings of the FAO World Symposium on Warm-Water Pond Fish Culture*, Rome, 18–25 May 1966. Vol. 3. Review and experience papers of meetings 2 and 3. *Food and Agriculture Organisation Fisheries Reports* (44), Vol. 3.

PILLAY, T. V. R. (1968a). (Ed.) *Proceedings of the FAO World Symposium on Warm-Water Pond Fish Culture*, Rome, 18–25 May 1966. Vol. 4. Review, experience and working papers of meetings 4, 5 and 6. *Food and Agriculture Organisation Fisheries Reports* (44), Vol. 4.

PILLAY, T. V. R. (1968b). (Ed.) *Proceedings of the FAO World Symposium on Warm-Water Pond Fish Culture*, Rome, 18–25 May 1966. Vol. 5. Review and experience papers of meetings 7, 8 and 9. *Food and Agriculture Organisation Fisheries Reports* (44), Vol. 5.

PINCHOT, G. B. (1970). Marine farming. *Scientific American*, 223, 15–21.

REGIER, H. A. (1966). A perspective on research on the dynamics of fish populations in the Great Lakes. *Progressive Fish Culturist*, 28, 3–18.

REGIER, H. A. (1972). Community transformation – some lessons from large lakes. In *Proceedings of the 50th Year Anniversary Symposium*, University of Washington, College of Fisheries. *University of Washington Publications in Fisheries*, No. 5, 35–40.

REGIER, H. A., APPLEGATE, V. C., & RYDER, R. A. (1969). Ecology and management of the walleye in western Lake Erie. *Technical Reports of the Great Lakes Fishery Commission*, No. 15.

REGIER, H. A., & COWELL, E. B. (1972). Applications of ecosystem theory: succession, diversity, stability, stress, conservation. *Journal of Biological Conservation*, 4, 83–8.

REGIER, H. A., & HENDERSON, H. F. (1973). Towards a broad ecological model of fish and fisheries. *Transactions of the American Fisheries Society*, 102, 56–72.

RICKER, W. E. (1958). Handbook of computations for biological statistics of fish populations. *Bulletin of the Fisheries Research Board of Canada*, No. 119.

RYDER, R. A. (1970). Major advances in fisheries management in North American glacial lakes. In *A Century of Fisheries in North America*, Ed. N. C. Benson,. Special Publications of the American Fisheries Society No. 7, 115–27.

RYTHER, J. H. (1969). Photosynthesis and fish production in the sea. *Science*, **166**, 72–6.
SCHAEFER, M. B. (1970). Management of the American Pacific tuna fishery. In *A Century of Fisheries in North America*, Ed. N. C. Benson. Special Publications of the American Fisheries Society No. 7, 237–48.
STANSBY, M. E. (1962). Proximate composition of fish. In *Fish in Nutrition*, pp. 55–60, Ed. E. Heen & R. Kreuzer. London: Fishing News (Books) Ltd.
SWINGLE, H. S., & SWINGLE, W. E. (1967). Problems in dynamics of fish populations in reservoirs. In *Reservoir Fishery Resources Symposium*, pp. 229–43. Athens, Georgia: American Fisheries Society, Reservoir Committee Southern Division.

DISCUSSION

By G. WILLIAMS

Department of Zoology, University of Reading

AND P. A. JEWELL

Department of Zoology, University College, London

Within the framework of Dr Phillipson's ecological setting, two major points emerged: the possibility of increasing net primary production and the possibility of channelling primary production through selected secondary producers.

Except for some agricultural systems and assuming light to be the ultimate limiting factor, realised net primary production is well below the theoretical expectation, and it is in this area that dramatic increases might be expected. The example was given of cocoa production in which removal of shade trees, combined with a management system which provided nutrients and water and which protected the trees from insect attack, increased production by a factor of 5. The average production of leaf matter in the UK was about 7×10^3 kg ha^{-1} per annum, but with irrigation and fertiliser application this could be doubled. Similarly, in the wet tropics, where water was not limiting, fertiliser treatment could treble the output to 40–50×10^3 kg ha^{-1} per annum. To justify the expenditure involved, intensive management was needed to direct a larger fraction of output to consumption by large mammals. This meant proportionally less for the invertebrates, particularly soil organisms, and there were instances in Great Britain where soil structure and the persistence of grasses was adversely affected by sustained high yields.

To exploit the existing net primary production, which is all consumed in one way or another, it would seem logical to exploit animals in relation to their efficiency as protein converters. This would give the sequence, invertebrates, poikilothermal vertebrates and, in the last resort, the homeothermal birds and mammals. Unfortunately the problem of harvesting and the palatability of the products to humans had reversed this sequence. Some felt that exploitation of unusual food resources, such as broth enriched with beetles to provide crunchiness, would only be contemplated when conventional protein sources became intolerably expensive. Others felt that this had already occurred where the human population pressure was very high and the time required to collect these small food items no longer presented any problem.

If production of animal proteins is to be by the warm-blooded vertebrates, a decision was needed on whether this was done more efficiently by mixed stocking schemes or by single-species systems. Direct evidence for the superiority of mixed stocking had already been put forward in the papers by Dr Phillipson and Professor Regier. The latter's dramatic figures were the result of exploiting the three-dimensional world of the aquatic environment and the difference between the rate of production by wild game and domestic cattle quoted by Phillipson seemed very small in relation to the uncertainties of some other flow rates quoted. Clearly further work was needed. The question was raised of whether a strong case could be made for the heavy expenditure that would be involved in carrying out proposals for such a comparison using African herbivores. It was pointed out that, without further evidence, there was a danger that the superiority of mixed stocking would become a part of ecological dogma. In a discussion of game ranching in East Africa, emphasis was placed on the contrast between the costs of providing a suitable environment for introduced cattle and those of exploiting by simpler methods the community of large mammals already well adapted to the difficult environment. Several speakers objected that the difficulties of shooting wild animals, of marketing the carcasses and of controlling transmissible disease reduced the apparent benefits. There was admittedly a big problem here but the answer could lie in the complete domestication of new species that would be capable of being herded and handled but would retain all their innate advantages for man's use. For example, the oryx can thrive on water requirements that are one-fifth of those of cattle and they can be tamed and herded with ease.

Having selected a species for exploitation there remained the final question of the system of management. The closely controlled methods of stall-feeding would seem to offer the most efficient way of converting plant material to animal protein but Professor Homb's results did not confirm this. Although a number of suggestions were put forward to explain this apparent paradox, it could well be that any superiority of zero-grazing would show better in other indices such as energy transformation and that protein conversion was not the relevant measure.

The ideas of ecological energetics had been infiltrating agriculture for the past ten years and the time was ripe to attempt to quantify all the processes at work in a given agricultural system. Despite the dominance of crop and animal production such a study should include the above-ground invertebrates and the soil fauna. Dr Phillipson's flow diagrams provided a model which would allow real comparisons to be made between natural and agricultural systems.

PART V

THE BIOLOGICAL EFFICIENCY OF
INDUSTRIAL SYSTEMS OF
PROTEIN PRODUCTION

17

CONVERSION OF AGRICULTURAL PRODUCE FOR USE AS HUMAN FOOD

By F. AYLWARD and B. J. F. HUDSON

Department of Food Science, University of Reading,
London Road, Reading, Berkshire

INTRODUCTION

The agricultural scientist concerned with agronomy or animal production has normally considered problems of biological efficiency in terms of the production of protein or other components in the whole crop, in some major part such as the wheat grain, in the whole animal or in some special product such as milk or eggs.

In practice most animal and plant products have to be subjected to some form of separation and processing procedure on the farm, or in the domestic or restaurant kitchen, or at some point in between – for example, in the flour mill and bakery, or in the abattoir and meat cannery. At these different points losses or diversion of the original plant or animal material may occur. Some of these losses are inevitable; some are avoidable and can be reduced through 'good housekeeping' in the home or restaurant, or through procedures for the utilisation of products and byproducts.

The factors for what can be broadly called 'waste' vary from commodity to commodity and also with time and place. One contribution of the food processing industries in this and other countries, is to reduce such waste to the minimum, by ensuring the maximum utilisation of the products of agriculture, fisheries and wild life. Before discussing the application of new methods to what are often described as novel proteins, a brief account will be given of the more traditional picture.

UTILISATION OF PROTEIN FROM TRADITIONAL SOURCES

Protein consumption in the United Kingdom

Table 1 summarises the present pattern of protein consumption in the United Kingdom; on average some 60 per cent of protein comes from animal produce and about 40 per cent from cereals and other plant foodstuffs. As indicated by the National Food Surveys, the pattern varies between income groups and has varied with time. During the present century there has been in the United Kingdom a decline in consumption of plant foodstuffs and an increase in consumption of animal produce (Hollingsworth, 1971; Aylward, 1969). This change is a reflection of the increasing affluence of our society and a narrowing of the gap between different economic groups.

Table 1. *Sources of protein consumed in the UK (% of total protein consumed)*

	1909–13	1968
Milk and dairy products	15	23
Carcass meat, poultry and meat products	25	28
Eggs and egg products	3	5
Fish and fish products	10	5
All animal products	53	61
Potatoes, other vegetables, nuts, etc.	10	13
Cereals, bread and flour products	38	26
All plant products	48	39

In several parts of Europe, people consume on average a much higher percentage of plant protein, for example from cereals or from potatoes. In Africa, Asia and South America, even in areas where there is no absolute protein deficiency, animal protein may represent only a small fraction of the total protein eaten.

Animal produce

Table 2 indicates the three main routes for the utilisation of animal products: (i) in the 'fresh' form usually following some form of cooking or other heat treatment, (ii) in some modified form following domestic or commercial processing, and (iii) as components of other foods. In routes (ii) and (iii) the catering and food manufacturing industries have played an increasingly important role in the past half-century, and in particular over the past two decades. The animal processing industry includes such traditional operations as cheesemaking and bacon and ham manufacture, formerly carried out on a domestic or farmhouse scale.

Plant produce

The protein contribution from plant produce in the United Kingdom is summarised in Table 3. Cereals occupy a central place and provide a variety of products such as bread, biscuits and flour confectionery, and, in more recent times, prepared breakfast foods. Wheat is the most important

Table 2. *Traditional utilisation of animal products*

	Utilisation		
Commodity	Fresh, heat-treated or cooked	Processed	Component of other foods
Milk	As milk	Cheese, yoghourt, dried milk	Fresh or processed milk in domestic or trade products
Carcass meat and poultry	As meat, etc.	Canned or frozen meat, etc.	Sausages, pies and other fabricated products
Eggs	As eggs	Dehydrated eggs	Confectionery, etc.
Fish	As fish	Canned or frozen fish	Fish products

Table 3. *Traditional utilisation of plant products*

	Utilisation		
Commodity	Fresh, heat-treated or cooked	Processed	Component of other foods
Potatoes	As potatoes	Canned, frozen or dehydrated	In mixed fabricated products
Other vegetables	Salads or cooked vegetables	Canned, frozen, dehydrated, etc.	In mixed fabricated products
Wheat and other cereals	Maize ('corn on the cob'), porridge oats	After milling – bread, flour, confectionery, breakfast cereals. Brewing	Miscellaneous products
Oilseeds	Nuts	Peanut butter	Soyaflour in bread

cereal for human consumption in Britain; in other countries cereals such as rye, maize (Indian corn) and rice are of greater importance. Potatoes and other vegetables make a significant contribution to protein supplies in Britain; in other countries such as Belgium with a very high *per capita* consumption of potatoes the protein intake from this source is even more important.

The protein contribution of oilseeds is at present relatively small in the United Kingdom, although for many years soya flour has been used as a bread additive. In the United States peanut butter is well known as a protein food.

EFFICIENCY OF UTILISATION

General considerations

In examining the utilisation of proteins we have to survey the whole area from field to consumer (see Table 4), namely losses on the farm, in the postharvest period, in the home or restaurant, and finally, physiological losses.

Table 4. *Protein losses*

Area	Stage
In field	In growing crops or animals. At harvesting
Postharvest or slaughter	Immediate postharvest treatment, storage and transport during processing
In home, restaurant or institution	Larder, or storage losses; losses in kitchen preparation or cooking; plate losses
After eating	Poor utilisation of food components by the body

In agricultural circles there is a natural emphasis on losses on the farm which decrease the efficiency of crop and animal production through disease or the action of pests or through inefficiency. There is no need to stress the obvious truism that there may be a great difference in efficiency of production on experimental plots and actual farm practice.

Two other areas of loss – during postharvest storage and processing, and in the restaurant or home – usually receive much less attention, even though they may be considerable. In connection with food wastage four points deserve special mention: (i) in general all animal produce deteriorates rapidly at environmental temperatures in temperate – and even more so in tropical – regions; (ii) with plant produce, losses, e.g. through pests, may be particularly high in protein-rich materials unless special precautions are taken; (iii) losses may be increased because of damage during harvesting with the induction or acceleration of enzymic or other biochemical changes; (iv) poor harvesting or postharvest storage conditions, e.g. in respect of moisture levels, may lead to fungal growth which in some cases, e.g. aflatoxin formation in groundnuts, may lead to the rejection of the crop for human consumption.

Much of the effort of the food manufacturer and distributor is directed to reducing to the minimum wastage in the postharvest phase, and many of the commercial processes such as canning and quick freezing are designed specifically to obtain the maximum utilisation of harvested material both through the preparation of the major product and the manufacture of byproducts.

Losses of protein and other nutrients through the action of pests, microorganisms and chemical changes can be reduced by good technical practices but there are other forms of wastage which may be inherent in processes and which vary from commodity to commodity.

Utilisation of animal protein

Carcass meat

The slaughtering, cutting and dressing processes involve losses of protein through blood, inedible or only partly edible offal and hides. In modern abattoir practice efforts are made to utilise all byproducts either as human food, as animal feed or for other purposes. Older domestic or butchery processes now form the basis of major enterprises to make soups, sausages, pies and other products which, while perfectly wholesome and good sources of protein, enable poor cuts of meat or other parts of the animal to be used as human food. The tradition of good housekeeping by the skilful farmer's wife, illustrated by the production of black puddings, haggis and tripe, has been carried over to the well-managed food enterprise. Nevertheless the losses of protein may be considerable unless rigid hygienic and other precautions are taken; rigorous veterinary inspection inevitably diverts some proportion of animal protein to animal feed or for use as fertiliser.

Booher (1948) in her review quotes the views of Haecker (1920) who estimated that 40–52 per cent of the total protein in veal calves (45–115 kg liveweight) were in the edible parts of the animal. For yearling beef calves (225–270 kg liveweight) her figures were 55–65 per cent, and for steers (545 kg liveweight) 62 per cent.

In contrast, the present-day production of pig meat for pork represents a comparatively efficient use of animal protein for edible purposes. According to privately communicated trade data, an average carcass of 60 kg yields 43 per cent of lean meat. If the protein content of head, rind, trotters, etc., is added, this figure is raised to about 50 per cent. Since 80 per cent of the carcass weight is retailed in the form of well-defined cuts which, between them, account for virtually all the lean meat, 80–90 per cent of the protein passes directly to edible use, and a substantial amount of the rest will also be consumed either as rind or as manufactured product, e.g. sausages.

Any consideration of meat protein must of course take into account the percentage efficiency of conversion of feed protein to food protein, which varies within the range 7–15 per cent, and is highest with pork and lowest with beef. On the other hand, the production of poultry, eggs and milk is 20–30 per cent efficient (Booher, 1948). In this context better use of meat protein, whilst economically desirable, will never allow it to compete in terms of overall efficiency with protein of plant or microbiological origin. For this reason an inherent economic advantage lies with textured vegetable protein, as will be seen later.

Fish

A similar pattern of partial utilisation of total protein is seen in the fishing industry. Problems are accentuated in deep-sea fisheries; a major proportion of protein-rich offal, together with fish unacceptable because of type or size, is discarded at sea or converted into fish meal for animal feed or fertiliser.

Eggs

The storage life of the whole egg even under the best conditions is short; techniques such as pasteurisation and dehydration enable egg protein to be kept in good condition for long periods. Freeze-drying, which because of its expense has been adopted for relatively few products, has been applied with success on a large scale to eggs for subsequent use in cake making.

Milk

Milk is one of the most sensitive food materials which even at low temperatures can deteriorate under the action of psychrophilic micro-organisms. Considerable research and development work today is being carried out on 'long-life' milk with special attention to the reduction of off-flavours, some of which are caused by the interaction of protein and other components.

The traditional method of storing milk protein was as cheese; fermented products such as yoghurt only have a reasonable shelf life at low temperature. The most important large-scale method of storing milk protein is in the dehydrated form; newer techniques enable the dried material to be readily reconstituted as liquid milk or for use as an ingredient in other foods. Significant losses of milk protein occur in the whey produced in cheese manufacture; there is now a considerable research effort directed to the economic processes for the recovery of whey protein.

Utilisation of plant protein

The edible parts of a crop usually represent only a fraction of the whole plant and in examining the picture of agricultural production this fact must be taken into account in calculations relating to actual yield in terms of use by the consumer. This can be illustrated in terms of cereals, vegetables and oilseeds.

Wheat and other cereals

Wheat, rye and other cereals after dehusking are usually subjected to some form of milling process as a preliminary to use as human food. 'Corn on the cob' (maize) is one of the relatively few examples of the use of a cereal as a vegetable; oats and rice can be used also in more or less unprocessed forms either as porridge or in other ways.

Table 5. *Components of wheat meal and flour (%)*
(from McCance & Widdowson, 1969)

	Whole meal 100 % extraction		Flour 85 % extraction		Flour 70 % extraction	
	English	Canadian	English	Canadian	English	Canadian
Protein	8.9	13.6	8.6	13.6	7.9	12.8
Available carbohydrate	73.4	69.1	79.1	74.9	81.9	76.9
Fat	2.2	2.5	1.5	1.7	1.0	1.2
Other	15.5	14.8	10.8	10.7	9.2	9.1

The wheat milling process can be used to produce a wholemeal containing all the protein and other components of the grain, but more usually the objective of milling is to produce a white flour of low fibre content. The rate of *extraction* was the subject of controversy before, during and after the Second World War when the question of protein (as well as total food supplies) became a matter of national importance (McCance & Widdowson, 1956). Table 5 illustrates some of the differences between whole wheat meal and flours of 85 and 70 per cent extraction rates; rates of extraction varying from 80 to 90 per cent were mandatory in the 1941–9 period but from 1950 onwards the 70 per cent extraction rate was provided for. Thus the protein produced in the grain must be reduced by a factor of 0.75 or more to take into account its actual use as human food in cereal products. It is true that some protein-containing byproducts of the wheat milling process such as the germ are used in food preparations, but the bulk of the residues go to animal feed. This 'recycling' must be taken into account in calculations of national food supplies and food balance sheets. In such calculations additional factors are necessary to take cognisance of the diversion of some proportion of the cereal crops either for animal feed or for other non-food uses such as malting and brewing.

Vegetables

The proportion of protein in the whole plant which is actually used varies from crop to crop. Anyone who has witnessed the garden pea harvest must be struck by the enormous agricultural effort to produce peas of the right quality, including degree of maturity, for canning, quick freezing or dehydration. The pea itself contains a small portion of the total protein produced, the pods and the rest of the vine are waste materials from the point of view of human use.

Reviews such as that of Aykroyd & Doughty (1966) have drawn attention to the importance of legumes as a source of protein in human nutrition. Many legumes, however, contain toxic factors and anti-nutrients closely associated with the protein (Liener, 1969); so the crop cannot be used directly by man.

Oilseeds

Nuts normally make only a small contribution to supplies of edible protein. Several of the commercially imported oilseeds (e.g. groundnuts and soya) produced by leguminous crops and high in protein find their use is normally limited by their content of fibre or by the presence of antinutrients (Altschul, 1958; Markley, 1950; Woodroof, 1966).

NEWER METHODS OF PROTEIN EXTRACTION AND UTILISATION

General considerations

Within the past decade several factors have combined to create a new trend in protein foods. The factors may be summarised under five headings: (i) nutrition, (ii) functional use in relation to food trends, (iii) economics, (iv) advances in polymer (including protein) science and (v) in processing techniques.

Growing consciousness of the serious shortage of protein in many parts of the world, and predictions of future population trends, have stimulated efforts to devise means of using current protein supplies more efficiently. On a more localised basis interest in dietetic, 'health' and 'slimming' foods has stimulated interest in plant foodstuffs and in plant proteins. Quite apart from nutritional considerations advances in food science have led to a recognition of the functional properties of proteins in different types of manufactured foodstuffs. Thus soya is added to bread formulations primarily for functional purposes (Learmonth, 1956; Learmonth & Wood, 1962).

The increasing cost of animal proteins has encouraged the food industries in many countries to seek substitutes for animal proteins, first as a replacement of the animal component in various fabricated foods and more recently as meat analogues.

Progress would have been impossible without the very considerable background of new knowledge of the basic structure and reactions of proteins which has been accumulated over the past quarter of a century, and without the parallel increase in our knowledge of polymer systems, derived in large part from the plastics and synthetic fibre research and development centres. Technological advances in relation to polymers have now been applied successfully to protein systems to produce extruded, moulded and spun products (Odell, 1969). Other advances in our knowledge of sensory evaluation techniques and of the properties and uses of flavouring materials are also being widely used by food manufacturers.

In summary, food processing technology, aided by improved knowledge of polymer science, has become sufficiently sophisticated to be able to develop textures and flavours at will. Much greater knowledge of consumer needs and preferences has given marketing organisations greater confidence in new product development. As a result we have developed means for the utilisation for food of various sources of protein not previously considered suitable. We have also developed various entirely new biochemical engineering operations for the production of single-cell protein from a variety of substrates.

The newer techniques have had an impact on many different commodity fields such as cereals and dairy products, as well as meats (Altschul, 1958). The wide range of developments is illustrated in the recent publication of the US Department of Agriculture entitled *Synthetics and Substitutes for Agricultural Produce* (1969).

Animal products (including fisheries)

When we talk about new or novel proteins, products of land animals are usually omitted from consideration. It should be noted, however (Table 6), that protein concentrates from milk, and in particular dried skimmed milk or isolated casein, are widely used as food components because of their functional properties or nutritional contribution. Much work has been done in recent years on the recovery of protein from the whey produced in large quantities as a byproduct in cheese making. Active research and development work is taking place in the use of milk proteins as components of foodstuffs in order to obtain new outlets for dairy products.

Another source of animal protein is the blood obtained as a byproduct from slaughterhouses; this is of course used as a constituent of animal feed but work was carried out during the war years, and more recently, to obtain materials suitable as food ingredients.

A very considerable research and development effort has been put into the preparation of fish meal in flours for human use. The main research investment has been in the United States, but there have been activities sponsored by private enterprise, by governments and by UN technical development agencies in several other countries. The objective has been to obtain good-quality protein from fisheries material, particularly material that would not normally be directly acceptable for human use. A colourless bland 80 per cent concentrate from fish protein has been produced and is available for use in food preparations.

Table 6. *Protein products from animals and micro-organisms*

Commodity	Protein for human use	Other major food or feed products	Non-food products
Milk	Dehydrated skimmed milk, casein, whey protein	Butter, cheese, etc., whey (as feed)	Fibres, mouldings
Fish	Fish meal or flour	Fish meal (feed)	Fertilisers
Blood	Purified protein	Animal feed	Fertilisers
Meat	Fresh meat and manufactured products	Petfoods and meat and bone meal (feed)	Tallow, hides, etc.
Eggs	Manufactured products, albumin	—	—
Fungi, bacteria	(Under development)	Feed concentrates	—

We would expect these proteins from land animals or fisheries sources to be more expensive than plant proteins, but much depends on the economy of the sector of industry and in particular on the nature of the main products and byproducts from the various operations. The contribution of the newer techniques of extraction and processing may lie in the better utilisation of protein now wasted or diverted from human use.

Soya protein preparations

Soya has been consumed for centuries in East Asia in both fermented and unfermented forms, and even in this country soya is no newcomer. It has been used in the form of full-fat soya flour or defatted soya flour in a very wide range of baked goods and convenience foods for around forty years. The flours confer water- and fat-binding properties on baked goods and meat products and impoved shelf-life in, for example, bread.

It is true, nevertheless, that until quite recently the huge world production of soyabeans was geared to the production of edible oil with defatted soyabean meal as a byproduct utilised primarily by the animal feeds industry. A new industry, based on the processing of soyabeans into a variety of new protein products, has now been created (Table 7).

Table 7. *Protein from soyabeans*

Successive products	Protein content (%)	Overall protein yield (%)
Soyabeans	43	100
Dehulled beans (+hulls)	40	86
Defatted meal (+soyabean oil)	50	86
Isolate (+fibre, soluble carbohydrate, protein and peptides)	95	60*
Fabricated products	40–60	60

* This implies a 70 % yield of protein isolate from the original defatted meal. By comparison, if the defatted meal is used as an animal feed it will produce 6 parts of beef – a yield of 7 % only.

Extraction of oil

The upper part of the table outlines the traditional process of extraction of edible oil from soya to give a defatted protein fibre residue containing about 50 per cent protein. If this is to be used as human food, as distinct from animal feed, the extraction process must be carefully controlled – for example to avoid excessive heat treatment which can lead to 'waste' through the deterioration of the biological value of the protein. The heat treatment must, however, be adequate for the destruction of anti-nutrients and enzymes. In general solvent extraction is preferred to the older expressing techniques; the solvents must be chosen with a view to their health safety and their ease of removal from the product.

Fractionation

The fractionation of the defatted material follows, on the whole, classical methods of separation worked out at the beginning of the century, but now refined and operated on an industrial scale. The purification procedures involve the removal of fibre and other insoluble and soluble carbohydrates and of some soluble proteins.

Processing

The soya protein flour, concentrate or isolates can be texturised by extrusion and other processes or turned into spun products. The texturised products are used as additives in a variety of plant and animal foodstuffs; spun products to be used as meat analogues are formulated into composite material which may include fat, vitamin and mineral components, together with flavouring and colouring materials.

The industrial production and use of soyabean concentrates, isolates and texturised forms of these is essentially a development of the past five years (Meyer, 1970). A whole range of fabricated meat-like, or cheese-like foods have become possible by exploiting techniques for the conferment of texture.

Proteins from other plant sources

An earlier paper in this symposium (Pirie, 1973) has reviewed in some detail the potentialities of various crops for protein production. It will therefore be sufficient to note here the current industrial position and likely developments (Table 8).

Table 8. *Protein products from plants*

Commodity	Protein for human use	Other major food or feed products	Non-food products
Oilseeds	Flours, concentrates, isolates	Edible oil, animal feeds	Drying oils, soap, fibres, phospholipids, pigments
Legumes (other than oilseeds)	Flours, concentrates, isolates	Animal feeds, fermentation products	Starches
Cereals	Protein-enriched cereals (air classification), extracts	Bread and flour products, fermentation products, animal feeds	Starches, adhesives
Chlorophyll-containing material (e.g. leaves)	Concentrates	High-fibre feed (ruminants), low-fibre feed (monogastric animals)	Fibres, pigments

Protein of adequate quality is also available from a variety of oilseed crops, and can be extracted from the defatted meal. Apart from soyabeans realistic possibilities include groundnut, cottonseed, sunflower, sesame and rapeseed. To a large extent the choice of oilseed suitable for processing in a particular country will depend on indigenous supplies. So far as the United Kingdom is concerned rapeseed may well prove to be a future standard commodity (as it is already in most of Northern Europe and in Canada) serving as raw material for protein processers (Aylward, 1971).

There is no need to confine our interests to conventional oilseed crops. Protein of good quality is also present in wheat and maize, in particular among the cereals, though levels are lower than in the oilseed group. Already there is a considerable business in the food industry in the extraction of protein from wheat (i.e. wheat gluten) and from maize in countries such as the USA.

Turning to the non-fatty legumes, a particularly interesting recent development is in the exploitation of field beans, with a rapidly growing acreage under cultivation in the United Kingdom. Protein is extracted from field beans, by processes similar to those used for extraction from defatted soya flour, and converted to spun fabricated material. The considerable carbohydrate residues can be employed as a substrate for a fermentation process which produces a second type of protein from a fungal micro-organism. The integration of the two processes represents a very elegant method of using to maximum advantage an agricultural crop readily produced in Britain.

Pirie (1973) has reviewed the work going on in various countries regarding the position of leaf protein. Much of the current activity has been stimulated by Pirie's earlier investigations; the older work was concerned with potentialities of leaf protein in developing countries and there is widening interest in this problem. Over the past two years there has been an alteration in the position. Several US universities and research centres have research programmes on alfalfa (lucerne) and other chlorophyll-containing material and over the past few months at least two firms in Britain have initiated research work.

The potential industrial sources of leaf protein include grasses, leaves generally and also green vegetable residues from other processes such as garden pea production. Commercial success in the use of leaf protein may well depend, as with oilseed and other systems, on the simultaneous production of several different food, feed or other materials.

EFFICIENCY OF CONVERSION

Utilisation of foodstuffs

In Table 4 attention was drawn to the various stages at which loss of protein can occur. If we are concerned primarily with protein as food for man some comments must be made on the reaction of man to the food available and the efficiency of utilisation within the body.

Food acceptability

It is a truism that no food is of nutritional value unless it is actually consumed. By this criterion a large proportion of the plant protein produced through agronomy and a significant proportion of the protein produced in domesticated animals is of no direct value. With such materials a major contribution of the food industries is to make available proteins that would otherwise be rejected by man, and used as animal feed or diverted to other purposes.

A parallel situation exists in respect of edible oils where a very large proportion of the fat of oilseeds is not normally consumed by man; it is consumed when, after oil milling or extraction, hydrogenation and other industrial procedures, it is made available as margarine and in the form of other edible fats. The margarine and related industries, during their growth over the past 100 years, have added greatly to national and international supplies of fats for direct human use and have provided alternatives to limited supplies of depot animal or milk fats.

Digestibility and utilisation in the body

A second nutritional point in relation to plant proteins is the efficiency of utilisation of proteins in the foods that are eaten. This problem is illustrated by the long-drawn out and continuing controversies about the digestibility of wholemeal bread as distinct from bread made from flours of different extraction rates. The complexity of this question is shown in the monograph by McCance & Widdowson (1956). From the many investigations carried out it can be concluded that whereas some proportion of roughage is an essential part of a normal diet, there are upper limits, which vary with individuals and with commodities, in the amounts that can be tolerated and there may be reductions in the utilisation of proteins and other nutrients. Thus within certain limits, and for certain commodities (including many legumes) some form of processing, over and above simple cooking procedures, may be important in increasing the efficiency of utilisation of the crop protein. And such identical procedures may be especially important with legumes and other crops which contain significant amounts of enzyme inhibitors or other toxic factors.

Protein quality

A third point of importance concerns the *quality* of the protein, in particular the quantities of the different essential amino acids present, and their proportions in relation to one another and to the non-essential amino acids. Many of the plant proteins are of good biological value but differ in their contents of essential amino acids such as lysine, cystine and methionine. It is well accepted that a mixture of plant protein from different sources may be better than a protein from any one source. The newer processing techniques enable such mixtures to be readily made to produce high-quality protein foods.

Theoretically supplementation can be carried out with the synthetic amino acids which have now become commercially available; this form of supplementation is used in animal feed, but there are differences of opinion about the application of this procedure to food for man.

When textured or other isolated plant proteins are to be used as components of other foods, questions of supplementation may not be of great importance. If, however, the commercially produced material is likely to become a major item in the diet as a food in its own right, or if the food is to be used for infants or for dietetic purposes, then formulations must take into account the need for supplementation with vitamins, minerals and accessory factors. In general this can readily be done by the food manufacturer on the basis of older experience with bread, margarine and other products. Provisional standards for various protein foods have been worked out recently in the United States on the basis of discussion between the food industries and the Food and Drug Administration. This and allied matters are now under review by the Food Standards Committee of the United Kingdom.

Overall efficiency of conversion

We can summarise the above section by saying that the newer commercial techniques can, if carried out efficiently, make a threefold contribution to protein utilisation from crop sources: (i) in making available crop protein that under normal circumstances cannot be directly available to man, (ii) in increasing the proportion of protein from some foods digested and used in the body, and (iii) in improving, by compounding and supplementative techniques, the quality of protein from any one source. These contributions can be expressed in mathematical and economic terms commodity by commodity and process by process.

There are few published figures available to show the actual efficiency of extraction of protein from different plant sources; the percentage of protein isolated will depend on the commodity, the process and the objectives. It is clearly in the interests of the processor to improve techniques to secure the maximum yield of the product desired for human use, with the knowledge that protein discarded during fractionation procedures will normally have a lower consumer value if used, for example, as animal feed. Thus turning to soya processing outlined in Table 7, it is noted that the defatted soya material contains almost all the protein in the original bean; with declining percentages as one descends the table to obtain more refined materials.

The processing enterprises will be concerned with both efficiency of extraction and also with the economics of the successive unit operations.

In some cases it may be worth while aiming not at high rates of extraction but at more simplified techniques to give a product for human food and a second product for animal feed. This appears to be the philosophy of the groups in the US currently engaged on the utilisation of alfalfa protein (Kinsella, 1970)

Some indication of the efficiency of the new soya processes is obtained from the current market prices, although many different factors are involved in fixing prices, in particular the distribution of costs between various feedstuffs and byproducts, e.g. between edible oil and edible protein from soya. From the evidence available it is clear that weight for weight, the cost of protein from soya is very much less than that of meat or milk protein. This is not surprising since the production of vegetable protein involves one step fewer in the 'food chain' than does that of meat protein. Dried extruded texturised products from soyabean protein (i.e. from defatted flour or from concentrate) are thus very economical in comparison with meat. Even after the elaborate additional processing involved in the production of spun protein products from isolate, and although the cost of the protein component is 2.5 to 3 times as high as that from the cheaper products, the spun products compete effectively with all but low-grade meats. In general we can expect prices to be reduced as greater volumes of material are processed and as the procedure becomes more efficient.

This review has been concerned with the utilisation of the raw materials of agronomy and, to a lesser degree, of animal husbandry. Many parallel considerations apply to products from unicellular organisms for which the efficiency of protein production must be considered in terms of the actual production and 'harvesting' of the raw materials, and of the preparation and processing of the materials for use as animal feed or by man. This point is made because a number of reviews on unicellular organisms have emphasised the high biological efficiency of production, but have failed to admit that at the preparation and processing stage the necessary purification procedures may, with some systems, present considerable difficulties which will add to the expense of the final products if designed for human use.

With agricultural systems, the overall efficiency of the processes must be calculated in terms of the complex of factors listed in this review. Practical decisions in any country will depend on agricultural resources available, on processing facilities and the types of products required. The success of the soya products points the way to the successful utilisation of other locally more important crops which should make a contribution to protein food supplies both in industrialised and in developing countries.

In this country the changing situation will involve close co-operation between the farm and the food manufacturer, and between agriculture and food scientists (Aylward, 1970, 1971).

REFERENCES

ALTSCHUL, A. M. (1958). *Processed Plant Protein Foodstuffs*. New York: Academic Press.
AYKROYD, W. R., & DOUGHTY, J. (1966). *Legumes in Human Nutrition*. Nutritional Study Number 19. Rome: Food and Agriculture Organisation.
AYLWARD, F. (1969). Synthesis, preparation and presentation in food. In *Chemistry and Industry in the 1990s*. London: Society of Chemical Industry.
AYLWARD, F. (1970). The food agriculture complex: interrelations between agriculture and the food industries. *Chemistry and Industry*, 1362–6.
AYLWARD, F. (1971). *The Influence of Food Science on the Future Pattern of Agriculture*. Seale-Hayne Agricultural College, Newton Abbot, Devon.
BOOHER, L. E. (1948). Economic aspects of food protein. In *Protein and Amino Acids in Nutrition*, Ed. Sahyun, M. New York: Reinhold Publishing Corporation.
HAECKER, T. L. (1920). University of Minnesota Agricultural Supplement Study Bulletin Number 193.
HOLLINGSWORTH, D. (1971). *Protein Needs and Demands in the United Kingdom*. British Nutrition Foundation Information Bulletin Number 6.
KINSELLA, J. E. (1970). Evaluation of leaf protein as a source of food protein. *Chemistry and Industry*, 550.
LEARMONTH, E. M. (1956). Soya in the field of nutrition. A review of existing knowledge. *Chemistry and Industry*, 360–7.
LEARMONTH, E. M., & WOOD, J. C. (1962). Some aspects of soya in food technology *Proceedings of 2nd International Congress of Food Science and Technology, London*.
LIENER, I. E. (1969). *Toxic Constituents of Plant Foodstuffs*. New York: Academic Press.
MCCANCE, R. A., & WIDDOWSON, E. M. (1956). *Breads White and Brown: Their Place in Thought and Social History*. London: Pitman Medical Publishing Company Limited.
MCCANCE, R. A., & WIDDOWSON, E. M. (1969). *The Composition of Foods*. London: HMSO.
MARKLEY, K. S. (1950). *Soybeans and Soybean Products*. New York: Interscience.
MEYER, E. W. (1970). Soya protein isolates for food. In *Proteins for Human Food*, Ed, Lawrie, R. A. London: Butterworths.
ODELL, A. D. (1969). Spun soy-protein fibre. *Chemical Engineering*, 1 Dec. 80–5.
PIRIE, N. W. (1973). Plants as sources of unconventional protein foods. *The Biological Efficiency of Protein Production*, Ed. Jones, J. G. W. London: Cambridge University Press.
UNITED STATES DEPARTMENT OF AGRICULTURE (1969). *Synthetics and Substitutes for Agricultural Products. A Compendium*. Economic Research Service Miscellaneous Publication Number 1141. Washington, DC Government Printing Office.
WOODROOF, J. G. (1969). *Peanuts: Production, Processing, Products*. Westport: Avi Publishing Company.

18

ASPECTS OF PROTEIN PRODUCTION BY UNICELLULAR ORGANISMS

By M. T. HEYDEMAN

Department of Microbiology, University of Reading

Unicellular organisms have had a part to play in man's nutrition since prehistoric time, being responsible for the changes in milk and fruit juices which lead to the formation of cheeses and the fermented drinks. In cheeses, however, the proportion of the dry weight actually due to microorganisms is small, and in the drinks there may be none at all. Yet it was yeast discarded as a waste product from the brewing industry which first suggested itself as a nutrient or food additive, first for animals and then for man, from the late nineteenth century onwards (Peppler, 1970). Surplus brewer's yeast is still used (Bunker, 1963), but in this symposium we shall discuss unicellular organisms as main food products rather than as by-products of another process. In World War I, when food was very short in Germany, dried, treated yeast in the form of a powder or little tasteless chips was added to broths to fortify them (Dumont & Rosier, 1969), and yeast was again recommended by the League of Nations in 1938 for distribution to the undernourished regions of the world. The outbreak of World War II stimulated further study of microbial proteins with a view to their use in emergency foods, not only yeasts (Strain, 1940) but also other micro-organisms being considered (Haehn, Gross & Glaubitz, 1940; Skinner & Müller, 1940). After the war, however, the lack of palatability and cost of preparation of unicellular organisms as food led to their relegation to a minor place, both in the rich countries which had no great need of them and in the poorer countries which lacked the means of production. In the mid-1960s unicellular organisms were deemed to have some potential, especially as by that time petroleum was being developed as a substrate, but in practice the rate of production was only a few million pounds annually in the US, from sulphite liquor of the paper industry, while larger scale production, even from very cheap molasses in the tropics, was considered too expensive for animal or human feeding (Scrimshaw, 1966). Discarded whey from the United States cheese industry, containing annually 204×10^6 kg sugar and 36×10^6 kg protein, was known to be usable for the growth of yeasts, but was not used on a major scale (Borgstrom,

1964). Since then, microbiologists have learned how to exploit more efficiently some of the special properties of certain unicellular organisms. Protein from unicellular organisms, termed 'single cell protein', is being produced in rapidly increasing amounts and many aspects have been reviewed (Blaxter, 1970; Bhattacharjee, 1970; Bunker, 1963; Enebo, 1970; Humphrey, 1970; Mateles & Tannenbaum, 1968; Peppler, 1970; Vincent, 1971; Wilkinson, 1971).

In this paper the biological efficiency of unicellular organisms in protein production will be discussed, considering first the organisms available and their particular advantages and utility, second the mechanism and energetics of protein biosynthesis in the organisms, third some problems and finally the biological efficiency of using protein from unicellular organisms.

UNICELLULAR ORGANISMS AVAILABLE FOR PROTEIN PRODUCTION

Four groups of unicellular organisms can be considered for protein production: algae, bacteria, protozoa and the single-celled fungi, the yeasts. The protozoa are nutritionally exacting and mechanically delicate, and seem never to have been considered for protein production. The other three groups each have their advantages.

Algae

The major advantage of algae appears to be their autotrophic nutrition; no organic carbon source need be supplied. However, light is required, so that the culture must be grown over large areas in lagoons. The cell concentration must be balanced between that which is biologically efficient in making use of the incident light and that which is economically efficient for harvesting. Centrifugal harvesting is most efficient with concentrated suspensions, the cost being related to the volume of culture handled rather than the mass of dry matter removed (Golueke & Oswald, 1965), but at high culture density the algal cells shade one another and there may be an increase in respiratory loss while the photosynthetic rate of the culture remains limited by light intensity. The great advantages unicellular algae have over higher plants are to be found in their growth kinetics and their protein content (Vincent, 1971). While the maximum rate of dry matter production is similar for higher plants (50 g m^{-2} per day; Verduin, 1953) and algae (50.4 g m^{-2} per day; Hindak & Pribl, 1968), the mass needed for this maximum is much less for algae, all cells of which photosynthesise, than it is for higher plants of complex structure: the algae thus have a higher specific growth rate. This enables more frequent repetition of the

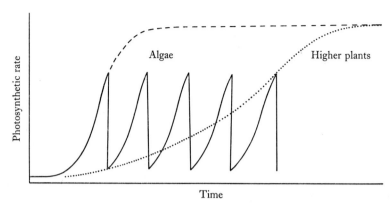

Figure 1. Effect of growth kinetics on seasonal yields of algae and higher plants. Algae are capable of attaining their maximum photosynthetic rate (-----) more rapidly than are higher plants (........), and can thus be harvested and regrown many times in a season. A possible pattern is shown (———), but that adopted would depend on the cost of labour and the importance of area productivity. (Modified from Vincent, 1971.)

harvesting–regrowth cycle of the algae in each season (Figure 1). Potential annual production of *Chlorella* in 5 per cent CO_2 was suggested by Dean (1958) to be 60 to 90×10^3 kg dry matter containing 50 per cent protein from each hectare of culture of mean doubling time 12 hours, comparing most favourably with a yield of 1.5×10^3 kg dried soybeans containing 33 per cent protein. It was suggested that closed systems would probably be needed for this, to maintain the CO_2 concentration, so it was not then seen as an economic possibility. CO_2 gas streams produced from the burning of cheap petroleum fractions can be used, however, in regions with much direct sunlight, warm temperatures and low land value, as in a process developed by the French Petroleum Institute using an algal culture, *Spirulina maxima*, used for centuries as human food in Chad and peculiarly easy to harvest (Tannenbaum & Mateles, 1968). McGarry (1971) has described a type of artificial pond, under construction in Thailand, fed with clarified sewage which could yield up to 360 kg ha^{-1} dried algae per day which, at an estimated 48 per cent protein content, would be over 30×10^3 kg protein annually. This protein content is within the range found by Hindak & Pribl (1968) for *Scenedesmus* and for various filamentous green algae. Oswald *et al.* (1964) found 8.5 per cent nitrogen, equivalent to over 50 per cent protein, in sewage-grown algae; Morimura & Tamiya (1954) made a powder of dried *Chlorella* containing 42 per cent protein. In contrast, even the foliage of higher plants normally contains only 20 to 35 per cent, exceptionally as high as 41 per cent protein

(FAO, 1970). However, the smallest of the Angiosperms, *Wolffia arrhiza*, has alga-like growth kinetics, with a harvesting–regrowth cycle time of 3–4 days, and is estimated by Bhanthumnavin & McGarry (1971) to be capable of yielding 8.4×10^3 kg ha^{-1} dry matter, 1.6×10^3 kg ha^{-1} protein, per nine month growing season of the edible form, which has the advantage of being a traditional food of Burmese, Laotians and Siamese.

Algal protein contains substantial amounts of the essential amino acids, the limiting component being the sulphur-containing amino acids with a chemical score (Mitchell & Block, 1946) referred to hen's egg of 41 for *Scenedesmus*, 49 for *Chlorella*, 32 for *Spirulina* (FAO, 1970) and 36 for a *Chlorella + Scenedesmus* mixture (calculated from McGarry, 1971). Another factor determining the value of protein is digestibility, found by Hindak & Pribl (1968) to depend very much on the species used and the method of preparation, but ultimately the proof of the protein is in the utilisation. This may be expressed by the protein efficiency ratio (PER) (FAO/WHO, 1965), and instanced by the studies of Cook, Lau & Bailey (1963) (Table 1). Uncooked algae had a PER of 1.61, while various mixtures of cereals, non-fat dried milk and algae were found to give PERs close to that of non-fat milk alone. This order of utilisability is substantiated by other workers (McGarry, 1971), while Lubitz (1963) gives a PER of 2.19 for *Chlorella pyrenoidosa* alone. Algae, especially *Scenedesmus*, of larger cell size than *Chlorella* and therefore more easily harvested, are already in use, grown in California, Czechoslovakia, Japan and the USSR in units of up to 10^6 litres (Bhattacharjee, 1970). Vincent (1971), however, adduces evidence that *Chlorella* is poorly digested and actually harmful in the diet of non-ruminants, including man.

Bacteria

From the point of view of this symposium, the apparently high protein content of bacteria is important. Protein as a percentage of dry weight ranges from 20 to 45 for moulds, 40 to 60 for yeasts, 30 to 60 for algae and 50 to 75 for bacteria. Such estimates of protein concentration are usually based on nitrogen determinations, sometimes on amino acid content, of the dried organisms, and can be upset by contributions to these values from non-protein sources. In bacteria a significant proportion of the nitrogen is in amino sugars and nucleic acids rather than amino acids, and some of the amino acids are combined not in protein but in the peptidoglycan of the bacterial cell wall where they may be inaccessible to animals. Accordingly one may conclude that the protein content of bacteria is only marginally, if at all, better than that of other unicellular organisms.

The biological efficiency of bacteria in nature is due to many factors, including their rapid growth rate under favourable conditions and their ability to use a wide variety of substrates. These are factors currently of interest in protein production. The ability of some bacteria to double their numbers or mass in only a few minutes is well known. In an early trial, *Bacillus megaterium* was grown with aeration for 6 hours in a molasses+inorganic salts medium; the yield of bacterial solids after centrifugal harvesting and drum drying was about 50 per cent of the weight of sugar used (Garibaldi et al. 1953). However, the high peptidoglycan and nucleic acid contents of bacteria make them less desirable than yeasts in conditions in which the latter can also thrive. Bacteria are also, because of their smaller size, more difficult than other unicellular organisms to harvest by centrifugation, though other methods such as flotation or flocculation may prove to be efficient.

Table 1. *Composition (% of diet) and protein efficiency ratios (g of gain per g of protein eaten) for various diets fed to rats (Cook et al. 1963)*

Diet number ...	2	3	4	5	6	7	8
Cooked algae	100	—	—	36	55	44	27
Nonfat dry milk	—	100	—	25	26	32	58
Cooked oatmeal	—	—	100	30	10	12	8
Cracked wheat	—	—	—	9	9	12	7
PER	1.85	2.64	2.35	2.45	2.51	2.55	2.75

On carbohydrate substrates, then, yeasts are usually the preferred organisms. Hydrocarbons, first reported to be used by moulds and bacteria, were later shown also to serve as carbon and energy source for some yeasts, which are now being industrially produced from *n*-alkanes. Higher animals are, indeed, also capable of oxidizing *n*-alkanes, but only unicellular organisms can convert them to non-lipid cell constituents, probably via the glyoxylate cycle (Kornberg & Madsen, 1957), the key enzyme of which has been shown to be induced in bacteria by their growth on alkane substrates (Heydeman, 1964). So much heat is produced during this growth on alkanes that an organism able to grow at elevated temperature, perhaps 60 °C, would present substantial advantages in cooling costs; such thermophiles are to be found among the bacteria (Mateles, Baruah & Tannenbaum, 1967) which may thus eventually prove to be the organisms of choice for full-scale production. Coal acids and coal tar aromatic hydrocarbons were found by Silverman, Gordon & Wender (1966) to support the growth only of bacteria; these substrates seem not to have been further investigated. Cheaper and more readily available than *n*-alkanes or coal derivatives, and without the toxicity of such substrates, is methane. Methane is so far known to be used only by bacteria which, although of unusual cytology, have compositions as nutritionally suitable as those of other bacteria (Wilkinson, 1971). There are problems in apparatus design and efficiency of substrate utilisation to be overcome, but these bacteria could then provide protein.

Another property peculiar to the prokaryotic organisms, bacteria and blue-green algae, is the possession of nitrogen fixing ability. The distribution and character of nitrogen fixation are reviewed by Postgate (1971), who lists many free-living organisms. There is some evidence for the ecological significance of these, while one strain appears to fix nitrogen actually in the intestines of animals and man, making good the apparent deficit found in the consistent negative nitrogen balances of the sweet-potato-eating people of the New Guinea highlands (Bergersen & Hipsley, 1970). During nitrogen fixation much substrate is diverted from biosynthesis, making commercial growth of free-living nitrogen fixers unattractive. The importance of nitrogen fixation in ecosystems and agriculture is discussed by Stewart (1966), but is outside the scope of this paper.

Fewer data are available for assessment of protein quality of bacteria than of algae or fungi. Analysis of bacteria shows some to have a high lysine content, suggesting their utility in supplementing cereal-based diets (Humphrey, 1970), but it must be noted that the presence of an amino acid does not guarantee its availability (Tannenbaum, Mateles & Capco, 1966). Comparatively high methionine levels are reported in some hydrocarbon-utilising bacteria (see Wilkinson, 1971). Feeding experiments seem to have been few, but Roberts (1953) showed that *Escherichia coli* grown aerobically in a simple medium could provide a useful protein supplement for rats and chicks, while Ambrose & De Eds (1954) fed dried *Bacillus megaterium* at up to 20 per cent of the diet to rats for 400 days without observing toxic effects.

Yeasts

As Pasteur showed over a century ago, anaerobically yeasts break down sugars more rapidly but grow more slowly than in the presence of oxygen; most of the sugar is converted to the fermentation products alcohol and CO_2. This anaerobic fermentation is, of course, the basis of alcohol production and bread making, but it is to be avoided if the organism itself is the desired product. Aeration is thus important in converting carbohydrate substrates to yeast protein. When n-alkanes are the substrate, a high degree of aeration is again necessary, not to avoid anaerobic fermentation, which is impossible with these substrates, but because of the enormous amount of oxygen required to oxidize them to the level of cell constituents. Long-chain alkanes are essentially $(CH_2)_n$, while carbohydrates are $(CH_2O)_n$: for every 14 g of alkane, 16 g oxygen will be needed. Complete oxidation of the alkane uses about $3\frac{1}{2}$ times its weight of oxygen. Accompanying these oxidations is an immense output of heat, and the agitation, aeration and cooling required for it are responsible for a substantial proportion of the production cost (Humphrey, 1970).

Yeasts are highly efficient producers of protein, as will be more fully discussed in the next two contributions to this symposium. They have the advantage over the bacteria of being more acceptable in the diet and having smaller numbers and proportions of undesirable components. The feeding of significant amounts of yeast is both practicable and biologically useful. Protein content is in the range 38.8 to 70.5 per cent of dry matter (FAO, 1970), the highest figures being given for yeast grown on petroleum hydrocarbons. In these, as in the algae, sulphur-containing amino acids show the lowest chemical score, but again with suitable mixing of the protein sources of the diet satisfactory results are obtainable, and even brewer's yeast alone has been shown to give a PER of 2.24.

PROTEIN BIOSYNTHESIS IN UNICELLULAR ORGANISMS

Our knowledge of the mechanisms of protein synthesis has been largely gained through studies of unicellular organisms, and it is possible to form an estimate of the biological efficiency of the process in the bacteria at least, with which this section primarily deals.

Mechanism

Protein biosynthesis not only requires amino acids, enzymes and energy, but also is absolutely dependent on the existence, and hence the prior synthesis, of another macromolecule: RNA. Production of this in turn depends on the presence of DNA, in the base sequence of which is carried the coded information which determines the sequence of amino acids and the length of each polypeptide chain of the proteins which the organism is capable of synthesising. The proportions of DNA and RNA and the total content of nucleic acid vary according to the rate and stage of growth, as well as from one organism to another. For bacteria, the DNA:RNA ratios given by Davidson (1965) (Table 2) are in the range of about 1:2 to 1:3, and the total nucleic acid ranges from 8.4 to over 20 per cent of dry weight. Baker's yeast, by contrast, has a DNA:RNA ratio of about 1:13 and a total nucleic acid content of only 4.26 per cent. Of the RNA, over

80 per cent is contained in the ribosomes, most of the remainder is in the form of transfer RNA (tRNA) of which there are about 40 species in the organism, and the rest is messenger RNA (mRNA). Protein is synthesised in the ribosomes, the other types of RNA also participating (Mahler & Cordes, 1971).

Table 2. *Nucleic acid contents of bacteria and baker's yeast*

Nucleic acid content as per cent dry weight (Davidson, 1965)

	total NA	RNA	DNA
Staphylococcus	11.57	8.75	2.82
Escherichia coli	12.37	8.59	3.78
	15.76	11.47	4.29
B. aertrycke	8.40	5.40	3.00
Strain S_1 (5 h)	21.40	15.30	6.10
(20 h)	13.10	8.70	4.40
Baker's yeast	4.26	3.95	0.31

An enzyme specific for both tRNA and amino acid joins the two to make amino-acyl-tRNA (aa-tRNA):

$$\text{amino acid} + \text{tRNA} + \text{ATP} = \text{aa-tRNA} + \text{AMP} + \text{PP}_i \quad (1)$$

$$\text{ATP} + \text{AMP} = 2\text{ADP}. \quad (1a)$$

Overall, two molecules of ATP are converted to ADP by this system.

mRNA is synthesised, as are the other varieties of RNA, on the surface of DNA. Each group of three bases, or codon, on the mRNA corresponds to a particular species of amino acid, selected by the formation of hydrogen bonds between the codon and a region, the anti-codon or nodoc, of the appropriate aa-tRNA. Special codons signal the initiation or termination of polypeptide synthesis.

Initiation of polypeptide synthesis involves the association of the smaller subunit of the ribosome with an appropriate region of mRNA, formylmethionyl-tRNA, GTP, at least three specific protein factors and, finally, the larger ribosomal subunit. During the process GTP breaks down to $\text{GDP} + P_i$. There then occur many cycles of elongation of the nascent polypeptide, each involving the sequence of operations shown in Figure 2, with the overall effect that for each amino acid added to the chain two molecules of GTP break down to $\text{GDP} + P_i$. When a termination codon of the mRNA reaches the ribosome the complete polypeptide chain is released. There is no evidence for any specific or energy-requiring process involved in the subsequent organisation of the polypeptide: it seems that

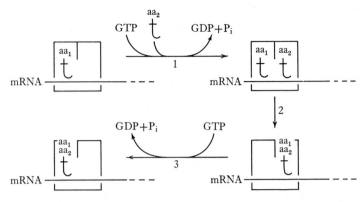

Figure 2. Schematic representation of the cycle of ribosomal events in peptide chain elongation: 1, uptake of the next aa-tRNA; 2, peptidyl transfer; 3, translocation of mRNA and peptidyl-tRNA with respect to the ribosome.

the primary structure of amino acid sequence spontaneously acquires the secondary, tertiary and quaternary structure characteristic of the native protein (Epstein, Goldberger & Anfinsen, 1963).

Energetics of protein biosynthesis

For growth of the polypeptide chain there is an energy requirement, in terms of ATP, of four ATP for each amino acid added: two in forming aa-tRNA (equations 1, 1a) and two for regeneration of GTP used in ribosomal events (Figure 2). If any of the machinery of protein synthesis should be labile and require frequent resynthesis, this too could contribute significantly to the running cost of the mechanism.

It is known that mRNA is labile, being hydrolysed to mononucleotides from which it must again be resynthesised. The energy requirement for this resynthesis is two ATP for each nucleotide added to the RNA chain, or six ATP per amino acid codon. If the mRNA is used n times before degradation, the additional cost for its turnover is $6/n$ ATP for each amino acid incorporated into protein. Watson (1970) estimates that the rate of polypeptide synthesis in rapidly growing *Escherichia coli* at 37 °C approaches 30 amino acids per second, and the interval between ribosomes is of the order of 80 nucleotides, or 27 codons, along the mRNA. Each codon thus participates in protein synthesis approximately once per second. Even in a messenger with as short a life as 75 s (Jacquet & Kepes, 1969) n is thus 75, and the energy cost of mRNA, at $6/75 = 0.08$ ATP, only 2 per cent of that for peptide chain elongation.

If we assume that unicellular organisms can make 38 ATP during the complete oxidation of one molecule of glucose (equation 4, below), then to form one mole of peptide bond, using four ATP, requires the burning of 19 g glucose.

PROBLEMS IN THE USE OF UNICELLULAR ORGANISMS FOR PROTEIN NUTRITION

Some features of unicellular organisms make them less than ideal as foods. All those considered above have a cell wall consisting largely of β1–4 linked carbohydrates in the form of cellulose, chitin or peptidoglycan. These are in themselves indigestible by the animals intended to be fed on them, and could impede the digestion and absorption of the protein within the microbial cell. It may be desirable to rupture mechanically the walls of the organisms. Tannenbaum, Mateles & Capco (1966) used a high-pressure homogenising technique to break *Bacillus megaterium*, obtaining a water-soluble fraction containing high-quality protein, as judged by chemical analysis of amino acids. Rogozin, Mamcis & Val'kovskij (1970) used a procedure involving acetone drying, grinding and extraction with sodium chloride solutions to obtain a protein said to be closely similar nutritionally to such proteins as actin and casein. Mitsuda *et al.* (1970b) found that 2 per cent NaOH extracted 54 per cent of the dry matter from alkane-grown *Candida*, including 75 per cent of the original protein which they then isolated by isotonic precipitation and lyophilisation. However, in order to remove odour and precipitate wall substances, they introduced a dialysis step, and found that further treatment with ethanol, calcium chloride and a second dialysis resulted in a purer protein (Mitsuda *et al.* 1970a). Shami, Syed & Shah (1970), after harvesting *Bacillus subtilis* grown on hydrocarbons, go so far as to make a cell hydrolysate which they show to be a source of fourteen amino acids. Such processing must, however, increase cost and reduce the efficiency of production. A promising line of investigation seems that of Davies (1971) who is studying mutants of *Chlamydomonas* which have lost their ability to synthesise cell walls, but can still be grown and harvested as easily as wild type cells.

The nucleic acid content of unicellular organisms tends to be high (Table 2), but varies with growth rate; this is further discussed by Worgan (1973). Man and the higher apes lack uric acid oxidase and therefore find deleterious the feeding of large amounts of purines. The safe level of nucleic acid intake by man has been suggested to be 2 g per day (Waslien, Calloway & Margen, 1968; Edozien *et al.* 1970). This corresponds to about 30 g per day of the yeast used by Edozien *et al.* (1970). Men fed 90 to 135 g per day developed unacceptably high uric acid levels. Two methods of avoiding this problem are apparent: by reducing the original nucleic acid content of the organisms or by removing it later from the harvested product. Some of the methods already mentioned for extracting protein will separate it from the nucleic acid, but Maul, Sinskey & Tannenbaum (1970) describe a simple three-step heating process which removes about 80 per cent of the nucleic acid from *Candida utilis* yeast, the final product containing 1.0–1.5 per cent nucleic acid and 50 per cent protein. A man should be able to tolerate eating this at over 130 g per day. The mechanism of the heat treatment process may be the disintegration of the ribosomes with activation of enzymes, followed by digestion of the nucleic acid by these enzymes and finally extraction of the products of nucleic acid hydrolysis.

Other problems in the acceptance of microbial foods for direct human consumption include the finding that dried algae, consumed at 150 g per day, gave oedema of face and hands in two out of five subjects tested, possibly an allergic reaction (Kondrat'ev *et al.* 1966). *Hydrogenomonas eutropa*, though tolerated by animals including chimpanzees, produced nausea and vomiting in most people tested, though the results varied from one test to another; there was also the possibility that resistance might be developed by the chronic eater, or that slight variation of cultural or preparative conditions might reduce the level of toxin in the bacteria (Waslien, Calloway & Margen, 1969).

At a still earlier stage of consumption one finds the problem that foods containing unicellular organisms tend to be unpalatable. Cook, Lau & Bailey (1963) found only small amounts of algae were tolerable, to a maximum of 8 per cent in peanut butter cookies, and Morimura & Tamiya (1954) found that up to 6.5 parts *Chlorella* powder could be added to 100 parts wheat flour without adversely affecting the taste to the Japanese palate. Note, however, that even these small supplements increase the protein content of the resultant foods by 25 and 20 per cent respectively. When algae have been harvested by the alum flocculation process, they may prove unpalatable even to animals (Oswald, 1964), but can still be disguised.

Other methods of harvesting the minute organisms are generally expensive, involving centrifugation of enormous masses, but *Spirulina*, because of its morphology, forms clumps or mats which are readily recovered using a simple filter, making it an economically more attractive proposition than the coccal algae (Enebo, 1970). The filamentous algae, in which the organisms are attached to one another in trichomes often of great length, would also be easy to harvest. They contain about as much protein as the coccal algae and are more digestible (Hindak & Pribl, 1968). The final problem is that of cultivation of the organisms, where there is always a risk, especially in large-scale work, that a contaminant organism of unknown effect may have entered the reaction vessel.

THE BIOLOGICAL EFFICIENCY OF USING PROTEIN FROM UNICELLULAR ORGANISMS

Man requires protein for his own food and that of his animals. The protein is digested again to amino acids, some of which will be more useful than others in supplementing the diet. Is not the elaboration of the protein a waste of energy? We can estimate from known pathways of metabolism (Mahler & Cordes, 1971) the amount of glucose which might be used in producing each amino acid. Two examples only are given here: tryptophan (MW = 204) with a long biosynthetic pathway, and alanine (MW = 89) with a comparatively short pathway. Figure 3 represents the biosynthetic pathway of tryptophan from glucose in oversimplified fashion, ignoring many other reactions linked with it but giving a good indication of the overall reaction summarised in equation (2).

Glucose requirement for amino acid synthesis

$$2G + 2NH_3 + 3NADPH + ATP \rightarrow Tryptophan \qquad (2)$$
$$G + 12NADP^+ + ATP + 5G\text{-}6\text{-}P \rightarrow 6Ru\text{-}5\text{-}P + 6CO_2 + 12NADPH \qquad (3)$$
$$38ADP + G + 6O_2 \rightarrow 6H_2O + 6CO_2 + 38ATP \qquad (4)$$
$$2.25G + 2NH_3 + 1.25ATP \rightarrow Tryptophan \quad \text{(from 2, 3)} \qquad (5)$$
$$\underset{410g}{2.28G} + 2NH_3 \rightarrow \underset{204g}{Tryptophan} \quad \text{(from 4, 5)} \qquad (6)$$
$$\underset{85g}{0.47G} + NH_3 \rightarrow \underset{89g}{Alanine} \qquad (7)$$

The main source of reduced NADP is the hexose-monophosphate pathway, represented by equation (3), while ATP production from glucose under aerobic conditions may be summarised as equation (4). Equations (5) and (6) are derived from equations (2) and (3) and from equations (4) and (5) respectively, showing that 410 g glucose will be used in making 1 mol, 204 g, tryptophan. Analogous but simpler operations show that 85 g glucose are needed to make 1 mol, 89 g, alanine. Compared with either of these amounts of glucose, the 19 g further required for peptide bond formation is significant, representing 4.6 and 20 per cent respectively of the requirements for tryptophan and alanine synthesis, and its elimination would represent an increase in production efficiency. Furthermore, if free amino acids are being produced, cell wall and nucleic acid material will not need either to be synthesised by the organism or to be disposed of by the

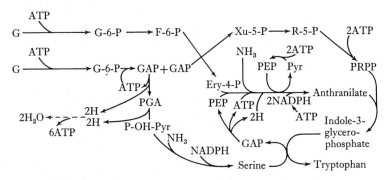

Figure 3. Biosynthesis of tryptophan from glucose and ammonia. Input and output substances are shown in heavy type; other substances are recycled or intermediate. GAP = glyceraldehyde-3-phosphate; Pyr = pyruvate; PEP = phospho-enol-pyruvate; PRPP = 5-phosphoribosyl-1-pyrophosphate; NADPH = reduced nicotinamide-adenine dinucleotide phosphate; P-OH-Pyr = phosphohydroxypyruvate; PGA = 3-phosphoglyceric acid; Xu = xylulose; R = ribose; F = fructose; G = glucose; Ery = erythrose.

consumer. Forrest (1970) has pointed out that the major consumption of energy by cells is in correct placement of molecules within the cells into organised groups or structures; if all the ATP available from catabolic processes were used strictly for chemical syntheses, up to ten times as much cellular material could be synthesised. Here is scope for improving biological efficiency for our own purposes. Finally, if amino acids rather than proteins are our products we need not be content with only a few per cent of an essential amino acid: it will usually be pure amino acid that we recover in good yield.

Production of certain amino acids by micro-organisms from carbohydrate substrates has been known for some years; 90 per cent of the 113×10^6 kg per annum monosodium glutamate (Demain, 1971) and virtually all the lysine (McPherson, 1966) is made by fermentation. The latter, being the limiting amino acid in most of the world's cereal crop, is particularly significant and can be produced by this method at less than 90 pence kg^{-1}, the yield being 25 per cent of consumed glucose (Demain, 1971). Extreme vegetarian diets are also liable to shortage of threonine and tryptophan. Threonine production by unicellular organisms may be economic in the near future. Here our knowledge of the fundamental principles of metabolic regulation in *Escherichia coli* is being put to practical use. Some stages in threonine production are inhibited or repressed by the metabolically related methionine and lysine (Figure 4), and a mutant unable to synthesise these amino acids will, when grown in culture in which they are limiting, overproduce threonine to the extent of 4 g l^{-1}. An alternative approach has been to produce a triple mutant in which, firstly, the normally threonine-inhibited homoserine dehydrogenase (step 3 in Figure 4) is no longer sensitive to threonine inhibition. Secondly, it is blocked at point *a* of Figure 4, so that isoleucine, a co-repressor of steps 1, 3, 4 and 5, is no longer made from threonine. This double mutant produced 4.7 g l^{-1} threonine. Finally the organism was mutated to methionine auxotrophy by blocking at point *b* of Figure 4, permitting production of over 6 g threonine l^{-1} (Shiio & Nakamori, 1969). A still more useful organism has now been obtained by Shiio & Nakamori (1970) – a single-step mutant of *Brevibacterium flavum* which produces up to 14 g l^{-1} threonine. The control mechanisms involved in threonine biosynthesis in this species are under investigation (Miyajima & Shiio, 1971).

Tryptophan production by direct fermentation is not yet developed, but yeast of greatly increased free tryptophan content has been obtained by Sobczak & Majchrzak (1970) by adding a key precursor, anthranilic acid (Figure 3), to the culture medium: 0.082 per cent addition increased free tryptophan five-fold, the total in the culture going up from 0.69 to 1.24 per cent of total dry weight recovered.

There are good prospects for the production of any or all of the amino acids directly from cheap substrates, including hydrocarbons. It is suggested, however, that there may be some nutritional advantage in supplementing with amino acids in peptide chains, as in proteins, rather than with the free amino acids (Altschul, 1965).

Another aspect of biological efficiency concerns the use of animals as intermediates in our use of protein from unicellular organisms. Not only is there a high percentage loss during this passage, but also we are introducing a slow step into the food-making process. It has been pointed out by many writers (e.g. Vincent, 1971; Bhattacharjee, 1970) that one of the advantages of micro-organisms is their high specific growth rate. If animals are to be used at all, it could be that the protozoa may yet have a part to play in the production of protein. These unicellular creatures have the composition of animals, lacking the cell wall of the algae, yeasts and bacteria. They find bacteria palatable, and seem able to digest them efficiently. A major problem in their use might be that aseptic techniques are needed to exclude contaminant organisms from protozoal cultures, but there seem to be possibilities worthy of investigation.

To be efficient, the protein must, when ready, be acceptable to the consumer and here there seem to be real problems, although the product from unicellular organisms does not differ fundamentally from fish and vegetable protein concentrates (Johnston & Greaves, 1969). Local customs and resources must be considered in developing particular solutions to

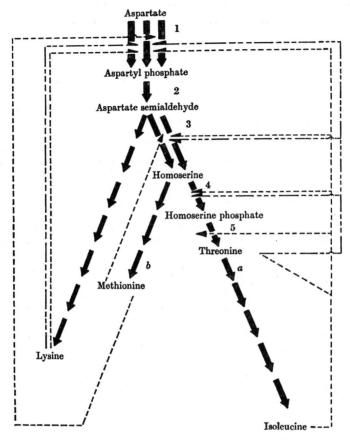

Figure 4. Biosynthesis of the aspartate family of amino acids in *Escherichia coli*. Feedback regulation of only the five reactions from aspartate to threonine is shown. Note that there are three isoenzymes for step 1 and two isoenzymes for step 3. — — — indicates repression; — · · — · · — feedback inhibition; *a* and *b* refer to auxotrophic blocks in a high threonine producer (Demain, 1971).

this problem on a regional or national basis, and Johnston & Greaves (1969) point out that the introduction of a new, protein-rich food as a commercial product for sale has been more successful than its free distribution.

In the long term, as Borgstrom (1964) has pointed out, heterotrophic micro-organisms cannot be depended on as a source of food; in the immediate future, as the next two contributions to this symposium will make clear, they can help us greatly. As our fundamental knowledge increases, we can expect increasingly to harness the biological efficiency of unicellular organisms, so making the most of our resources not only in space travel (Oswald, Golueke & Horning, 1965), where these are obviously limited, but also on the spaceship Earth which we inhabit.

REFERENCES

ALTSCHUL, A. M. (1965). *Proteins: Their Chemistry and Politics*, pp. 315–16. London: Chapman & Hall.

AMBROSE, A. M., & DE EDS, F. (1954). Comparative studies on the toxicity of washed cells of *Bacillus megaterium* and brewer's yeast. *Journal of the American Pharmaceutical Association* **43**, 185–7.

BERGERSEN, F. J., & HIPSLEY, E. H. (1970). The presence of N_2-fixing bacteria in the intestines of man and animals. *Journal of General Microbiology* **60**, 61–5.

BHANTHUMNAVIN, K., & McGARRY, M. G. (1971). *Wolffia arrhiza* as a possible source of inexpensive protein. *Nature, London* **232**, 495.

BHATTACHARJEE, J. K. (1970). Microorganisms as potential sources of food. *Advances in Applied Microbiology* **13**, 139–61.

BLAXTER, K. L. (1970). Animals and micro-organisms as food producers. *New Horizons for Chemistry and Industry in the 1990s*, pp. 67–72. London: Society for Chemistry and Industry.

BORGSTROM, G. (1964). The human biosphere and its biological and chemical limitations. *Starr* (1964), 130–63.

BUNKER, H. (1963). Microbial food. In *Biochemistry of Industrial Micro-organisms*, ed. C. Rainbow & A. G. Rose, pp. 34–67. New York and London: Academic Press.

COOK, B. B., LAU, E. W., & BAILEY, B. M. (1963). The protein quality of waste-grown green algae. I. Quality of protein in mixtures of algae, nonfat milk, and cereals. *Journal of Nutrition* **81**, 23–9.

DAVIDSON, J. N. (1965). *The Biochemistry of the Nucleic Acids*, 5th ed. London: Methuen.

DAVIES, D. R. (1971). Single cell protein and the exploitation of mutant algae lacking cell walls. *Nature, London* **233**, 143–4.

DEAN, R. F. A. (1958). Use of processed plant proteins as human food. *Processed Plant Protein Foodstuffs*, Ed. A. M. Altschul, pp. 205–47. New York and London: Academic Press.

DEMAIN, A. L. (1971). Microbial production of food additives. *Microbes and Biological Productivity*, 21st Symposium of the Society for General Microbiology, Ed. D. E. Hughes & A. H. Rose, pp. 77–101. London: Cambridge University Press.

DUMONT, R. & ROSIER, B. (1969). *The Hungry Future*. London: André Deutsch.

EDOZIEN, J. C., UDO, U. U., YOUNG, V. R., & SCRIMSHAW, N. S. (1970). Effects of high levels of yeast feeding on uric acid metabolism of young men. *Nature, London* **228**, 180.

ENEBO, L. (1970). Single-cell protein. *Evaluation of Novel Protein Products, Wenner-Gren Center International Symposium Series* **14**, 93–103. Oxford, New York, Toronto, Sydney and Braunschweig: Pergamon Press.

EPSTEIN, C. J., GOLDBERGER, R. F., & ANFINSEN, C. B. (1963). The genetic control of tertiary protein structure: studies with model systems. *Cold Spring Harbor Symposia on Quantitative Biology* **28**, 439–49.

FAO (1970). *Amino-acid Content of Foods and Biological Data on Proteins. Nutritional Studies No. 24*. Rome: FAO.

FAO/WHO (1965). *Protein Requirements, Nutrition Meetings Report Series No. 37*. Rome: FAO.

FORREST, W. W. (1970). Entropy of microbial growth. *Nature, London* **225**, 1165–6.
GARIBALDI, J. A., IJICHI, K., SNELL, N. S., & LEWIS, J. C. (1953). *Bacillus megatherium* for biosynthesis of cobalamin. *Industrial and Engineering Chemistry* **45**, 838–46.
GOLUEKE, C. G., & OSWALD, W. J. (1965). Harvesting and processing sewage-grown planktonic algae. *Journal of the Water Pollution Control Federation*, 471.
HAEHN, H., GROSS, W., & GLAUBITZ, M. (1940). Über Versuche neue Mikroorganismen für die technische Eiweissgewinnung dienstbar zu machen, I. *Vorratspflege und Lebensmittelforschung* **3**, no. 1/2, 47–52.
HEYDEMAN, M. T. (1964). Isocitrate lyase in a paraffin-utilizing coryneform bacterium. *VIth International Congress of Biochemistry, Abstracts*, p. 511. Washington, DC: Secretariat, VIth International Congress of Biochemistry.
HINDAK, F., & PRIBL, S. (1968). Chemical composition, protein digestibility and heat of combustion of filamentous green algae. *Biologia Plantarum* **10** (3), 234–44.
HUMPHREY, A. E. (1970). Microbial protein from petroleum. *Process Biochemistry*, June 1970, 19–22.
JACQUET, M., & KEPES, A. (1969). The step sensitive to catabolite repression and its reversal by 3′-5′ cyclic AMP during induced synthesis of β-galactosidase in *E. coli*. *Biochemical and Biophysical Research Communications* **36**, 84–92.
JOHNSTON, B. F., & GREAVES, J. P. (1969). *Manual on Food and Nutrition Policy, Nutritional Studies No. 22*. Rome: FAO.
KONDRAT'EV, J. I. et al. (1966). (Translated title.) Use of 150 g dry material of unicellular algae in human diets. *Voprosy Pitaniya* **25** (6), 14–19. (*Nutrition Abstracts and Reviews* **37**, no. 4475 (1967).)
KORNBERG, H. L., & MADSEN, N. B. (1957). Formation of C^{14}-dicarboxylic acids from acetate by *Pseudomonas* KB1. *Biochemical Journal* **65**, 13–14P.
LUBITZ, J. A. (1963). The protein quality, digestibility, and composition of algae, *Chlorella* 71105. *Journal of Food Science* **28**, 229–32.
McGARRY, M. G. (1971). Water and protein reclamation from sewage. *Process Biochemistry*, January 1971, 50–3.
McPHERSON, A. T. (1966). Production of lysine and methionine. *World Protein Resources, Advances in Chemistry Series* **57**, 65–74. Washington, DC: American Chemical Society.
MAHLER, H. R., & CORDES, E. H. (1971). *Biological Chemistry*, 2nd Ed. New York, Evanston, San Francisco and London: Harper & Row.
MATELES, R. I., BARUAH, J. N., & TANNENBAUM, S. R. (1967). Growth of a thermophilic bacterium on hydrocarbons; a new source of single-cell protein. *Science, New York* **157**, 1322–3.
MATELES, R. I., & TANNENBAUM, S. R. (1968) (Eds.). *Single-Cell Protein*. Cambridge, Mass: MIT Press.
MAUL, S. B., SINSKEY, A. J., & TANNENBAUM, S. R. (1970). New process for reducing the nucleic acid content of yeast. *Nature, London* **228**, 181.
MITCHELL, H. H., & BLOCK, R. J. (1946). Some relationships between the amino acid content of proteins and their nutritive value for the rat. *Journal of Biological Chemistry* **163**, 599–620.
MITSUDA, H., SUGUIRA, M., TONOMURA, B., & YASUMOTO, K. (1970a). Studies on protein foods. 5. On purification of protein isolates from a hydrocarbon assimilating yeast. *Journal of the Japanese Society of Food and Nutrition* **23**, 66–70. (*Nutrition Abstracts and Reviews* **41**, no. 195 (1971).)

MITSUDA, H., SUGUIRA, M., YASUMOTO, K., & TONOMURA, B. (1970b). Studies on protein foods. 4. Protein isolates from a hydrocarbon assimilating yeast and its nutritive values. *Journal of the Japanese Society of Food and Nutrition* **23**, 62–5. (*Nutrition Abstracts and Reviews* **41**, no. 195 (1971).)

MIYAJIMA, R. & SHIIO, I. (1971). Regulation of aspartate family amino acid biosynthesis in *Brevibacterium flavum*. Part IV. Repression of the enzymes in threonine biosynthesis. *Agricultural and Biological Chemistry* **35** (3), 424–30.

MORIMURA, Y., & TAMIYA, N. (1954). Preliminary experiments in the use of *Chlorella* as human food. *Food Technology* **8**, 179.

OSWALD, W. J. (1964). Production of algae for animal feed. *Solar Aeolian Energy, Proceedings of International Seminar, Sounion, Greece, 1961*, 377–83.

OSWALD, W. J. et al. (1964). *Nutritional and Disease Transmitting Potential of Sewage Grown Algae.* SERL Report no. 64–6. Berkeley, Cal.: University of California. Cit. McGarry (1971).

OSWALD, W. J., GOLUEKE, C. G., & HORNING, D. O. (1965). Closed ecological systems. *Journal of the Sanitary Engineering Division, American Society of Civil Engineers* **91**, 23–46.

PEPPLER, H. J. (1970). Food yeasts. *The Yeasts*, Ed. A. H. Rose & J. S. Harrison, vol. 3, pp. 421–62. New York and London: Academic Press.

POSTGATE, J. R. (1971). Relevant aspects of the physiological chemistry of nitrogen fixation. *Microbes and Biological Productivity, 21st Symposium of the Society for General Microbiology*, Ed. D. E. Hughes & A. H. Rose, pp. 287–307. London: Cambridge University Press.

ROBERTS, L. P. (1953). *Bact. coli* as a food supplement. *Nature, London* **172**, 351–2.

ROGOZIN, S. V., MAMCIS, A. M., & VAL'KOVSKIJ, D. G. (1970). (Translated title.) Isolation of nucleic acids and total protein from yeasts. *Prikladnaja Biokhimija i Mikrobiologija* **6**, 638–43. (*Nutrition Abstracts and Reviews* **41**, no. 4849 (1971).)

SCRIMSHAW, N. S. (1966). Meeting world food needs. *Prospects of the World Food Supply*, 48–55. Washington, DC: National Academy of Sciences.

SHAMI, M. D., SYED, B. H., & SHAH, F. H. (1970). Biosynthesis of amino acids from hydrocarbons. *Pakistan Journal of Scientific and Industrial Research* **12**, 370–2. (*Nutrition Abstracts and Reviews* **41**, no. 2317 (1971).)

SHIIO, I., & NAKAMORI, S. (1969). Microbial production of L-threonine. Part I. Production of *Escherichia coli* mutant resistant to α-amino-β-hydroxyvaleric acid. *Agricultural and Biological Chemistry* **33**, 1152.

SHIIO, I., & NAKAMORI, S. (1970). Microbial production of L-threonine. Part II. Production by α-amino-β-hydroxyvaleric acid resistant mutants of glutamate producing bacteria. *Agricultural and Biological Chemistry* **34**, 448–56.

SILVERMAN, M. P., GORDON, J. N., & WENDER, I. (1966). Microbial synthesis of food from coal-derived material. *World Protein Resources, Advances in Chemistry Series* **57**, 269–79. Washington, DC: American Chemical Society.

SKINNER, C. E., & MÜLLER, A. E. (1940). Cystine and methionine deficiency in mould proteins. *Journal of Nutrition* **19**, 333–44.

SOBCZAK, E., & MAJCHRZAK, R. (1970). (Translated title.) Optimum conditions for the biosynthesis of free tryptophan in the cells of *Torulopsis utilis* yeast. *Przemysł fermentacyjny rolny* **14** (10), 11–15. (*Nutrition Abstracts and Reviews* **41**, no. 4851 (1971).)

STARR, M. P. (1964) (Ed.). *Global Impacts of Applied Microbiology*. Stockholm/Göteborg/Uppsala: Almqvist & Wiksell; New York/London/Sydney: John Wiley & Sons, Inc.

STEWART, W. D. P. (1966). *Nitrogen Fixation in Plants*. London: Athlone Press.

STRAIN, H. H. (1940). Extraction of proteins and proteolytic enzymes from yeast. *Compte rendu des travaux du Laboratoire de Carlsberg, Série Chimique* **23**, 149–62.

TANNENBAUM, S. R., & MATELES, R. I. (1968). Single cell protein. *Science Journal*, May 1968, 87–92.

TANNENBAUM, S. R., MATELES, R. I., & CAPCO, G. R. (1966). Processing of bacteria for production of protein concentrates. *World Protein Resources, Advances in Chemistry Series* **57**, 254–60. Washington, DC: American Chemical Society.

VERDUIN, J. (1953). A table of photosynthetic rates under various conditions. *American Journal of Botany* **40**, 675–9.

VINCENT, W. A. (1971). Algae and lithotrophic bacteria as food sources. *Microbes and Biological Productivity, 21st Symposium of the Society for General Microbiology*, Ed. D. E. Hughes & A. H. Rose, pp. 47–76. London: Cambridge University Press.

WASLIEN, C. I., CALLOWAY, D. H., & MARGEN, S. (1968). Uric acid production of men fed graded amounts of egg protein and yeast nucleic acid. *American Journal of Clinical Nutrition* **21**, 892–7.

WASLIEN, C. I., CALLOWAY, D. H., & MARGEN, S. (1969). Human intolerance to bacteria as food. *Nature, London* **221**, 84–5.

WATSON, J. D. (1970). *Molecular Biology of the Gene*, 2nd Ed. New York: Benjamin.

WILKINSON, J. F. (1971). Hydrocarbons as a source of single-cell protein. *Microbes and Biological Productivity, 21st Symposium of the Society for General Microbiology*, Ed. D. E. Hughes & A. H. Rose, pp. 15–46. London: Cambridge University Press.

WORGAN, J. T. (1973). Protein production by micro-organisms from carbohydrate substrates. *The Biological Efficiency of Protein Production*, Ed. J. G. W. Jones. pp. 339–61. London: Cambridge University Press.

19
PROTEIN PRODUCTION BY UNICELLULAR ORGANISMS FROM HYDROCARBON SUBSTRATES

By T. WALKER

BP Proteins Ltd, London

This paper refers only to recent published reports of research into protein production by the fermentation of hydrocarbons. The use of petrochemicals as substrates for protein production has not been considered although the potential of such compounds as methanol, acetic acid and ethanol has been assessed in a number of laboratories.

HYDROCARBON MICROBIOLOGY

The first published report that hydrocarbons were subject to microbial degradation appears to be that of the Japanese botanist Miyoshi in 1895. He observed that certain paraffins were attacked by the fungus *Botrytis cinerea*. The first reports of microbial oxidation of methane, the simplest hydrocarbon, were made about ten years later by Kaserer (1905 a, b) and Söhngen (1905).

Recently Quayle (1967) has reviewed the literature on hydrocarbon biochemistry and he postulated that, for any representative microbial culture collection, about a quarter of the organisms would be capable of utilising hydrocarbons. Moreover, the hydrocarbon-utilising fraction would include bacteria, mycelial bacteria and fungi (including yeasts). Quayle found the literature on hydrocarbon substrate specificity to be generally inadequate, but, on the basis of the data of Fuhs (1961) and Lukins (quoted by Forster, 1962), he made the following tentative comments:

(i) Growth on aromatic, alicyclic and branched-chain hydrocarbons appeared to be less common than on aliphatic normal (*n*-) paraffins.

(ii) The ability to grow on methane seemed to be a comparatively restricted property.

(iii) Observations on two genera, *Pseudomonas* and *Mycobacterium*, both of which have been studied extensively with respect to hydrocarbon substrate specificity, indicated a strong preference for *n*-paraffins containing 9 to 16 carbon atoms.

Evans (1969) stated that whereas micro-organisms could be found to utilise almost any hydrocarbon, those utilising C_{10} to C_{20} n-paraffins showed the greatest potential for industrial development. According to Evans, n-paraffins of C_2 to C_{10} are less interesting to the industrialist on account of their being less liable to microbial attack and also because the specificity of micro-organisms tends to be very great in this range. n-paraffins of C_{20} and higher are acceptable from the purely microbiological aspect but since they are solids at normal fermentation temperatures of 30–35 °C they could not be used directly.

Methane is also considered to have potential as a substrate for industrial protein production. Until recently comparatively few obligate methane-utilising micro-organisms had been isolated (Ribbons, 1967), but Whittenbury, Phillips & Wilkinson (1970) have now isolated more than 100 methane-utilising strains, all of which were bacteria. Moreover, there is other evidence of a general advance in knowledge of methane fermentation.

In practice the two types of hydrocarbon which have received most attention from the point of view of the industrial production of protein, are liquid C_{10} to C_{20} n-paraffins and methane, and these form the basis of this paper.

THE SUPPLY OF HYDROCARBONS

Estimates of the supply and reserves of hydrocarbon substrates vary, but all indicate very large quantities of apparently suitable materials. Lainé (1970) has estimated that on average crude oil contains about 2 per cent weight for weight (w/w) of the appropriate n-paraffins. With world production of crude oil exceeding 2300 million tons in 1970 the current annual production of n-paraffin substrate appears to be about 50 million tons. Assuming a conversion of hydrocarbon to protein of 50 per cent (actual yield data are discussed later) the potential production of protein from n-paraffins alone exceeds 20 million tons per annum. Moreover, with 'Published Proved' oil reserves at the end of 1970 exceeding 84000 million tons the supply of n-paraffin fermentation substrates seems assured for several decades to come.

The supply of methane is more difficult to define. The methane content of natural gas varies from less than 50 per cent (Kapuni field, New Zealand) to more than 95 per cent (West Sole field, North Sea). Vast, unmeasured quantities of natural gas are currently wasted in locations where they cannot be utilised economically. Total utilisation of natural gas in Western Europe, North America, Japan and Australasia amounted to over 600 million tons of crude oil equivalent in 1970. World reserves of natural gas were estimated to be equivalent to about 28000 million tons

UNICELLULAR ORGANISMS–HYDROCARBON SUBSTRATES 325

of crude oil in 1970. Assuming a conversion of methane to protein of 50 per cent and a methane content of, say, 75 per cent in natural gas, the potential production of protein from methane appears to be many times greater than that from the liquid n-paraffins.

It is the enormous potential of hydrocarbons as sources of protein which has attracted so much attention from research workers in a variety of scientific disciplines.

INDUSTRIAL PROCESSES FOR PROTEIN PRODUCTION FROM HYDROCARBONS

Processes utilising liquid n-*paraffins*

The first suggestion of an industrial process for the cultivation of microorganisms on hydrocarbons as a source of protein was made by Champagnat, Vernet, Lainé & Filosa (1963). This group, from the Société Française des Pétroles BP, had been studying microbial growth on certain refinery effluents and also the microbial desulphurisation of oil products. Since then the development by British Petroleum (BP) of two processes for the production of proteins from hydrocarbons has been widely disclosed and the reports by Llewelyn (1967), Evans (1969), Bennett, Hondermarck & Todd (1969) and Bennett & Knights (1970) describe many of the technical principles of the processes.

Both the BP processes employ yeasts. Certain other groups (Malek, 1967; Ivengar, 1967) have also used yeasts but Ko & Yu (1967) have described work with bacteria. According to Wilkinson (1971) the use of bacteria is more difficult, except under aseptic conditions, because of their liability to attack from phage and other lytic predators. Harvesting yeasts is considered to be simpler and less costly than harvesting bacteria which are generally smaller (Wang, 1967). A limiting factor with yeasts is that very few thermophilic strains have been isolated and the normal fermentation temperature for yeasts is 30–35 °C which would necessitate expensive refrigeration of the fermentors in some locations. Thermophilic strains of n-paraffin-utilising bacteria are, however, fairly common and, as an example, Mateles, Baruah & Tannenbaum (1967) described a strain of *Bacillus stearothermophilus* which grew in the range 60–70 °C. Differences between yeasts and bacteria in their composition and nutritional value are described below.

At the present time the two BP processes are probably at the most advanced stage of development. Figure 1, taken from Evans (1969), shows the principal stages in the two processes. The first process uses a substrate

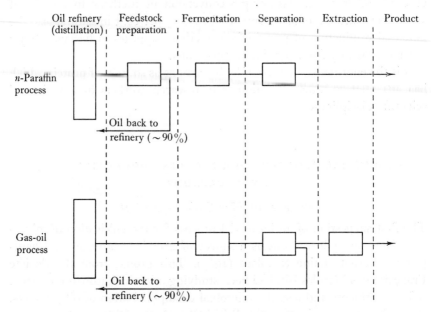

Figure 1. Comparative process flow diagrams.

of high-purity n-paraffins in the range C_{11} to C_{18}, and these are almost completely consumed during fermentation. The second process uses standard refinery heavy gas-oil (i.e. c. 300–380 °C TBP) as substrate from which the n-paraffins are preferentially consumed. In this case only about 10 per cent of the substrate is utilised and the remainder, dewaxed and with reduced cloud and pour points, is recovered for re-use as a component of a range of refinery products. In addition the gas-oil process employs a solvent extraction step to remove all traces of the gas-oil trapped in the yeast cells.

Both processes are truly continuous. Continuous fermentation offers very considerable advantages over batch operation in terms of both costs and close control of the process (Mateles et al. 1967).

Production plants for each of the two BP processes have been built. The first plant using pure n-paraffins is at Grangemouth, Scotland, and has a capacity of 4000 tons of product per annum. The first gas-oil unit has been built at Lavera, France, and has a capacity of 16 000 tons of product per annum. Bennett et al. (1969) made an interesting comparison of manufacturing costs for the two processes and this is shown in Table 1. Broadly speaking, it can be said that the low substrate cost in the gas-oil version of the process is counter-balanced in the other process by the absence of costs for gas-oil separation and solvent extraction of the product. Clearly these values were very preliminary estimates and more reliable data on the costs of manufacture will come from the large-scale production units.

The essential difference between liquid n-paraffins and carbohydrate fermentation is that the latter furnishes some of the elements essential to cell growth, carbon (C), hydrogen (H) and oxygen (O) in a soluble form, whereas a hydrocarbon supplies only C and H in a form which is virtually insoluble in water. Oxygen must then be supplied in greatly increased quantities and in practice this is likely to be from air blown into the fermentor. Other nutrients essential to both substrates are the cations NH_4^+, K^+, Mg^{2+}, Fe^{2+} and Zn^{2+}, and the anions SO_4^{2-} and PO_4^{3-}. Specific growth factors may also be needed to achieve maximum performance. Bennett et al. (1969) expressed the overall conversion of hydrocarbons to yeast cells as follows (in kg mol):

$2nCH_2 + 2nO_2 + 0.19n\ NH_4^+ +$ other essential elements (P, K, S, etc.)
$\longrightarrow n\ (CH_{1.7}\ O_{0.5}\ N_{0.19}\ Ash) + nCO_2 + 1.5n\ H_2O + 200\,000n$ kcal.

Table 1. *A comparison of the estimated process costs for the two BP processes for the production of dried yeasts from hydrocarbons (from Bennett, Hondermarck & Todd, 1969)*

	Process	
Item	n-paraffins	Gas-oil
Substrate (%)	40	13
Minerals and chemicals (%)	18	30
Utilities (%)	18	25
Personnel, maintenance, overheads, depreciation (%)	24	32
Total manufacturing cost (%)	100	100

This may be compared with a similar equation for a carbohydrate feedstock, using the same empirical formula for the cells:

$1.8nCH_2O + 0.8nO_2 + 0.19nNH_4^+ +$ other essential elements (P, K, S, etc.)
$\longrightarrow n(CH_{1.7}\ O_{0.5}\ N_{0.19}\ Ash) + 0.8nCO_2 + 1.3n\ H_2O + 80\,000n$ kcal.

The heat releases correspond to about 7600 and 3000 kcal respectively per kg cell dry weight.

Thus the use of n-paraffins instead of carbohydrates involves the supply of about 2.5 times as much oxygen and the removal of about 2.5 times as much heat. Also, because the two liquid reaction phases are virtually immiscible, sufficient agitation must be provided to disperse the smaller volume hydrocarbon phase thoroughly into the larger volume aqueous phase in order to ensure efficient hydrocarbon mass transfer. The transfer mechanism of the hydrocarbon to the cell has been the subject of much study and speculation. Evans (1969) indicated, from a knowledge of the productivity of liquid hydrocarbon fermentation systems, that a combination of intermediate solution of the hydrocarbon in water, and of direct contact between the droplet of hydrocarbon and the cell, was responsible. In practice, according to Bennett & Knights (1970), oxygen transfer is more likely to be the first limiting step in productivity.

Table 2. *Changes in oxygen requirement and heat release of a micro-organism (yeast) with change in yield factor on hexadecane substrate (after Evans, 1969)*

Yield factor	Oxygen requirement mol O_2/kg dry weight yeast	Heat release kJ/kg dry weight yeast
1.1	50	25 000
0.3	250	120 000

Evans (1969) has also explained the considerable importance of optimising process conditions and, in particular, of achieving a high yield factor, i.e. dry cell weight:weight substrate assimilated. In an optimised system a yield factor of 1.1 can be achieved but this can fall to 0.3 when the supply of nutrients is not optimal. This has a direct effect on substrate costs but, for the engineer, inefficient operation has two other serious consequences, an increased demand for oxygen and an increased heat output. This is shown in Table 2.

Process utilising methane

It is probably true to say that no one has yet produced a definite outline of an industrial process for the fermentation of methane. Nevertheless there is considerable and world-wide interest in this possibility and no general consideration of hydrocarbon fermentation would be complete without some mention of it. Ribbons (1967) has listed the following industrially attractive features of methane fermentation:

(i) Methane is volatile and should leave no undesirable residues in the products of fermentation.

(ii) The solubility of methane in water is similar to that of oxygen.

(iii) Methane is available in large quantities in many of those areas of the world in great need of an improved protein supply.

According to Whittenbury (1969) no yeasts able to utilise methane have been identified and all the published work refers to bacteria. Until recently a barrier to the industrial development of methane fermentation has been the poor growth rates and hence low productivity of methane-utilising bacteria. Since productivity is the prime determinant of operating costs the aim, as in all hydrocarbon fermentation processes, is for strains which will permit high cell densities with short holding times, in continuous culture. Whittenbury (1969) claimed to have isolated strains which were capable of growing at 5.0 g dry weight per litre with generation times of 6–8 hours in batch culture. More recently Hamer & Norris (1971) have achieved cell densities of up to 15.0 g dry weight per litre with pure cultures of methane-utilising micro-organisms. Harwood (1970) reported generation times of 4.33 hours in batch culture and stable continuous operation with a dilution rate of 0.14 h^{-1} (i.e. a holding time of about 7.0 hours). Sheehan & Johnson (1971), working with a mixed methane-oxidising bacterial culture in continuous culture, achieved a fermentor productivity of 2.39 g (dry weight) of cells per hour per litre from a cell concentration of 12.8 g (dry weight) of cells per litre at a dilution rate of 0.187 h^{-1}. The limit on productivity was identified as the mass transfer rate of oxygen in the fermentor. Moreover this culture grew at 45 °C which, for the reasons already discussed, could make methane fermentation particularly interesting in some parts of the world. The maximum specific growth rate (μ_{max}) of the Sheehan & Johnson culture was 0.303 h^{-1} (generation time 2.29 h). For a series of 15 fermentations Sheehan & Johnson reported average yields of 0.62 g (dry weight) cells per g methane used and 0.22 g (dry weight) cells per g oxygen used. Other workers have reported more efficient conversion of methane into bacterial cells and, as a recent example, Wilkinson (1971) achieved yields of 1.1 g (dry weight) cells per g methane consumed which coincides with the cell yield claimed by Evans (1969) for yeasts on liquid n-paraffins. Despite this confusion on the question of cell yield from methane and also of oxygen requirement, there is evidence of considerable recent progress in methane fermentation. In their recent review of the subject Hamer & Norris (1971) concluded that yields and productivities now justify serious consideration of industrial development.

THE PRODUCTS OF HYDROCARBON FERMENTATION

Chemical composition and nutritional value

Workers at the Rowett Research Institute (Woodham & Deans, 1971; Palmer & Smith, 1971) have recently studied six samples of single cell proteins. In these reports it was implied but not stated that all six samples were grown on hydrocarbons. Assuming that this was the case this collection is interesting in that it included both yeasts and bacteria. Table 3

Table 3. *Some chemical components of a collection of single-cell proteins (from Palmer & Smith, 1971, and Sheehan & Johnson, 1971)*

Constituent	Palmer & Smith Series						Sheehan & Johnson
	Y2	Y3	Y4	Y5	B6	B7	MB
Nitrogen							
% of dry matter	10.5	10.4	9.7	8.7	14.2	13.6	12.1
Nucleic acid nitrogen							
g per 100 g total N	9.0	12.3	14.2	15.0	18.4	7.5	23.7
Total lysine							
g per 16 g N	7.6	7.8	7.9	7.5	5.9	6.0	4.3
% of dry matter	5.0	5.1	4.8	4.1	5.2	5.1	3.2
Total methionine							
g per 16 g N	1.8	1.6	1.9	1.7	1.8	1.8	3.0
% of dry matter	1.2	1.0	1.2	0.9	1.6	1.5	2.3
Total S amino acids							
g per 16 g N	2.8	2.5	2.6	2.4	2.8	2.8	3.4
% of dry matter	1.8	1.6	1.6	1.3	2.5	2.4	2.5

Y = Yeast; B = Bacteria; MB = Mixed bacterial culture

shows some of the data of Palmer & Smith and those of Sheehan & Johnson (1971) for their methane-oxidising mixed bacterial culture.

From the data in the table and also that from other sources (Ko & Yu, 1967; Yamada *et al.* 1967) it appears that nitrogen concentrations are higher in bacteria than in yeasts. Non-protein nitrogen in the form of nucleic acids also appears to be higher in bacteria although it cannot be said with certainty that all the products shown in Table 3 still had their full complement of nucleic acids at the time of analysis. Furthermore it is known that for many micro-organisms the nucleic acid content can be varied widely by varying the growth rate of the culture. For these reasons comparisons of nucleic acid contents would be valid only if the conditions of culture were similar, which was probably not the case with the data of Palmer & Smith. Apparently the lysine content of cells can also be varied by varying the culture conditions (Yamada *et al.* 1967) and so comparisons are again difficult. From the samples shown in Table 3, the hydrocarbon-grown yeast proteins contained substantially higher concentrations of total lysine than the bacterial proteins but, by virtue of the higher crude protein contents of the bacteria, the yeast and bacterial products themselves contained similar amounts of lysine. The exception seems to be the sample of Sheehan & Johnson which had a much lower lysine content.

Even lower lysine concentrations (2.6 and 2.2 per cent of dry weight) were recorded for two pure methane utilisers by Ribbons (1967). This is insufficient evidence to conclude that methane-utilising bacteria are poorer with respect to lysine content than other hydrocarbon-grown micro-organisms but, so far, no one appears to have described a high-lysine methane utiliser. Lysine is often the first limiting amino acid in cereal-based diets for both farm animals and human beings and, for this reason, a great deal of attention is given to it. Synthetic *l*-lysine is an established commodity but at current world prices of about £1.00 per kg the cost of correcting the Sheehan sample, to, say, the lysine concentration of the Y_2, Y_3, B_6 and B_7 samples of the Rowett series, would be almost £20 per ton, which would be prohibitive for some animal production applications. Yeast and bacterial proteins are generally poor in the sulphur amino acids although bacteria appear to be better than yeasts, again by virtue of their higher crude protein content. Deficiency of methionine is not so serious as that of lysine because synthetic *dl*-methionine is widely available and is comparatively inexpensive. The current world price for *dl*-methionine is about £0.50 per kg and on this basis the difference in value between the best and the poorest samples in Table 3 is only about £6.00 per ton.

Determinations of the nutritive value of micro-organisms grown on hydrocarbons have been reported recently by Shacklady (1970), working with the two BP products, and by Palmer & Smith (1971); both investigations used microbial proteins as the sole source of protein for rats. In anticipation of deficiencies of the S amino acids in both investigations the various proteins were fed with and without methionine supplementation. A representative selection of results is shown in Table 4.

Shacklady's observations showed that methionine is the first limiting amino acid. Shacklady also observed that when this deficiency was corrected the biological value of the two BP yeasts is raised to that of dried skim milk or dried whole egg protein, thus showing the very high potential value of these particular yeast samples. Palmer & Smith also demonstrated a response to methionine although the biological values of their supplemented samples, particularly those for bacteria, were poor by comparison with those reported by Shacklady. The digestibility of the yeasts appeared to be marginally higher than that of the two samples of bacteria.

As with other proteins which are subjected to industrial processing the nutritional quality of single-cell proteins will depend on the techniques used in their preparation. The nutritional availability of lysine is particularly sensitive to harvesting conditions. Several recent and independent reports have indicated the lysine availability of the two BP yeasts to be

Table 4. *A comparison of the nutritive value of various hydrocarbon-grown microbial proteins with that of two conventional sources of protein*

Report	Product		Criteria of nutritive value		
			NPU	TD	BV
Shacklady (1970)	(a)	Yeast from *n*-paraffins	59	96	61
,,	(b)	(a) + 0.3 % *dl*-methionine	88	96	91
,,	(c)	Yeast from gas-oil	50	94	54
,,	(d)	(c) + 0.3 % *dl*-methionine	91	95	96
,,	(e)	Soya protein isolate	42	100	42
,,	(f)	(e) + 0.3 % *dl*-methionine	64	99	65
,,	(g)	Dried whole egg	90	100	90
,,	(h)	(g) + 0.3 % *dl*-methionine	97	100	97
Palmer & Smith (1971)	(a)	Yeast (Y2 from Table 3)	56	94	60
,,	(b)	Y2 + 0.9 % *l*-methionine	74	94	79
,,	(c)	Yeast (Y3 from Table 3)	58	93	62
,,	(d)	Y3 + 0.9 % *l*-methionine	78	93	84
,,	(e)	Bacterium (B6 from Table 3)	38	90	42
,,	(f)	B6 + 0.9 % *l*-methionine	63	90	70
,,	(g)	Bacterium (B7 from Table 3)	46	91	51
,,	(h)	B7 + 0.9 % *l*-methionine	68	90	75

NPU = Net protein utilisation.
TD = Total digestibility.
BV = Biological value.

very high. For example Shannon & McNab (1971), using colostomised adult hens, found a lysine digestibility of 95 per cent; Lewis & Boorman (1971), using growing chickens, found the lysine availability of the two BP yeasts to be between 98 and 103 per cent. Chemical determinations of lysine availability using the Roach, Sanderson & Williams (1967) procedure on many samples in BP's own laboratories have shown values in excess of 95 per cent. Independent laboratories have also obtained similar chemically determined available lysine values. Microbiological assays for the methionine availability of the BP yeasts have also indicated very high values (quoted by Barber *et al.* 1971). Overall there appears to be a growing body of evidence that the principal amino acids in correctly harvested hydrocarbon-grown yeasts are readily available. There is no published information on the nutritional availability of the amino acids in hydrocarbon-grown bacteria or, indeed, in yeasts other than those used by BP.

The uses of hydrocarbon-grown single-cell proteins

The first requirement for a single-cell protein is that it should be proved to be safe to use. Because of the nature of the substrates employed it is particularly important that the products should be shown to be devoid of any tendency to induce cancer. In addition the products must be proved to be free from all toxic effects. In different countries the law is not always very definite in its requirements for products that will form a major source of nutrients in food or feed, as distinct from minor additives over which it is usually quite decisive. Nevertheless it is anticipated that all single-cell products will have to undergo stringent tests as to their safety in use. It seems very likely that the safety of these products will be assessed on the basis of their behaviour, in comparison with materials generally regarded as safe, when subjected to standard toxicological tests with a variety of animals. The species of animals used in these tests will depend on the intended application for the materials.

Details of the programme of toxicological tests to which the two BP products have been subjected have been published by Llewelyn (1967) and Shacklady (1970). These products, which are destined to be used only in animal feeds, came through a stringent testing programme without reservation on the part of the independent authorities who performed the tests (de Groot, Til & Feron, 1970 a, b).

Uses in animal feeding

Because very little has been published on the projected uses of hydrocarbon proteins other than the BP products, these will be used as the only example in this section. The BP yeasts are destined to be used in animal feeds and it is interesting to compare their characteristics as animal feedingstuffs with those of the two most important feed protein commodities, fishmeal and soyabean meal. Table 5 shows such a comparison.

With the exception of the S amino acids, where differences in nutritional availability between the yeasts and fishmeal probably partially correct the apparently lower concentrations in the former, the yeasts are very similar to fishmeals. The lower energy values of the gas-oil grown yeast are due to the removal of the yeast lipids during solvent extraction of the residual gas-oil. In practical calculations of the market value of n-paraffin-grown yeast the higher energy value compared with fishmeal cancels the effect of the lower S amino acid content of the yeast. For this reason the market value of the n-paraffin-grown yeast for use in pig and poultry diets is likely to be similar to that of fishmeals of the type described in Table 5. A comprehensive series of experiments demonstrating the successful use

Table 5. *The principal characteristics of BP dried yeasts as animal feedingstuffs compared with fishmeal and soyabean meal*

	BP yeasts			
	n-paraffin process	Gas-oil process	Fishmeal*	Soyabean meal
Crude protein				
$N \times 6.25$ % as received	63	66	62–65	45
Metabolisable energy				
kcal/kg determined in:				
(a) Poultry	3050	2550	(2750)	(2200)
(b) Pigs	3900	3500	—	—
Essential amino acids				
g/16 g N				
Iso-leucine	5.0	5.3	4.6	5.4
Leucine	7.4	7.8	7.3	7.7
Phenylalanine	4.3	4.8	4.0	5.1
Tyrosine	3.6	4.0	2.9	2.7
Threonine	4.9	5.4	4.2	4.0
Tryptophan	1.4	1.3	1.2	1.5
Valine	5.8	5.8	5.2	5.0
Arginine	5.1	5.0	5.0	7.7
Histidine	2.1	2.1	2.3	2.4
Lysine	7.4	7.8	(7.0–7.7)	6.5
Available lysine	7.2†	7.8†	(6.3–7.0)	(6.5)
Cystine	1.1	0.9	1.0	1.4
Methionine	1.8	1.6	2.6	1.4

* Shacklady (1967). † Lewis & Boorman (1971).

Values in brackets are 'table' figures commonly used in the animal feedingstuffs industry.

of the BP yeasts either as complete or partial replacements for fishmeal in a variety of diets for pigs and poultry have been described elsewhere (Shacklady, 1970; van der Wal, Shacklady & van Weerden, 1969; van Weerden, Shacklady & van der Wal, 1969).

Uses in diets for human beings

BP have given first priority to the development of products suitable for feeding to animals. In biological terms, which are the terms of this symposium, the use of animals as intermediates in the use of high-quality proteins involves a considerable loss of efficiency. We are frequently asked why we did not develop products suitable for direct feeding to human beings immediately. The answer is comparatively simple. Although it must be acknowledged that animal production is comparatively inefficient biologically, it is a sure method of maintaining and improving protein supply. Despite the enormous variations in eating habits of communities a growing demand for animal products is evident in almost every part of the world regardless of the stage of development.

Moreover, because our two products have been proved to be suitable for feeding to animals in Europe it seems certain that they will be accepted for that purpose elsewhere. This is entirely different from the situation in human foods. There is no guarantee that a food which is enjoyed by the people of one country will even be accepted by those of other countries. The question of acceptability is at the core of the problem of human protein nutrition because the biological value or efficiency of proteins which are not eaten, for whatever reason, is zero.

This does not mean that BP has discounted the possibility of feeding single-cell proteins directly to human beings. In fact we believe that direct feeding of certain derivatives of hydrocarbon-grown yeasts to human beings will develop in the future. The yeasts can be produced in the form of bland, white powders which could be incorporated into confectionery and bread. Yeast protein isolates can be coloured, flavoured and textured and it is entirely feasible that they could eventually replace conventional proteins in human diets. In the meantime, however, we are giving priority to animal production.

Other groups working in the field of microbial protein production have opted to concentrate immediately on the development of products suitable for direct feeding to human beings. Clearly opinions vary as to the best avenue of approach. However, most people would agree that the successful development of high-protein human foodstuffs will depend on competent research of the markets involved.

For a description of some of the processing techniques currently being investigated to make single-cell proteins 'look and taste like food' readers are referred to the paper by Tannenbaum (1967).

CONCLUSION

The story of protein from hydrocarbons is still one of potential rather than achievement. However, the construction of two industrial production units by BP could be the first stage of large-scale production during the next decade and thereafter.

The first European processes will employ yeasts but the development of thermophilic strains of bacteria is a possibility for the future. Similarly the research currently being applied to methane fermentation could lead to the development of industrial processes in the foreseeable future.

Initially the products of hydrocarbon fermentation will be fed to animals, particularly pigs and poultry. However, the development of derivatives of these products suitable for direct feeding to human beings is predicted for the future.

Permission to publish this paper has been given by the British Petroleum Company Limited.

REFERENCES

BARBER, R. S., BRAUDE, R., MITCHELL, K. G., & MYRES, A. W. (1971). *Br. J. Nutr.* 25, 285.
BENNETT, I. C., HONDERMARCK, J. C., & TODD, J. R. (1969). *Hydrocarbon Processing* 48 (3), 104.
BENNETT, I. C., & KNIGHTS, D. L. (1970). AIChE 67th National Meeting, Atlanta, Georgia, USA, February 1970.
CHAMPAGNAT, A., VERNET, C., LAINÉ, B. M., & FILOSA, J. (1963). *Nature* 187, 13.
DE GROOT, A. P., TIL, H. P., & FERON, V. J. (1970a). *Fd Cosmet. Toxicol.* 8, 267.
DE GROOT, A. P., TIL, H. P., & FERON, V. J. (1970b). *Fd Cosmet. Toxicol.* 8, 499.
EVANS, G. H. (1969). 4th Petroleum Symposium, UN Economic Commission for Asia and the Far East.
FORSTER, J. W. (1962). *J. Microbial. Serol.* 28, 241.
FUHS, G. W. (1961). *Arch. Mikrobiol.* 39, 374.
HAMER, G., & NORRIS, J. R. (1971). Proceedings of the 8th World Petroleum Congress, Moscow.
HARWOOD, J. H. (1970). Ph.D. Thesis, University of London, September 1970.
IVENGAR, M. S. (1967). In *Single-Cell Protein*, p. 263, Ed. R. I. Mateles & S. R. Tannenbaum. Cambridge, Massachusetts: MIT Press.
KASERER, H. (1905a). *Z. Landw. Versuchsw. Oesterreich* 8, 789.
KASERER, H. (1905b). *Centr. Bakt. Parasitenk. Abt.* 2 15, 573.
KO, P. C., & YU, Y. (1967). In *Single-Cell Protein*, p. 255, Ed. R. I. Mateles & S. R. Tannenbaum. Cambridge, Massachusetts: MIT Press.
LAINÉ, B. (1970). Personal communication.
LEWIS, D., & BOORMAN, K. R. (1971). *Proceedings of the Xth International Congress of Animal Production*, Versailles, July 1971.
LLEWELYN, D. A. B. (1967). Symposium on Microbiology, London, September 1967. Proceedings published by Institute of Petroleum, London, 1968.
MALEK, I. (1967). In *Single-Cell Protein*, p. 268, Ed. R. I. Mateles & S. R. Tannenbaum. Cambridge, Massachusetts: MIT Press.
MATALES, R. I., BARUAH, J. N., & TANNENBAUM, S. R. (1967). *Science, NY* 157, 1322.
MIYOSHI, M. (1895). *Jb. wiss. Bot.* 28, 269.
PALMER, R., & SMITH, R. H. (1971). Proceedings of 230th Meeting of the Nutrition Society, Glasgow, February 1971.
QUAYLE, J. R. (1967). Symposium on Microbiology, London, September 1967. Proceedings published by Institute of Petroleum, London, 1968.
RIBBONS, D. W. (1967). Symposium on Microbiology, London, September 1967. Proceedings published by Institute of Petroleum, London, 1968.
ROACH, A. G., SANDERSON, P., & WILLIAMS, D. R. (1967). *J. Sci. Fd Agric.* 18, 274.

SHACKLADY, C. A. (1967). 2nd Int. Conf. on Global Impacts of Applied Microbiology, Addis Ababa, November 1967.
SHACKLADY, C. A. (1970). 3rd Int. Congress of Food Science and Technology, Washington, DC, August 1970.
SHANNON, D. W. F., & MCNAB, J. M. (1971). Personal communication.
SHEEHAN, B. T., & JOHNSON, M. J. (1971). *Appl. Microbiol.* **21**, 511.
SÖHNGEN, N. L. (1905). *Centr. Bakt. Parasitenk. Abt.* 2 **15**, 513.
TANNENBAUM, S. R. (1967). In *Single-Cell Protein*, p. 343. Ed. R. I. Mateles & S. R. Tannenbaum. Cambridge, Massachusetts: MIT Press.
VAN DER WAL, P., SHACKLADY, C. A., & VAN WEERDEN, E. J. (1969). 8th Int. Nutrition Congress, Prague, September 1969.
VAN WEERDEN, E. J., SHACKLADY, C. A., & VAN DER WAL, P. (1969). 8th Int. Nutrition Congress, Prague, September 1969.
WANG, D. I. C. (1967). In *Single-Cell Protein*, p. 217, Ed. R. I. Mateles & S. R. Tannenbaum. Cambridge, Massachusetts: MIT Press.
WHITTENBURY, R. (1969). *Proc. Biochem.* **4** (i), 51.
WHITTENBURY, R., PHILLIPS, K. C., & WILKINSON, J. F. (1970). *J. Gen. Microbiol.* **61**, 205.
WILKINSON, J. F. (1971). In *Symposia of the Society for General Microbiology.* Number 21, *Microbes and Biological Productivity*, p. 15.
WOODHAM, A. A., & DEANS, P. S. (1971). Proceedings of 230th Meeting of the Nutrition Society, Glasgow, February 1971.
YAMADA, K., TAKAHASHI, J., KAWABATA, Y., OKADA, T., & ONIHARA, T. (1967). In *Single-Cell Protein*, p. 193. Ed. R. I. Mateles & S. R. Tannenbaum. Cambridge, Massachusetts: MIT Press.

20
PROTEIN PRODUCTION BY MICRO-ORGANISMS FROM CARBOHYDRATE SUBSTRATES

By J. T. WORGAN

National College of Food Technology (University of Reading),
St George's Avenue, Weybridge, Surrey

The carbohydrates are the largest replenishable source of carbon compounds which are available for conversion by micro-organisms into biomass containing protein. If the primary aim of this conversion is to provide protein of high nutritional quality for human consumption, then the following aspects are relevant to the biological efficiency of the process:

(a) the yield of protein per unit weight of carbohydrate,
(b) the rate of conversion to protein,
(c) the nutritional quality of the biomass as a source of protein in the human diet,
(d) the value of the carbohydrate in the human food chain. The conversion to protein of cellulose or pentose sugars, for example, is considered to be more biologically efficient than the conversion of glucose, a sugar which is directly available as human food.

Aspects (a), (b) and (c) may be measured quantitatively. The relative values of sources of carbohydrate in the human food chain are more difficult to express in numerical terms. Before discussing the above aspects in relation to actual or potential microbiological processes based on carbohydrate substrates, a few general comments on micro-organisms as sources of food are of interest.

Foods in the diet of modern man have been selected by a process of trial and error which has continued throughout the history of mankind. With a few minor exceptions the biomass of micro-organisms has never been available during this process of selection. The production of sufficient quantities to provide new sources of food has become a possibility by the application of modern technology and selection by the lengthy process of trial and error can be replaced by our knowledge of nutrition and toxicology. Since a high proportion of microbial biomass may consist of protein, the term Single-Cell Protein (SCP) has become widely accepted to describe all sources which may be utilised as human or animal food.

Although tentative conclusions may be drawn from investigations which have already been reported, there remains a wealth of experimental material from which to select the most efficient micro-organisms to produce protein foods from a given substrate. There are probably 100000 species of fungi and yeasts and 1500 species of bacteria. Within each species there are numerous naturally occurring strains and further strains may be produced artificially by adaption or mutation. These strains may differ from the parent strain in their toxicological properties and in their ability to utilise carbohydrate substrates. Differences may also occur between strains in the nutritional value of the cell protein. The environment in which the cells are grown also has a considerable influence on chemical composition. Statements in the literature that certain species yield cell biomass with high protein contents therefore have little meaning.

In most industrial microbiological processes the strains used have been specifically developed for their efficiency in producing the required metabolic product. Since the maximum production of a metabolite implies an imbalance in the efficient growth of the cell, neither the strains used in industrial processes nor the conditions of growth adopted are directly applicable to the production of SCP. Although several predictions have been made of the potential efficiency of microbiological processes for the production of protein, the actual number of production units is still small and only three species of yeast, all of which are grown on carbohydrate substrates, are being produced on an industrial scale. The oldest established process is that for the production of baker's yeast (*Saccharomyces cerevisiae*) which is grown on a molasses substrate with the addition of ammonium hydroxide or sulphate as the nitrogen source. The other species of yeast produced on a commercial scale are food yeast (*Candida utilis*, formerly *Torula utilis*) and *Saccharomyces fragilis*. Food yeast has the advantage over baker's yeast that it is able to utilise pentose sugars and may therefore be grown on a wider variety of substrates. Waste sulphite liquor (from paper pulp manufacture) is used in Russia, Canada and the USA; wood hydrolysates are used in Russia and Czechoslovakia; molasses in Taiwan, South Africa, Cuba and the Philippines. Only a few small units produce *S. fragilis*, a yeast which can utilise lactose and is grown on cheese whey. The efficiency of protein production in this process is difficult to assess because protein from the whey is included in the harvested product.

Since the production of food yeast is based on the process for manufacturing baker's yeast, the latter process will be described and its biological efficiency discussed. Although *S. cerevisiae* is primarily produced for its baking properties, it does nevertheless contain 50 per cent protein and is an example of the microbiological production of protein on a large scale.

PRODUCTION OF YEAST PROTEIN ON AN INDUSTRIAL SCALE

In the production of baker's yeast the process is initiated by a single yeast cell which is allowed to develop into a yeast culture in the laboratory. After incubation the whole of this culture is used as inoculum for a larger volume of culture medium and this procedure is repeated through several stages of increasing size, first in the laboratory, then in stainless steel vessels in the factory. Strict aseptic conditions are maintained throughout this procedure. From these initial stages of the process sufficient seed yeast is produced to inoculate the main fermenter vessels of $45-230 \times 10^3$ l capacity. Although precautions are taken to minimise infection during the final stage of growth it is not economical to maintain completely aseptic conditions. The fermenter vessels are vigorously aerated throughout the final stage; temperature and pH are controlled. Higher yields of yeast are obtained by the operation of the Zulauf or Differential process in which the carbohydrate source in the form of molasses is initially provided in dilute concentration. As the growth of yeast in the fermenter proceeds, more molasses are added to keep pace with the growth of the yeast. The quantities added ensure that there is sufficient but never a large excess of carbohydrate in the growth medium. At the end of this procedure the yeast, which has now reached a mass of 100 tons, is harvested by centrifugation.

The whole process from single yeast cell to 100 tons of yeast biomass occupies a period of 300 hours and involves 60 yeast generations. The average generation time is therefore 5 hours. The yeast cell mass increases from 10^{-10} to 10^8 g. Based on the sugar content of the molasses the yield of yeast is 50 per cent (Dawson, 1952). Since the yeast cells will contain an average of 50 per cent protein the rate of protein synthesis will be comparable to that of the yeast cell mass. The rate of protein synthesis can therefore be expressed as a protein doubling time of 5 hours.

In the Distillers Company plant at Dovercourt the process has been improved by replacing the final stage with a continuous system which is maintained for a period of approximately 80 hours. The production of seed yeast up to the 4500 l stage follows the procedure described above. The seed yeast is then fed into a system of eight fermenters in series into which molasses, other nutrients and water are introduced to keep pace with the increasing mass of yeast cells. This procedure gives the same proportionate yield of yeast from molasses and increases the productivity of the process by 33 per cent (Olsen, 1960). A protein doubling time of 3.75 hours is estimated for this process.

YIELD OF MICROBIAL PROTEIN FROM CARBOHYDRATES

In the process for the production of baker's yeast the average generation time can be reduced from 5 to 1.5 hours by increasing the temperature of incubation from 25 °C to 30 °C. At the higher growth rate oxygen becomes limiting in the fermenter and a change from aerobic to the less efficient anaerobic metabolism occurs. Thus the yield of yeast from the carbohydrate supplied decreases. Reducing the growth rate by lowering the temperature below 25 °C does not necessarily give higher yields of yeast since the maintenance energy required by the cell biomass becomes significant. For *Aerobacter cloacae* the sugar required to provide energy for maintenance has been determined to be about 0.1 g sugar per g dry matter per hour (Pirt, 1965). A similar requirement has been proposed for yeast cells (Harrison, 1967). Thus the generation time of 5 hours maintained in the batch process for the production of yeast is optimal for the conversion of carbohydrate to yeast cell mass and hence of carbohydrate to yeast protein.

Calculations of the maximum yield of yeast from sugar have been found useful to the yeast manufacturer and are based on the Gay-Lussac equation for the fermentation of sugar to alcohol. Under anaerobic conditions sugar is converted almost quantitatively to alcohol and carbon dioxide with a negligible increase in yeast cell mass. When oxygen is introduced into the system the alcohol is almost completely utilised for yeast cell growth.

$$C_6H_{12}O_6 = 2CO_2 + 2C_2H_5OH.$$

The assumption is made that two C atoms are used to form CO_2 and four C atoms are used to form yeast cell mass. From 180 g glucose, therefore, 48 g C are available for cell synthesis. Since the average yeast cell contains 47 per cent carbon the yield of cells is 102.1 g, or 57.5 per cent of the weight of glucose used (White, 1954).

It has been proposed that the biological energy required for the synthesis of cell biomass is similar for all micro-organisms. Experimentally the amount of energy has been determined as 1 mol ATP per 10.5 g cell mass (Senez, 1962). Assuming that, during growth in aerobic conditions, all of the energy required for synthesis is derived from the complete metabolism of the substrate to CO_2 and water, the maximum yield of biomass of any micro-organism from a given substrate may be calculated from equation (1).

Maximum yield of microbial biomass from 1 mole substrate (Y_m)

$$= \frac{10.5\,M \times C_s}{M/E \times C_s + 10.5\,C_m}, \tag{1}$$

where M = molecular weight of substrate,
C_s = carbon content (per cent) of substrate,
C_m = carbon content (per cent) of microbial biomass,
E = number of moles of ATP produced by the aerobic metabolism of 1 mole of substrate.

Applying equation (1) to the production of *S. cerevisiae* cells from glucose,

Molecular weight of glucose (M) = 180 g,
Carbon content of glucose (C_s) = 40 per cent,
Carbon content of yeast cells (C_m) = 47 per cent (White, 1954),
Number of moles ATP per mole of glucose (E) = 38,

∴ Yield of yeast cells (Y_m) = 111 g.

Thus there will be sufficient provision of C atoms for 111 g yeast from 130 g glucose, leaving 50 g glucose to provide the 10.6 mol ATP required for the synthesis. The maximum yield of yeast from glucose will therefore be $111/180 \times 100 = 61.7$ per cent.

```
180 g glucose → 130 g glucose → 111 g yeast
      ↓                                │
  50 g glucose                         │
      ↓                                ↓
    yields                          requires
50 × 38/180 = 10.6 mol ATP    111/10.5 = 10.6 mol ATP
```

Although the yield of 10 to 10.5 g cells per mol ATP has been confirmed for a number of micro-organisms, there is some evidence from anaerobic growth experiments that higher yields may be attained (Hobson, 1965). The method of calculation given above may therefore give an underestimate of potential yields. That the yield of 50 g yeast from 100 g sugar is approaching the maximum which might be achieved on a production scale seems probable from the maximum yields quoted for laboratory experiments. Yields from 100 g glucose of 53.3 g *S. cerevisiae* (Olson & Johnson, 1949), 53.3 g *Arthrobacter globiformis* and 52.2 g *Escherichia coli* (Morris, 1960) are some of the highest values reported. Higher yields than those for yeast or bacteria have been reported for fungal mycelium from carbohydrates. *Aspergillus niger*, for example, has produced 61 g mycelium from 100 g sucrose (Perlman, 1965). Since the C content of mycelium may vary from 40 to 63 per cent (Lilley, 1965) the higher yield may be due to the different cell composition. Thus the maximum yield from glucose of microbial biomass containing 40 per cent C calculated by applying equation (1) would be 70 per cent. For mycelium containing 50 per cent protein it may be assumed that the C content will be similar to that of yeast and the maximum yield of biomass from carbohydrate will therefore be similar.

Yields of microbial biomass from most carbohydrates will be of the same order as those from glucose since the hydrolysis of polysaccharides to monosaccharides produces comparatively little biological energy. Somewhat lower yields may be anticipated from pentose sugars. An estimate of the maximum possible yields of protein from carbohydrates may be made if the maximum true protein contents of yeasts and fungi are assumed to be 50 per cent and of bacterial cells to be 70 per cent. The significance of these values is discussed below. Thus from carbohydrate substrates the maximum yield of true protein will be $60.5 \times 50/100 = 30.2$ per cent for yeasts and fungi and for bacteria will be $60.5 \times 70/100 = 42.3$ per cent. On a production scale the maximum yields will probably be $50 \times 50/100 = 25$ per cent for yeasts and fungi and for bacteria $50 \times 70/100 = 35$ per cent.

RATE OF CONVERSION OF CARBOHYDRATE TO PROTEIN

A mean protein doubling time of 5 hours has been shown to be the optimum rate for the large-scale batch production of yeast protein. Bacteria when grown in the laboratory are known to have a shorter generation time than yeasts. On a large scale the efficiency of utilisation of substrate will be limited by the rate at which oxygen can be supplied to the growing cells. Since it is this aspect which limits the rate of yeast growth on a large scale, it is doubtful whether bacteria could be produced more rapidly than yeast without reducing the efficiency of the conversion of carbohydrate to protein. Fungi are reported to grow more slowly than bacteria or yeasts. This generalisation does not apply to all species. A protein doubling time of 2 hours has been reported for *Neurospora crassa* (Zalokar, 1959) and the same rate of protein production has been maintained by the growth of fungal mycelium in a 3000 l fermenter (Banks, Hunston & Worgan, 1971).

Continuous culture has been advocated as a method of increasing the productivity of microbiological processes. The pattern of growth in a normal batch culture is illustrated in Figure 1. When a micro-organism is inoculated into a culture medium a lag period follows before cell num-

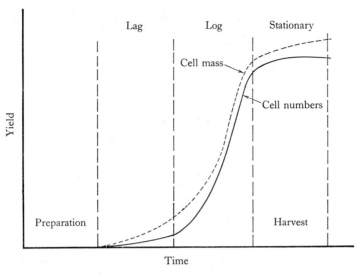

Figure 1. Changes in the cell mass and cell numbers during growth of a micro-organism in batch culture.

bers begin to increase. Growth, as measured by increase in cell mass, does take place during this lag phase. After a period of time which depends upon the state of the inoculum, rapid cell division begins and the culture enters into the log phase. Eventually the rate of production of new cells decreases and the culture passes into the stationary phase. For cell biomass the time of harvest in a batch culture will be at the beginning of this stationary phase. The aim of a continuous culture system is to establish an equilibrium in the log phase of growth by continually withdrawing a proportion of the culture and replacing it with fresh medium. In practice it is difficult to achieve efficient substrate utilisation and maintain growth at the maximum logarithmic rate which has been determined by growth in a batch culture (Luedeking, 1967). Nevertheless it is possible to maintain a higher mean protein doubling time in continuous culture than in a batch process. Baker's yeast has been produced continuously in a pilot plant at a yield of 52 per cent based on the weight of molasses sugar. The doubling time was between 3 and 4 hours (Hospodka, 1966).

A further limitation on the rate of production of SCP in both batch and continuous culture is the nucleic acid content of the cells which increases at higher growth rates (see Table 1). The amount of nucleic acid limits the use of SCP in the diet and this aspect is discussed below. In continuous culture at a fixed growth rate the nucleic acid content of cells can be varied (Table 1) by adjusting the temperature of incubation (Tempest & Dicks, 1967). Whether a reduction in nucleic acid content could be achieved on a production scale without reducing the efficiency of conversion of substrate to protein has still to be determined.

Thus the rates of conversion of carbohydrates to protein by micro-organisms on a production scale are limited by other factors than the maximum growth rate which can be achieved in laboratory culture. Processes for producing SCP will therefore probably operate at the same rate whether bacteria, fungi or yeasts are used. The maximum mean protein doubling time will probably be about 5 hours in batch processes and about 3 hours in continuous culture.

Table 1. *Effect of growth rate and temperature on the protein and nucleic acid contents of micro-organisms (after Elsworth et al. 1968; Tempest & Dicks, 1967)*

Micro-organism	Doubling time (hours)	Composition	
		Protein (%)	Nucleic acid (%)
C. utilis	4.7	55	10.1
	1.7	61	11.9
A. aerogenes	4.7	62	12.3
	1.7	60	14.2
	0.8	56	18.9
A. aerogenes			
25 °C	3.5	68.9	17.7
40 °C	3.5	69.0	12.3

TYPES OF CARBOHYDRATE SUBSTRATES

With the exception of cellulose and lactose, most of the carbohydrates likely to be available as substrates for the growth of micro-organisms on a large scale can be utilised by several bacterial species and by most species of fungi. Yeasts are less versatile in the range of carbohydrates they can use as C sources. *S. cerevisiae* for example will not use starch, pentose sugars or lactose for growth. *C. utilis* will grow on a wider range of substrates since it does utilise pentose sugars.

The efficiency of utilisation by fungal species or strains is about the same for all monosaccharides provided growth does take place. Where apparently less efficient utilisation of disaccharides and polysaccharides occurs, a longer period of incubation may give yields as high as those from glucose (Cochrane, 1958). This ability to adapt to the substrate probably also applies to bacteria and yeasts. Poor yields from a given carbohydrate source should not, therefore, be regarded as being indicative of the potential value of the carbohydrate as a substrate for a process. The rate of growth on the substrate can usually be improved by adaptation. A new strain may therefore be developed which has the ability to rapidly utilise the new substrate. This property may be retained if the stock culture of the new strain is maintained on a culture medium containing the relevant carbohydrate. Strain specificity in the utilisation of carbohydrate sources by fungi is not uncommon and mutants of *Aspergillus niger* and *Penicillium chrysogenum* are known which have different carbohydrate utilisation patterns from the parent strains. Some mutants are able to grow on C sources which are not available to the parent strain. Polysaccharides are sometimes better C sources than the sugars obtained from them by hydrolysis. Starch, for example, is often a better C source for fungal growth than glucose or maltose (Cochrane, 1958). This effect may be due to the gradual hydrolytic release of the sugar. Similar conditions are therefore created to those which apply during the incremental feed of nutrients in the process for producing baker's yeast.

Yeasts, although less versatile in the range of carbohydrates they are able to utilise, may also be adapted to new carbohydrate substrates. *Rhodotorula glutinis*, for example, has been adapted to growth on xylose. Several yeasts from seven different genera have also been adapted to utilise seaweed carbohydrates such as laminarin and sodium alginate (Morris, 1958), compounds which do not occur in the natural surroundings of the yeasts. In the 'Symba' yeast process which has been developed to use starch wastes as substrate, two yeasts are grown in association. *Endomycopsis fibuliger* will grow on starch as the C source and in doing so produces excess amylase which hydrolyses the starch to sugar. *C. utilis*, which is not able to utilise starch, grows on the sugar produced. When both yeasts are inoculated into a fermenter vessel rapid growth can be maintained and the harvested product consists of about 9 parts *C. utilis* to 1 part *E. fibuliger* (Tveit, 1967).

Starch and sucrose are the two carbohydrates which can be produced in the highest yield by the growth of crop plants. Both are biologically valuable since they are used directly as energy sources in the human diet. Cellulose and hemicelluloses (mainly polymers of uronic acids and pentose sugars) are the most abundant sources of carbohydrates and vast quantities are produced and harvested inadvertently with many food crops. In some cases they exceed the quantity of the edible portion of the crop three or fourfold. Since these sources of carbohydrate have negligible value as human food, it would be a more biologically valuable process to convert them to protein than it would be to convert starch or sucrose. The annual quantities available of some agricultural byproducts are listed in Table 2.

Although these byproducts are often termed cellulosic materials only about one-third of many of them consist of cellulose. As measured by the crude fibre content, for example, wheat, barley and oat straws contain less than 36 per cent cellulose.

Table 2. *Annual production of agricultural byproducts containing carbohydrates (after Woodward, 1952; Development Commission, 1953)*

Byproduct	Country of origin	Annual production ($kg \times 10^6$ dry weight)
Oat husk	UK	70
Potato haulm	UK	1250
Wheat straw	UK	2073
Wheat straw	India	8000
Sugar-cane bagasse	India	8000

In the main bulk of agricultural and forestry byproducts the carbohydrates are present in ligno-cellulosic plant tissue. This material is resistant to rapid degradation by micro-organisms due to the protective effect of the lignin polymer which forms a three-dimensional cage-like structure throughout the tissue and thus prevents the penetration of enzymes (Pew & Weyna, 1962). The partially crystalline structure of cellulose also prevents contact between enzyme molecules and the substrate. Polysaccharides, polyuronides and even the smaller molecules of sugars, which are occluded in the ligno-cellulose matrix, may also be protected from the enzyme activity of micro-organisms. There are therefore several stages in the biological degradation of ligno-cellulosic materials.

The resistance of ligno-cellulose to degradation differs within different plant tissues. The lignin of some annual plants for example is soluble in 1.5 per cent NaOH at 20 °C, whereas temperatures of 160–180 °C may be necessary to dissolve the lignins of resistant woody tissue (Brauns, 1952). The crystalline structure of cellulose is also more extensive in some plant materials which are therefore more resistant to chemical reagents and solvents. These differences in resistance to chemical reagents are reflected in the susceptibility of the plant materials to biological degradation. The degradation of ligno-cellulosic material is illustrated in Figure 2.

Only the wood-rotting fungi are capable of carrying out reaction L on the more resistant ligno-cellulose of wood. A wider range of fungi and some bacteria are able to degrade the ligno-cellulose of less resistant materials such as that of cereal straws and husks. Other species of fungi

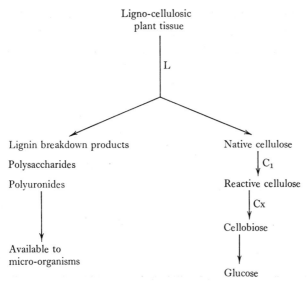

Figure 2. The degradation of ligno-cellulosic plant tissue.

and bacteria, which are not able to carry out reaction L, do have cellulase (C_1) activity. A much larger number of species of fungi and bacteria have cellulase (Cx) activity and are able to hydrolyse cellulose once it is in the reactive form. Experimentally neither reactions L nor C_1 have been found to take place at a rate which compares with the growth rate of micro-organisms. If the enzyme systems involved could be prepared in an active form and their optimum conditions for activity established, huge quantities of substrates could be made readily available for the production of SCP. Separation of the processes into two or three stages may be necessary since the optimum conditions for microbial growth and for L, C_1 and Cx activity may differ. The conversion of reactive cellulose to glucose in high yield has been shown to be possible with an enzyme preparation from *Trichoderma viride* (Katz & Reese, 1968).

Acid hydrolysis of the carbohydrates to sugars has been investigated as a method for replacing reactions L, C_1 and Cx. Due to the resistance of cellulose to hydrolysis, the sugars which are produced in the initial stages of the reaction are decomposed. Careful control of the hydrolysis conditions is therefore essential to avoid considerable destruction of carbohydrates. In practice two types of processes have been developed to minimise these losses. Concentrated acids at normal temperatures were used in the Bergius process and dilute acids at high temperatures (160–180 °C) in the Scholler process. The Scholler process was operated on a large scale in Germany during World War II. About 6.8 kg yeast protein were obtained from 100 kg (dry weight) wood (Lock, Saeman & Dickerman, 1945). In pilot plant studies of the Madison Wood Sugar process, which was a modification of the Scholler method, the maximum yield of sugar from wood was 56 per cent and of yeast from sugars 45 per cent (United States Forest Service, 1964). At the maximum protein content of yeast reported (56 per cent) the yield of yeast protein from 100 kg (dry weight) wood is estimated to have been 14.4 kg. The limits on the efficiency of producing microbial protein from the hydrolysates of ligno-cellulosic materials will be the same as those already discussed for other carbohydrate substrates. Any difference in the efficiency of the process will therefore be due to the losses of carbohydrate which may occur during acid hydrolysis. From the results of pilot plant studies these losses are of the order of 20–25 per cent (Gilbert, Hobbs & Levine, 1952).

In Russia the process for the production of *C. utilis* on wood hydrolysates is considered to be sufficiently efficient to justify production on a large scale and an annual output of 900×10^6 kg yeast is envisaged (Bunker, 1968).

Enzymic reactions L and C_1 can be replaced by mechanical or chemical treatments. The prolonged grinding of wood in a ball mill, for example, will produce material which can be degraded by cellulases which only have Cx activity. The energy required for the mechanical griding is considerable and the process is not economically viable. Alternative treatments of ligno-cellulosic materials with acid, alkali or urea reagents at temperatures below 100 °C may be used to prepare culture media suitable for the growth of several fungal and bacterial species (Worgan, 1967). *Actinomucor elegans*, grown on a culture medium prepared from 100 g oat husk by this method, gave a yield of 12.6 g mycelial protein (Delaney & Worgan, 1970). Sugar-cane bagasse degraded by oxidation in alkali conditions at 149 °C has been used as a C source for the growth of a *Cellulomonas* species. In a pilot plant this bacterium has been grown in a culture medium containing the degraded bagasse with a protein doubling time of 3.2–3.7 hours (Han, Dunlop & Callihan, 1971).

NUTRITIONAL VALUE OF MICROBIAL SOURCES OF PROTEIN

The proportion of protein in the biomass of a given species of microorganism varies with the environment in which it is grown. There do, however, appear to be different maximum limits for bacteria, yeasts and fungi. Bacteria have the highest protein contents and 87 per cent of the cell has been reported to be protein in some of the samples analysed (Bunker, 1963). These values refer to the crude protein content ($N \times 6.25$) and therefore include non-protein nitrogen compounds. The nucleic acid content varies with growth rate. At growth rates which would be used in production processes (5 hours doubling time) the average nucleic acid content of yeast is about 10 per cent. Values quoted for commercially produced baker's and food yeasts vary between 8 and 12 per cent.

Bacteria produced at the same growth rate will probably contain a higher proportion of nucleic acid. *Salmonella typhimurium* grown at a 5 hours doubling time contained 83 per cent protein (N × 6.25) and 16 per cent nucleic acid (Elsworth *et al.* 1968). The maximum proportion of true protein in yeast cells is therefore estimated to be 50 per cent and that of bacterial cells 70 per cent.

Fungi are reported to produce biomass with lower protein contents than bacteria or yeasts. Where protein yield and protein content are used as the criteria for determining the optimum conditions for their culture, the mycelium of a number of fungal species can be produced with protein contents as high as those of yeasts. Mycelium of *Fusarium semitectum* containing 60 per cent protein and of *Rhizopus oryzae* containing 59.6 per cent have been obtained by growth in submerged liquid culture. When grown at a mean protein doubling time of less than 5 hours, the nucleic acid content of the mycelium of *F. semitectum* was 5 per cent (Delaney & Worgan, 1970). Whether it is a general characteristic of fungi that, when produced at the same growth rate, they contain less nucleic acid than yeasts and bacteria has still to be established.

In addition to the reduction in the reported protein contents of microorganisms, the nucleic acid limits the use of SCP in the diet. From results of recent feeding trials on human beings it appears that the amount of nucleic acid in the human diet should be limited to 2 g per day (Edozien *et al.* 1970). At a level of 10 per cent nucleic acid in yeast cells this would limit the daily allowance to 20 g yeast per day, i.e. 10 g yeast protein. Processing of the cells may reduce the nucleic acid content. By subjecting the cells of *C. utilis* to a heat shock followed by successive periods of incubation at 50 °C and 60 °C the nucleic acid content has been reduced to 1 per cent. The heat processing does not result in a loss of protein from cells (Maul, Sinskey & Tannenbaum, 1970). If similar processes are applicable to other sources of SCP the nucleic acid content of harvested microbial biomass need not limit its use in the diet.

Except for a small quantity of *S. fragilis*, food yeast (*C. utilis*) is the only example of a microbial source of protein which is currently produced for use as human food. The choice of *C. utilis* was made because of its ability to grow on a wider range of substrates than baker's yeast and was not based primarily on the nutritional quality of the cell protein. Large-scale production and human feeding trials were investigated in Germany during World War I and laid the foundation for the output of 15×10^6 kg per annum during World War II. The yeast produced was used as a supplement to protein food supplies (Lock et al. 1945). C. utilis has therefore acquired an established reputation as a safe food, and it is primarily for this reason that it has been selected for production on an industrial scale.

Table 3. *Comparison of the nutritive value of* C. utilis *cells and separated cell protein (from Mitsuda et al. 1969)*

Protein source	Biological value (%)	Digestibility (%)	NPU (%)
Cells	36.6	64.6	23.6
Protein	47.3	95.8	45.3
Protein + methionine	61.1	94.5	57.7

The results of feeding trials show that *C. utilis* cells are not nutritionally equivalent to sources of animal protein. Values of 38 per cent (Miller, 1963) and 23.6 per cent (Mitsuda, Yasumoto & Nakamura, 1969) have been found for the Net Protein Utilisation (NPU). The digestibility of the cells is part of the reason for the low NPU values. Isolated *C. utilis* protein gave an NPU value of 45.3 per cent (Table 3), which increased to 57.7 per cent after supplementation with methionine (Mitsuda et al. 1969). Food yeast is therefore only suitable for use as a protein supplement. Equal parts of maize and yeast protein for example are complementary in their amino acid patterns and the mixture has greater nutritional value than the separate components (Bressani, 1968). Only a few NPU values have been reported in the literature for other species of micro-organisms. Examples of species with higher values than *C. utilis* are given in Table 4.

The quantitative essential amino acid composition of the proteins of several micro-organisms have been reported. Most yeasts (Nelson et al. 1960) and fungi (Rhodes et al. 1961) are deficient in the sulphur amino acids, cystine and methionine. Although several bacterial species do not lack methionine these same species are deficient in the proportion of tryptophan (Anderson et al. 1958). A recent report emphasises that it is difficult to obtain reliable values for the quantities of the amino acids, cystine and methionine, in proteins (Porter, Westgarth & Williams, 1968). That methionine is probably the only essential amino acid in which *F. semitectum* is deficient has been determined by Smith (1971) by chemical analysis and by feeding tests (Table 5).

Table 4. *NPU values of microbial sources of protein*

Micro-organism	NPU (%)	Reference
Candida utilis	38	Miller (1963)
Escherichia coli	51	Miller (1963)
Boletus edulis	64.9	Rafalski *et al.* (1966)
Aspergillus niger	58.4	Mitrakos *et al.* (1970)
Fusarium semitectum (Laboratory fermenter sample)	60	Miller (1968)

Table 5. *Nutritional value of the protein of* Fusarium semitectum

Sample	Methionine content of protein (g/100 g)	NPU (%)	Digestibility (%)
Laboratory fermenter (Miller, 1968)	2.1	60	—
Pilot plant (Smith, 1971)	1.7	52	80
+0.9 % methionine	—	66	—
+1.8 % methionine	—	76	—

Table 6. *The effects of strain, substrate and pH on the amino acid composition of protein from micro-organisms*

Micro-organism	Substrate	Amino acid content (g/100 g protein)		Reference
		Lysine	Tryptophan	
Candida tropicalis	Glucose	4.4	0.7	Lipinsky & Litchfield (1970)
	Xylose	10.0	0.5	
Candida utilis	Waste sulphite liquor	6.7	1.2	Peppler (1970)
	Molasses	10.7	0.5	
Escherichia coli				
B210		7.9	0.3	Yamada *et al.* (1968)
B766		3.7	0.2	
		% of cell mass		
Pseudomonas aeruginosa	Glucose	9.7	0.3	Anderson *et al.* (1958)
	Xylose	4.4	0.5	
	Hexadecane			
	pH 7	2.9	0.4	
	pH 8	6.4	0.4	

Variations may occur in the essential amino acid composition of the cell protein of a given species of micro-organism. The term cell protein usually includes free amino acids in the cell in addition to those in protein molecules. The differences may be due to the strain, the growth media or the environment (Table 6). It has been found that the proportions of some amino acids in microbial cells can be increased by the addition of precursors. The lysine content of yeast, for example, is increased by the inclusion of 2-oxo-adipic acid in the growth medium (Jensen & Shu, 1961). Similar attempts to increase the methionine content of yeasts were not effective (Chiao & Peterson, 1953).

Thus there are several micro-organisms which are known to be better nutritional sources of protein than food yeast. Many other species and strains remain to be investigated. From a study of factors which determine the amino acid composition of microbial biomass, methods for improving the nutritional value might be developed.

PALATABILITY, TOXICITY AND PROCESSING ASPECTS OF MICROBIAL PROTEINS

Food yeast is one of the few examples of a micro-organism specifically produced as a food. Many other harmless micro-organisms occur in foods and are consumed in significant quantities. Cheese, the oriental foods such as miso and tempeh, and the numerous types of fermented liquors all contain bacteria, fungi or yeasts. More than 2000 edible species of the larger fungi (mushrooms) have been reported in the literature and it is feasible that the mycelium of some of them could be produced on an industrial scale (Worgan, 1968). There are therefore numerous species which do not appear to have any toxic effects.

Although some bacteria and fungi are known to produce toxins they form a relatively small proportion of the total number of species. Micro-organisms which produce acute toxic effects can be eliminated by biochemical tests and by animal feeding trials. Very little is known about the effects of the remaining species if they were to be regularly consumed in the diet. The difficulty of establishing the safety of new sources of food is one of the main obstacles to the possibility of their use in the immediate future. Neither the genera nor even the species can be used as an infallible guide to the selection of a micro-organism. *Amanita phalloides* is one of the most toxic species of fungi; *Amanita fulva* is a popular edible species.

The biomass from potential industrial processes for producing microbial foods should be regarded as raw material corresponding to a harvested crop. Some established foods contain toxins which are destroyed by processing. Soya is an example, and eggs contain a toxin which is inhibited by a short period of cooking. Simple processing methods such as mild heat treatment are known to destroy some of the toxins which may occur in micro-organisms. More extensive processing techniques for the extraction of proteins may also remove toxins from the cell mass.

From bacteria and yeasts it is possible that textured, flavoured products will be developed similar to those prepared from soya protein. The preparation of a spun protein fibre from a bacterium has already been reported (Heden et al. 1971). Fungi may yield mycelium which without further processing has an acceptable flavour and texture.

ASSESSMENT OF THE BIOLOGICAL EFFICIENCY OF PROTEIN PRODUCTION

It has been shown that microbial sources of protein of good nutritional quality can be produced on an industrial scale from carbohydrate substrates. The rate of protein synthesis is limited to a protein doubling time of 5 hours in batch culture and 3 hours in continuous culture. The maximum yield of protein is about 25 per cent for yeasts and fungi and 35 per cent for bacteria. Established processes for the production of food yeast are probably near the limits of efficiency. On theoretical grounds for batch processes, neither the rate of protein production nor the yield of protein are capable of any considerable increases. An improvement in productivity will be obtained by the adoption of continuous culture systems. There is scope for improvement in the nutritional value of the product. C. utilis cells do not compare nutritionally with animal sources of protein, although processing or extraction of protein from cells will improve the digestibility. Other species of micro-organism are known, and others may be discovered, in which the protein is of higher nutritional quality.

The following equation is suggested for comparing the biological efficiency of these processes for the production of protein for use as human food:

$$\text{Biological efficiency of protein production (BEP)} = \frac{W \times R \times (S \times A)}{N \times 1000},$$

where W = g carbohydrate to yield 100 g protein,
  N = NPU,
  R = protein doubling time (hours),
  S = starch equivalent (SE),
  A = 1 unless otherwise specified.

The greater the value of the BEP index the lower the efficiency. For ligno-cellulosic woody tissue the SE is assumed equal to 1. The factor A is a discretionary term to allow for the fact that SE does not always measure the differences in the biological value of carbohydrates as sources of human food. For pentosans and pentose sugars, for example, A may be assigned the value of 0.01 when comparing the efficiency of different microbiological processes. Fibrous sources of carbohydrate will be represented by the SE, since the extent to which they are utilised by ruminants is a measure of their potential value in the human food chain.

Table 7. *Comparison between the biological efficiency of protein production (BEP) by micro-organisms and beef cattle using sucrose as the carbohydrate source ($SE = 100$)*

	Carbohydrate to yield 100 g protein (W) (g)	NPU* (N) (%)	Protein doubling time (R) (hours)	BEP Index
Candida utilis (batch culture)	400	38	5	5.3
Candida utilis (continuous culture)	400	38	3	3.2
Fusarium semitectum	400	65	5	3.1
Escherichia coli	286	51	5	2.8
Beef cattle	1900	76	2800	7000

* Micro-organisms: Table 4. Beef cattle: Food and Agriculture Organisation (1964).

Calculations of values for the BEP index for a carbohydrate substrate such as sucrose are given in Table 7. The calculations on the production of food yeast protein are based on the actual results of large-scale processes; those for *F. semitectum* and *E. coli* on the probable values which have been discussed above. An accurate value representative of the BEP index for beef protein production is difficult to determine, since neither the rate of protein production nor the efficiency of energy utilisation for protein synthesis is constant throughout the life cycle of the animal. If the rearing of a 400 kg bullock over a period of 2 years is taken as an example, data available for the 250 kg stage of growth (Evans, 1960) may be considered representative of the overall production cycle.

Liveweight gain (LWG) for 250 kg bullock = 900 g per day.

A 250 kg bullock requires 1.83 kg SE for maintenance and 1.62 kg SE for production, making a total SE requirement of 3.4 kg SE or 3.8 kg SE per kg LWG.

Each kg of LWG contains 200 g protein.

Thus 3.8 kg SE will yield 200 g protein and 1900 g SE will yield 100 g protein

$\therefore W = 1900$ g.

At 900 g LWG per 24 h, rate of protein production = 180 g per 24 h. Assuming that 60 per cent of the liveweight is carcass, 70 per cent of the carcass is meat and 20 per cent of the meat is protein, then weight of protein in 250 kg bullock

$$= 250 \times 60/100 \times 70/100 \times 20/100 = 21 \text{ kg}.$$

Thus 21 kg protein increases by 180 g in 24 h.

Protein Doubling Time $(R) = 21\,000/180 \times 24$ h $= 2800$ h.

Calculations of the BEP index from data available for other stages during the growth cycle show that, although the value of the index varies, it is of the same order of magnitude as that for the 5 cwt bullock when comparisons are made with values calculated for microbiological processes.

Table 8. *Comparison between the biological efficiency of protein production (BEP) by micro-organisms and beef cattle using oat husk as the carbohydrate source (SE = 20.6)*

	Carbohydrate to yield 100 g protein (W) (g)	NPU* (N) (%)	Protein doubling time (R) (hours)	BEP index
Candida utilis	666	38	5	1.8
Fusarium semitectum	910	65	5	1.4
Beef cattle	9224	76	2800	7000

* Micro-organisms: Table 4. Beef cattle: Food and Agriculture Organisatio (1964).

The results of the calculations for the BEP indices which are given in Table 7 show that the industrial process for the conversion of sugar to protein by food yeast is about 1300 times as efficient as the conversion by beef cattle. The production of *E. coli* as a protein source would be almost twice as efficient as the food yeast process. Comparisons are made in Table 8 of the BEP indices for the conversion of fibrous sources of carbohydrate to protein. The conversion factor for food yeast is based on the assumption that a 60 per cent yield of sugars would be obtained by the acid hydrolysis of oat husk (Dunning & Lathrop, 1945). For *F. semitectum* the conversion factor is based on the experimental yield of 11 g mycelial protein per 100 g oat husk (Delaney & Worgan, 1970). The process for fungal mycelium would be approximately 5000 times and the yeast process 4000 times more efficient than the conversion of fibrous sources of carbohydrate by the ruminant.

In Table 9 hydrocarbons are compared with carbohydrates as C sources for conversion to protein by micro-organisms. On a weight for weight basis higher yields are obtained from hydrocarbons although the efficiency of energy conversion to protein is less than in the carbohydrate process.

Table 9. *Comparison between hydrocarbons and carbohydrates as substrates for protein production by micro-organisms*

Substrate	Quantity to produce 100 g protein		
	Weight (g)	Energy (kcal)	Weight of carbon (g)
Carbohydrate	400	1492	160
Hydrocarbons			
CH_4	200	2656	150
C_8H_{18}	200	2240	170

Yields of protein assumed: from carbohydrates 25 %; from hydrocarbons 50 %

Land use is an alternative method of comparing the efficiency of protein production.

Land use for food yeast production

Quantity of sugar to yield 1 ton yeast protein = 4 tons.

Yield of beet sugar = 5.6 tons per hectare per year (British Sugar Corporation Ltd, 1961).

Number of hectares to yield 1 ton yeast protein = 4/5.6 = 0.71.

An equal weight of yeast protein would probably be obtained by utilising the beet tops, beet pulp and the molasses.

Thus approximately 0.36 hectares would yield 1 ton yeast protein per year, or yield of yeast protein = 2800 kg per hectare per year.

Land use for beef cattle production

Assuming that a 2-year-old bullock weighs 400 kg and contains 34 kg protein,
Annual protein production per bullock = 17 kg.
Number of bullocks to yield 1 ton protein per year = 1000/17 = 59.

Each bullock requires 0.40 hectare of land (Dudley Stamp, 1960).
Thus 23.6 hectares would yield 1 ton meat protein per year,
or yield of beef protein = 42 kg per hectare per year.

The conversion of carbohydrates to protein by micro-organisms is therefore a more biologically efficient process than conversion by ruminant livestock. A bonus in overall biological terms will be obtained if the sources of carbohydrate used for these processes are wastes or industrial effluents which create pollution problems. Serious consideration should therefore be given to the possibility of food production by microbiological processes. Products which are equivalent to meat in flavour, texture and nutritive value are capable of being developed in the future. Although it is probable that few micro-organisms are toxic it is a difficult and time-consuming procedure to establish that a new food is safe for human consumption. It is probably this latter consideration which will prevent the introduction of microbial foods in the immediate future.

REFERENCES

ANDERSON, R. F., RHODES, R. A., NELSON, G. E. N., SHEKLETON, M. C., BARRETS, A., & ARNOLD, M. (1958). Lysine, methionine and tryptophan content of micro-organisms. *Journal of Bacteriology* **76**, 131–5.

BANKS, G. T., HUNSTON, M. J., & WORGAN, J. T. (1971). Unpublished results.

BRAUNS, F. E. (1952). In *Wood Chemistry*, Ed. L. E. Wise & E. C. Jahn, Vol. 1, chapter 11. New York: Reinhold.

BRESSANI, R. (1968). Use of yeast in human foods. In *Single-Cell Protein*, Ed. R. I. Mateles & S. R. Tannenbaum, pp. 90–121. London: MIT Press.

BRITISH SUGAR CORPORATION LTD (1961). *Home-grown Sugar*. London: British Sugar Corporation.

BUNKER, H. J. (1963). Microbial food. In *Biochemistry of Industrial Micro-organisms*, Ed. C. Rainbow & A. H. Rose, pp. 34–67. New York and London: Academic Press.

BUNKER, H. J. (1968). Sources of SCP. In *Single-Cell Protein*, Ed. R. I. Mateles and S. R. Tannenbaum, pp. 67–78. London: MIT Press.

CHIAO, J. S., & PETERSON, W. H. (1953). Yeasts, methionine and cystine contents. *Agricultural and Food Chemistry* **1**, 1005–8.

COCHRANE, V. W. (1958). *Physiology of Fungi*. New York: John Wiley & Sons Inc.

DAWSON, E. R. (1952). The cultivation and propagation of Baker's Yeast. *Chemistry & Industry*, 793–7.

DELANEY, R. A. M., & WORGAN, J. T. (1970). The production of protein foods by the mycelial growth of fungi in culture media prepared from plant waste products. *3rd International Congress of Food Science and Technology*. Washington, DC, Abstract 113.

DEVELOPMENT COMMISSION REPORT (1953). *A Survey of Agricultural Forestry and Fishery Products in the UK*. London: HMSO.

DUDLEY STAMP, L. (1960). *Our Developing World*, p. 119. London: Faber & Faber.

DUNNING, J. W., & LATHROP, E. C. (1945). The saccharification of agricultural residues. *Industrial & Engineering Chemistry* **37**, 24–9.

EDOZIEN, J. C., UDO, U. U., YOUNG, V. R., & SCRIMSHAW, N. S. (1970). Effects of high levels of yeast feeding on uric acid metabolism of young men. *Nature, London* **228**, 180.

ELSWORTH, R., MILLER, G. A., WHITAKER, A. R., KITCHING, D., & SAYER, P. D. (1968). Production of *Escherichia coli* as a source of nucleic acids. *Journal of Applied Chemistry* **17**, 157–66.

EVANS, R. E. (1960). *Rations for Livestock*. MAFF Bulletin Number 48, pp. 47–51. London: HMSO.

FOOD AND AGRICULTURE ORGANISATION (1964). Protein. *World Food Problems*, Number 5, p. 13. Rome: FAO.

GILBERT, N., HOBBS, I. A., & LEVINE, J. D. (1952). Industrial and Engineering Chemistry **44**, 1712–20.

HAN, Y. W., DUNLOP, C. E., & CALLIHAN, C. D. (1971). Single-cell protein from cellulosic wastes. *Food Technology* **25**, 130–3, 154.

HARRISON, J. S. (1967). Aspects of commercial yeast production. *Process Biochemistry* **2**, 41–5.

HEDEN, C. G., MOLIN, N., OLSSON, U., & RUPPRECHT, A. (1971). Preliminary experiments on spinning bacterial proteins into fibres. *Biotechnology and Bioengineering* **13**, 147–50.

HOBSON, P. N. (1965). Continuous culture of some anaerobic and facultatively anaerobic rumen bacteria. *Journal of General Microbiology* **38**, 167–80.

HOSPODKA, J. (1966). Industrial application of continuous fermentation. In *Continuous Culture of Micro-organisms*, Ed. I. Malek & Z. Fenel, pp. 493–645.

JENSEN, A. L., & SHU, P. (1961). Production of lysine-rich yeast. *Applied Microbiology* **9**, 12–15.

KATZ, M., & REESE, E. T. (1968). Production of glucose by enzymic hydrolysis of cellulose. *Applied Microbiology* **16**, 419–20.

LILLEY, V. G. (1965). Chemical constituents of the fungal cell. In *The Fungi*, Ed. G. C. Ainsworth & A. S. Sussman, Volume 1, pp. 163–75. London: Academic Press.

LIPINSKY, E. S., & LITCHFIELD, J. H. (1970). Algae, bacteria and yeasts as food or feed. *Critical Reviews in Food Technology* **1**, 581–618.

LOCK, E. G., SAEMAN, J. F., & DICKERMAN, D. K. (1945). The production of wood sugar in Germany and its conversion to yeast and alcohol. Fiat Final Report, Number 499. London: HMSO.

LUEDEKING, R. (1967). Fermentation process kinetics. In *Biochemical and Biological Engineering Science*, Ed. N. Blakebrough, Volume 1, pp. 181–243. New York and London: Academic Press.

MAUL, S. B., SINSKEY, A. J., & TANNENBAUM, S. R. (1970). New process for reducing the nucleic acid content of yeast. *Nature, London* **228**, 181.

MILLER, D. S. (1963), Some nutritional problems in the utilisation of non-conventional protein for human feeding. In *Recent Advances in Food Science*, Ed. J. Muil Leitch & D. N. Rhodes, Volume 3. London: Butterworths.

MILLER, D. S. (1968). Queen Elizabeth College, London, unpublished results.

MITRAKOS, K., SEKERI, K., DROULISKOS, N., & GEORGI, M. (1970). Microbial protein from carob. *3rd International Congress of Food Science and Technology*, Washington, DC, Abstract 110.

MITSUDA, H., YASUMOTO, K., & NAKAMURA, H. (1969). A new method for obtaining protein isolates from Chlorella alga, Torula yeasts and other microbial cells. *Chemical Engineering Progress Symposium Series* **65**, no. 93, 93–103.

MORRIS, E. O. (1958). Yeast growth. In *The Chemistry and Biology of Yeasts*, Ed. A. H. Cook, pp. 251–321. London: Academic Press.

MORRIS, J. G. (1960). Studies on the metabolism of *Arthrobacter globiformis*. *Journal of General Microbiology* **22**, 564–82.

NELSON, G. E. N., ANDERSON, R. F., RHODES, R. A., SHEKLETON, M. C., & HALL, H. H. (1960). Lysine, methionine and tryptophan content of micro-organisms. II. Yeasts. *Applied Microbiology* **8**, 179–82.

OLSEN, A. J. C. (1960). Manufacture of Baker's Yeast by continuous fermentation. *Chemistry & Industry*, 416–25.

OLSON, B. H., & JOHNSON, M. J. (1949). Factors producing high yeast yields in synthetic media. *Journal of Bacteriology* **57**, 235.

PEPPLER, H. J. (1970). Food yeasts. In *The Yeasts*, Ed. A. H. Rose & J. S. Harrison, Volume 3, pp. 421–60.

PERLMAN, D. (1965). The chemical environment for fungal growth. In *The Fungi*, Ed. G. C. Ainsworth & A. S. Sussman, Volume 1, pp. 479–88. London: Academic Press.

PEW, J. C., & WEYNA, P. (1962). Fine grinding, enzyme digestion and the lignin–cellulose bond in wood. *Tappi* **45**, 247.

PIRT, S. J. (1965). The maintenance energy of bacteria in growing cultures. *Proceedings of the Royal Society* (B) **163**, 224.

PORTER, J. W. G., WESTGARTH, D. R., & WILLIAMS, A. P. (1968). A collaborative test of ion-exchange chromatographic methods for determining amino acids. *British Journal of Nutrition* **22**, 437–50.

RAFALSKI, J., SALM, J., KLUSZCYNSKA, Z., & SWITONICK, T. (1966). *2nd International Congress of Food Science & Technology*, Warsaw, A. 1.20.

RHODES, R. A., HALL, H. H., ANDERSON, R. F., NELSON, G. E. N., SHEKLETON, M. C., & JACKSON, R. W. (1961). Lysine, methionine and tryptophan content of micro-organisms. III. Moulds. *Applied Microbiology* **9**, 181–4.

SENEZ, J. C. (1962). Some considerations of the energetics of bacterial growth. *Bacteriological Reviews* **26**, 95–107.

SMITH, R. H. (1971). Rowett Research Institute, Aberdeen, unpublished results.

TEMPEST, D. W., & DICKS, J. W. (1967). Inter-relationships between potassium, magnesium, phosphorus and ribonucleic acid in the growth of *Aerobacter aerogenes* in a chemostat. In *Microbial Physiology and Continuous Culture*, Ed. E. O. Powell, C. G. T. Evans, R. E. Strange & D. W. Tempest, 3rd International Symposium, pp. 140–54.

TREIT, M. (1967). Industrial aspects of microbial food production with special reference to the potential of the new Symba-yeast process. In *Biology and the Manufacturing Industries*, Ed. M. Brook. Institute of Biology Symposia, Number 16, pp. 3–10.

UNITED STATES FOREST SERVICE (1964). Food yeast production from wood-processing by-products. Research note FPL-065. US Department of Agriculture.

WHITE, J. (1954). Baker's Yeast. In *Yeast Technology*, pp. 27–41. London: Chapman & Hall.

WOODWARD, F. N. (1952). Industrial utilisation of agricultural wastes. *Chemistry & Industry* **71**, 844.

WORGAN, J. T. (1967). Production of fermentation and cultivation media. British Patent Specification 1220807.

WORGAN, J. T. (1968). Culture of the higher fungi. In *Progress in Industrial Microbiology*, Ed. D. J. D. Hockenhull, Volume 8, pp. 74–139. London: Churchill.

YAMADA, K., TAKAHASHI, J., KAWABATA, Y., OKADA, T., & ONIHARA, T. (1968). SCP from yeast and bacteria grown on hydrocarbons. In *Single-Cell Protein*, Ed. R. I. Mateles & S. R. Tannenbaum, pp. 192–207.

ZALOKAR, M. (1959). Enzyme activity and cell differentiation in *Neurospora*. *American Journal of Botany* **46**, 555–9.

DISCUSSION

By E. J. ROLFE

National College of Food Technology (University of Reading),
St George's Avenue, Weybridge, Surrey

AND A. SPICER

Lord Rank Research Centre, Lincoln Road, High Wycombe, Bucks.

Food production in the world is keeping slightly ahead of population growth, due in a considerable measure to the expansion in cereal production over recent years. In this respect the new high-yielding cereals have made an important contribution. The rate of growth of fish production has also been striking in some areas, particularly South America, but far too much of the production is of poor quality and used for fertiliser or animal feeding. There is a marked need of protein for human consumption in these areas and work is imperative to enable utilisation of a greater proportion of the catch for human consumption, not only as fresh but in a processed form or as a concentrate.

In the UK a typical diet provides about 3150 kcal per day, including a total protein intake of about 90 g. Food intake in developing countries is considerably less, e.g. in India 1810 kcal per day and a total protein intake of 46 g. Assuming that the biological value of the proteins in the foods eaten in the developing countries is about 60, then the desirable intake for an adult would be about 53 g. Thus it is apparent that the problem is one of adequate food; if the people received an adequate amount of the food they are presently eating their protein deficiency would be solved. Assuming inadequate calories and that the protein intake of the people were increased the problem of protein malnutrition would still remain as the additional protein would be largely used for energy production.

In some developing countries the staples are low-protein foods, e.g. cassava or plantain, and there is need for supplementation by protein-rich foods. There are also the so-called vulnerable groups of the population that have an enhanced requirement for protein, e.g. infants, and expectant and nursing mothers, and in many countries the protein content of the diet of these groups is below a satisfactory level. During recent years it has been shown that undernutrition at certain critical stages of brain growth is associated with a subsequent irreversible impairment of mental function. The vulnerable period occurs in humans at the time when the brain is developing most rapidly, i.e. it begins about three-quarters of the way through pregnancy and continues through the first $1\frac{1}{2}$ years after birth. In the developed countries the protein problem is of quite a different kind. Traditional sources of protein, particularly meat, have become expensive. Animals are poor converters and a high price must be paid to have vegetable proteins converted into the more desirable meat proteins. Interest is therefore centred on the conversion by application of food technology of vegetable and other forms of protein into more desirable forms. This then in brief is the protein problem.

Significant nutritional improvement of cereal grains is possible both by increasing the protein content and by improving the quality of the protein. Genes for high protein content in wheat have been identified at the University of Nebraska and the stability and potential usefulness of these genes in the major wheat-producing areas of the world is being studied. However, the high-protein genes do not affect significantly the amino acid composition of the protein. In both wheat and rice, the mainstays of many diets in the world, the proteins are deficient in essential amino acids. The gene in maize to increase the lysine content has been identified and could lead to significant improvements in the protein quality of diets of maize-eating populations. Such success has stimulated the search for genes with similar effects in other cereals, e.g. barley, sorghum, rice and wheat. It should be possible in the future to tailor plant varieties to meet man's protein and other nutritional needs much better than at present. In the meantime the improvement of cereal-based diets by the addition of small amounts of protein concentrates, or by fortification with the limiting essential amino acids, is a means of compensating for the qualitative deficiencies of cereal protein.

The adoption of cereals with improved protein value is not easy; the industrialised countries do not feel the need. In the developing countries where there is the major need for cheap protein the problem is to persuade the farmers to grow the protein-improved varieties. The farmers require to be convinced that to grow them is profitable, which implies a quality-dependent differential payment. To introduce the latter presents problems not only of an administrative nature, but would also present problems of sampling and provision of analytical laboratories to produce the results on which quality payments could be based.

Protein deficiencies may also be relieved by utilisation of oil seed proteins for direct human consumption and also by reducing the substantial food losses which at present occur during harvesting and storage.

It is against this background that the usefulness of single-celled proteins

(SCP) must be assessed and their potential contribution to human nutrition evaluated. For the sake of completeness SCP is taken to include algae and microfungi. Their major advantage is that they can utilise waste, surplus or cheap carbon sources, e.g. sewage effluent, molasses or petroleum derivatives, for the conversion of inorganic nitrogen into protein. Organisms which have been proposed and studied as potential protein sources include:

Bacteria

In particular they have tolerance to relatively high temperatures which confers important technological advantages. Difficulties arise in harvesting the cells.

Yeasts

Production on a commercial scale has been operated for a large number of years for the baking industry and also food yeast has been produced on a commercial scale. During the 1939–45 war, surplus molasses in Jamaica was used for protein production. The yeast cells were separated by centrifuging, washed and dried on a drum drier.

Microfungi

Less specific in substrate requirement than yeasts and the filamentous nature leads to simplication of harvesting techniques.

Algae

Being green, algae utilise sunlight as energy and CO_2 as carbon source. Unfortunately the green colour also limits acceptability, and there are difficulties in devising an economic production process.

Available substrates vary widely in nature from hydrocarbons (including gaseous methane, which can be utilised by certain bacteria, and which, being gaseous, cannot contaminate the product and avoids the necessity for solvent extraction to remove contaminants, and also normal paraffins and gas oil), various carbohydrates (starch, sugar, molasses, and also, following chemical treatment, such waste agricultural materials as bagasse, oat husk, etc.) and petrochemicals, particularly methanol and ethanol. Different substrates for microbial conversion were discussed at a PAG meeting in Moscow in 1971; methanol, ethanol and carbohydrates were accepted as the three substrate nutrients so far acceptable for microbial conversion for food suitable for human consumption. It was recommended that, for the present, hydrocarbon-grown material should be used only for animal feeding.

SCP production may be carried out more simply as a batch process, but a commercial operation demands a continuous process. Though offering advantages of several kinds, including costs and close control, it is more difficult to devise. However, processes are operating successfully based on hydrocarbons and on starch.

In selecting the organism for culture a considerable number of factors and characteristics must be taken into account, e.g. growth temperature (a high growth temperature can be expected to enhance rate of protein production and reduce the need for refrigeration of the culture vessel), ease of harvesting, yield and composition of the protein, absence of toxic compounds, etc., but in the end it amounts to the selection of a strain that produces high-quality protein at optimum yield most economically. Having selected and developed a strain to the production stage, it will be difficult and expensive to take up another strain which subsequently may appear to offer economic advantages. An important factor causing this is that any new SCP must pass careful scrutiny, and every new strain of organism must be submitted to expensive and time-consuming tests for toxicity and other properties. There have been detailed descriptions of the procedures which should be followed as an essential prerequisite to evaluate a new protein-rich food on human subjects (PAG Guideline No. 6, 1970).

Before they may be used as human food sources, new foods must be assessed for the quality of their protein content and their safety for use. In addition to toxicity tests on animals, extensive microbiological examination is needed of the product from its production through its many-step evolution to the ultimate package for the consumer. Following this work, studies of the new protein's nutritional value should be carried out using animal feeding tests and chemical tests of amino acid content and availability. Such testing applies especially to new foods developed by isolation from conventional sources by unusual techniques, and to microbial proteins. Processes involving the use of solvent extractions and unusual heating conditions, or the utilisation of food additives in a variety of combinations, may result in changes in digestibility, absorption, metabolism, or safety of the food in question. In the case of SCP particular attention must be directed to the composition of the media from the viewpoint of the possible presence of chemical components regarded as hazardous to health.

Novel foods successfully completing such a series of tests need then to be evaluated for their suitability for human nutrition (PAG Guideline No. 7, 1970). The objective is to use with maximum safety the fewest subjects possible to obtain significant data. Human testing in the scheme falls into four main categories:

(1) Determination of product acceptability and physiological tolerance.
(2) Growth tests – measurement of body weight and height.

(3) Determination of nitrogen balance, Net Protein Utilisation.

(4) Other criteria, e.g. measurement of changes in amount of serum albumin, plasma amino acid and enzyme levels.

During actual production there are hazards which the producer must consistently guard against, viz.:

(1) If using bacteria as a source of SCP care must be taken to prevent attack from phage or other lytic predators which could effectively stop protein production.

(2) The difficulties of operating large-scale plant under sterile and aseptic conditions must be overcome in order to avoid danger of infection with a toxic or pathogenic species.

(3) How great is the hazard of the spontaneous development of a mutant with toxic or other undesirable properties? To date yeasts have proved extremely stable in this respect, and experience covers a large number of years where yeasts have been grown for brewing and baking. It is evident, however, that safety depends on an extremely careful control of all factors involved in cell production and can be ensured only by a rigid programme of frequent microbiological and biochemical analytical control and periodic biological re-evaluation of the material (PAG Statement No. 4, 1970).

After harvesting the cells there are still some problems of utilisation for human consumption. The cell wall of the micro-organism is indigestible and the nutrients will become available only provided the cell wall is ruptured either mechanically or by the action of appropriate enzymes. On an industrial scale this is an additional operation and adds to the cost. In the case of microfungi a more fragile cell wall is obtained by harvesting young cultures. This problem may be overcome by isolation of suitable mutants, e.g. it has been reported recently that, following treatment with a chemical mutagen, a large number of mutants with various defects in cell wall biogenesis in a green alga, *Chlamydomonas reinhardi*, were isolated; a few produced no detectable quantities of cell wall. In spite of this structural defect the cells retain ample mechanical strength to allow aeration, stirring, and harvesting by centrifugal separation. The quality and composition of the cell protein is essentially unchanged (Davies, 1971).

The microbial cells in general exhibit a high nucleic acid content which limits the intake of SCP. Toxicity to humans arises through lack of uric acid oxidase and consequently the total amount of nucleic acid in the diet should be limited in order to avoid excessive concentrations of uric acid in blood and urine. Present information suggests there should be a limit of 2 g per day on the amount of nucleic acid introduced by SCP into the diet of an adult and correspondingly less by weight for children. This is the amount contained in approximately 10–30 g of ordinary cells harvested in the rapidly growing phase. Nucleic acid removal therefore becomes obligatory if SCP is to be used in large quantities as a human food. A process has been developed to achieve this and it is relatively cheap. Even so it must increase the cost of the protein but it is anticipated that these novel foods will remain competitive and supply to the market proteins of high biological value at reasonable cost.

The nutritive value of the SCP for human food will depend on the digestibility and biological value of the protein. All SCPs appear to be rich in the essential amino acids. The bacterial and yeast proteins suffer from a shortage of sulphur amino acids which does not occur in microfungi. The latter are able to produce a protein profile very similar to the ideal prescribed by FAO.

All microbial protein whether for human or animal nutrition must be competitive in cost with existing or available alternative protein materials such as fish meal, vegetable protein concentrates or oil seed cakes. An important item in cost will be the cost of the substrate, and care must be taken that a process based on a cheap waste material is not priced out of the market subsequently by rising costs of the substrate following a rising demand for it. It seems doubtful at the moment that economy of production for animal feeding has been achieved so far, but with sustained progress during the next two or three years there can be expected to appear an economical conversion of hydrocarbon substrate into microbial protein for animal use.

Before a new food such as SCP can be used in any country for human consumption it must be approved by the appropriate legislative authority. Such approval is given only following the submission of evidence to show suitability as a human food and freedom from toxic or other materials deleterious to health. In the UK the Food Standards Committee is reviewing the use of all unconventional proteins as food, including those of vegetable and microbial origin. Aspects to be considered are safety and provisions regarding composition, description and labelling. The utilisation of vegetable proteins has been practised for many years, e.g. soya sausages during World War II, and soya is commonly added to bread. New products as textured vegetable protein (TVP) may be added to a meat product and make it more nutritious but also more expensive, as at the moment the TVP cannot legally replace any of the statutory meat requirement.

Extensive testing and feeding trials of SCP have been in progress for many years with favourable results and it can be expected that in due course SCP will find application in human foods.

With any new food the producer is faced with the problem of persuading the customer to eat it. Perhaps the simplest method is to disguise it, which may follow either of two procedures:

(1) Fortification of traditional foods, e.g. bread, soups, etc.

(2) Simulation of the conventional foods, e.g. meat analogues, milk substitutes.

In both procedures the application of SCPs is possible because they are in general colourless, odourless and tasteless. They can therefore be coloured and flavoured to simulate a chosen product or added as a bland material to a traditional dish.

Multicellular organisms such as microfungi will have the advantage of possessing textural qualities which are not inherent in yeasts and bacteria and therefore appear to be more suitable for conversion into food possessing some organisation or structure. Texturisation of yeast or bacterial biomass would be an additional and expensive operation. It is obvious that it will be necessary to produce formulations which would give to each area products which are acceptable to the indigenous population, and texture and flavour will be the predominant factors.

When considering the production of novel proteins it must be appreciated that the needs of developed and developing countries may differ. It is readily apparent that efficiency of production is important to both, but striking differences can emerge when the economy of individual countries is taken into account, e.g. labour utilisation, availability of technical manpower, average income of the working population and capital investment.

No one single material will have the monopoly of application for all the world but the availability of substrate material in each country, its advance in technology and other factors as mentioned above will ultimately decide whether, for example, it is the hydrocarbon-based conversion or some other substrate conversion which will be most economical. It will be necessary to produce formulations which would give to each area products which are acceptable to the people living in those countries, bearing in mind that texture and flavour are predominant factors in the introduction of novel foods. It would seem that the ideal situation would be to have the new foods manufactured and marketed by private companies on a commercial basis but with the full co-operation of governments. In such circumstances the aim should be to make bulk sales of the novel food as a welfare food with little or no profit, and the normal retail outlet should provide the profit-making part of the project. Success would depend on the product becoming a low-profit, high-volume commercial venture, and herein lies the problem. It is very difficult to persuade people to eat highly nutritious mixtures resulting from the collaborative work of the nutritionist and the food technologist. Tradition, religious beliefs and customs are as important in controlling food selection as is palatability and price. If the food differs from familiar foods there may be reluctance to accept it. If it resembles a well-liked staple food the consumer will ask why it should be more expensive. If the new food is given away it may become stigmatised as the poor man's food.

The problem must therefore be attacked in all its aspects, technological, economic and, not least important, social and political. This multidisciplinary approach is essential if any degree of success is to be assured.

REFERENCES

DAVIES, D. R. (1971). Single cell protein and the exploitation of mutant algae lacking cell walls. *Nature, London* **233**, 143–4.

PAG GUIDELINE No. 6, FAO/WHO/UNICEF Protein Advisory Group, 'Guideline for preclinical testing of novel sources of protein', 13 March 1970.

PAG GUIDELINE No. 7, FAO/WHO/UNICEF Protein Advisory Group, 'Guideline for human testing of supplementary food mixtures', 10 June 1970.

PAG STATEMENT No. 4, FAO/WHO/UNICEF Protein Advisory Group, 'Control and periodic biological re-evaluation of the material on single cell protein', 5 June 1970.

LIST OF PARTICIPANTS

Dr Th. Alberda
Instituut voor Biologisch en Scheikundig Onderzoek van Landbouwgewassen (IBS), Bornsesteeg 65–67, Wageningen, Netherlands.

Dr E. F. Annison
Animal Research Division, Colworth/Welwyn Laboratory, Unilever Ltd, Colworth House, Sharnbrook, Bedford.

Professor F. Aylward
Department of Food Science, University of Reading, London Road, Reading, RG1 5AQ.

Dr C. C. Balch
National Institute for Research in Dairying, Shinfield, Reading, Berks.

W. H. Beaumont
Wellcome Veterinary Research Station, Frant, Tunbridge Wells, Kent.

Dr D. E. Beever
Grassland Research Institute, Hurley, Maidenhead, Berks.

E. Berner, Jr
Botanical Institute, 1432 Vollebekk, Norway.

Dr R. Bickerstaff
Animal Research Division, Colworth/Welwyn Laboratory, Unilever Ltd, Colworth House, Sharnbrook, Bedford.

Dr K. L. Blaxter, FRS
The Rowett Research Institute, Bucksburn, Aberdeen, AB2 9SB.

M. A. B. Boddington
School of Rural Economics and Related Studies, Wye College, Ashford, Kent.

Professor J. C. Bowman
Department of Agriculture, University of Reading, Earley Gate, Reading, RG6 2AT.

Dr R. Braude
National Institute for Research in Dairying, Shinfield, Reading, Berks.

N. R. Brockington
Grassland Research Institute, Hurley, Maidenhead, Berks.

Dr A. W. Broome
ICI (Pharmaceuticals) Ltd, Biology Department, Mereside, Alderley Park, Macclesfield, Cheshire.

Dr W. H. Broster
National Institute for Research in Dairying, Shinfield, Reading, Berks.

LIST OF PARTICIPANTS

Dr P. J. Buttery
Department of Applied Biochemistry & Nutrition, University of Nottingham, School of Agriculture, Sutton Bonington, Loughborough, Leics.

R. Clark-Monks
Nutrition Department, Nitrovit Ltd, Dalton, Thirsk, Yorks

Dr K. E. Cockshull
Glasshouse Crops Research Institute, Worthing Road, Rustington, Littlehampton, Sussex.

Dr J. P. Cooper
Welsh Plant Breeding Station, Aberystwyth, Cards.

D. W. Cowling
Grassland Research Institute, Hurley, Maidenhead, Berks.

Professor C. M. Donald
Department of Agronomy, University of Adelaide, Waite Agricultural Research Institute, Glen Osmond., S. Australia.

Professor A. N. Duckham, CBE
Studio Cottage, Didcot Road, Blewbury, Berks.

Dr J. E. Duckworth
Meat & Livestock Commission, P.O. Box 44, Queensway House, Bletchley, Bucks.

Dr I. F. Duthie
Lord Rank Research Centre, Lincoln Road, High Wycombe, Bucks.

Dr V. R. Fowler
Rowett Research Insitute, Bucksburn, Aberdeen, AB2 9SB.

Miss F. E. Gunnell
Beechcroft, Dukes Road, Fontwell, nr Arundel, Sussex.

Dr F. J. Harte
Grange, Dunsany, Co. Meath, Ireland.

Dr S. B. Heath
Department of Agriculture, University of Reading, Earley Gate, Reading, RG6 2AT.

M. R. Heslehurst
39 Norman Street, Coorparoo, 4151, Brisbane, Australia

M. T. Heydeman
Department of Microbiology, University of Reading, London Road, Reading, RG1 5AQ.

R. W. Hodge
Grassland Research Institute, Hurley, Maidenhead, Berks.

Professor T. Homb
Norges Landbrukshøgskole, Post Box 25, 1432 Ås-NLH, Norway.

LIST OF PARTICIPANTS 373

Dr B. J. F. Hudson
Department of Food Science, University of Reading, London Road, Reading, RG1 5AQ.

Dr A. P. Hughes
Plant Environment Laboratory, University of Reading, Shinfield Grange, Shinfield, Reading, Berks.

K. E. Hunt
Institute of Agricultural Economics, Dartington House, Little Clarendon Street, Oxford.

J. Hutchinson
Protein Food Development Group, Nutrition Division, FAO, Via delle Terme di Caracalla, 00100 Rome.

Professor H. Hvidsten
Agricultural College of Norway, Vollebekk, Norway.

Dr P. A. Jewell
Department of Zoology, University College, Gower Street, London, W.C.1.

Dr O. R. Jewiss
Grassland Research Institute, Hurley, Maidenhead, Berks.

Dr A. S. Jones
Rowett Research Institute, Bucksburn, Aberdeen, AB2 9SB.

Dr J. G. W. Jones
Department of Agriculture, University of Reading, Earley Gate, Reading, RG6 2AT.

Dr L. H. P. Jones
Grassland Research Institute, Hurley, Maidenhead, Berks.

Dr D. C. Joshi
Norges Landbrukshøgskole, Post Box 25, 1432 Ås-NLH, Norway.

Dr A. J. Keys
Rothamsted Experimental Station, Harpenden, Herts.

R. V. Large
Grassland Research Institute, Hurley, Maidenhead, Berks.

Dr B. M. Laws
Pauls & Whites Foods Ltd, Hall-Marke House, New Cut West, Ipswich, IP2 8HP, Suffolk.

Dr E. L. Leafe
Grassland Research Institute, Hurley, Maidenhead, Berks.

W. Little
ARC Institute for Research on Animal Diseases, Compton, nr Newbury, Berks.

LIST OF PARTICIPANTS

G. R. de Lucia
Centro de Investigaciones Agricolas, 'Alberto Boerger', La Estanzuela, Colonia, Uruguay, South America.

J. C. McKenzie
Food & Drink Research Ltd, Centre House, 114-116 Charing Cross Road, London, WC2H 0JR.

Dr J. M. McNab
Poultry Research Centre, Kings Buildings, West Mains Road, Edinburgh, EH9 3JS.

Professor A. J. Matty
Department of Biological Sciences, University of Aston in Birmingham, Birmingham 4.

W. S. Miller
Spillers Ltd, Kennett Nutritional Centre, Kennett, Newmarket, Suffolk.

Dr T. R. Morris
Department of Agriculture, University of Reading, Earley Gate, Reading RG6 2AT.

J. Morrison
Grassland Research Institute, Hurley, Maidenhead, Berks.

R. F. A. Murfitt
National College of Agricultural Engineering, Silsoe, Beds.

Dr E. Owen
Department of Agriculture, University of Reading, Earley Gate, Reading, RG6 2AT.

Ir P. J. J. Philipsen
Instituut voor Bewaring en Verwerking van Landbouwproducten (IBVL), 59, Bornsesteeg, Wageningen, Netherlands.

Dr J. Phillipson
Animal Ecology Research Group, Department of Zoology, Oxford University, South Parks Road, Oxford.

Professor N. W. Pirie, FRS
Rothamsted Experimental Station, Harpenden, Herts.

W. Pooswang
Department of Horticulture, University of Reading, Shinfield Grange, Shinfield, Reading, Berks.

Professor G. S. Puri
Liverpool Polytechnic, Byron Street, Liverpool 3.

Dr H. A. Regier
Department of Zoology, University of Toronto, Ramsey Wright Zoological Laboratories, 25 Harbord Street, Toronto 5, Ontario, Canada.

LIST OF PARTICIPANTS

Professor E. H. Roberts
Department of Agriculture, University of Reading, Earley Gate, Reading, RG6 2AT.

Dr R. Roberts
J. Bibby Agriculture Ltd, Richmond House, 1 Rumford Place, Liverpool, L3 9QQ.

A. A. Robinson
ICI (Pharmaceuticals) Ltd, Biology Department, Mereside, Alderley Park, Macclesfield, Cheshire.

Professor E. J. Rolfe
National College of Food Technology, St George's Avenue, Weybridge, Surrey.

Dr B. A. Rolls
National Institute for Research in Dairying, Shinfield, Reading, Berks.

Professor J. A. F. Rook
The Hannah Dairy Research Institute, Ayr, Scotland.

Dr J. Rothwell
Department of Food Science, University of Reading, London Road, Reading, RG1 5AQ.

Dr R. H. Smith
National Institute for Research in Dairying, Shinfield, Reading, Berks.

J. J. Soulsby
White Fish Authority, 2/3 Cursitor Street, London E.C.4.

Professor C. R. W. Spedding
Department of Agriculture, University of Reading, Earley Gate, Reading, RG6 2AT.

Professor A. Spicer
The Lord Rank Research Centre, Lincoln Road, High Wycombe, Bucks.

Dr J. E. Storry
National Institute for Research in Dairying, Shinfield, Reading, Berks.

Dr D. A. Stringer
ICI (Agricultural Division) Ltd, Jealotts Hill Research Station, Bracknell, Berks.

Dr D. Swaine
Brooke Bond Liebig Research Centre, Blounts Court, Sonning Common, Reading, Berks.

Dr J. C. Tayler
Grassland Research Institute, Hurley, Maidenhead, Berks.

Dr D. J. Thomson
Grassland Research Institute, Hurley, Maidenhead, Berks.

Dr J. H. M. Thornley
Glasshouse Crops Research Institute, Worthing Road, Rustington, Littlehampton, Sussex.

Dr R. J. Treacher
ARC Institute for Research on Animal Diseases, Compton, nr Newbury, Berks.

Dr T. Walker
BP Proteins Ltd, Britannic House, Moor Lane, London, E.C.2.

Miss J. M. Walsingham
Grassland Research Institute, Hurley, Maidenhead, Berks.

Professor P. F. Wareing
Department of Botany, University College of Wales, Aberystwyth, Cards.

Dr N. Watchorn
ICI (Agricultural Division) Ltd, Jealotts Hill Research Station, Bracknell, Berks.

R. N. H. Whitehouse
Plant Breeding Institute, Maris Lane, Trumpington, Cambridge.

A. P. Williams
National Institute for Research in Dairying, Shinfield, Reading, Berks.

Professor G. Williams
Department of Zoology, University of Reading, Whiteknights Park, Reading, Berks.

T. E. Williams
Grassland Research Institute, Hurley, Maidenhead, Berks.

Dr P. N. Wilson
BOCM Silcock Ltd, Basingstoke, Hants.

Dr A. A. Woodham, The Rowett Research Institute, Bucksburn, Aberdeen, AB2 9SB.

Dr J. T. Worgan,
National College of Food Technology, St George's Avenue, Weybridge, Surrey.

INDEX

acetate, as source of energy for ruminants, 150
Actinomyces elegans, as source of protein, 350
Aerobacter aerogenes, cell composition of, 346
Aerobacter cloaceae, energy required for maintenance of, 342
aflatoxin, in groundnuts, 288
alanine: glucose requirement for synthesis of, 314; incorporation of ammonia into, in plants, 75, 77
alanine tRNA, structure of, 144
albumen, in different cereals, 88, 89
alfalfa, leaf protein of, 297, 300
algae: allergic reactions in human consumers of, 313; filamentous, compared with coccal, 313; potential usefulness of, as human food, 56, 365; protein production by, 304–6; *see also individual species*
alginate of seaweed, as substrate for yeast, 347
Amanita, edible and toxic species of, 354
amino acids: activation of, in protein synthesis, 144–5, 148, 149; essential, 7–8; imbalance of, 141, 152–5; possible replacement of protein foods by, 3, 314-15; production of, by micro-organisms, 315–16; protein synthesis from, by animals, 145–8, and by plants, 78–9; in proteins of algae, 306, of animals and plants, 34–5, 113, of bacteria, 162, 308, of cereals, 34, 84–6, of fish meal and soya beans, 334, of fungi, 368, of protozoa, 162, and of yeasts, 334, 352, 353, 354; specific dynamic action of, 141, 155–60, 161; synthesis of, in plants, 75–8, 134; supplementation of proteins by, 53, 299, 364; *see also individual amino acids*
amino-acyl bond, energy of, 145, 148
amino-acyl transfer RNA synthase, 145
ammonia: dilution of high-quality protein by, 7; fed to dairy cattle (as citrate), 243; reduction of nitrate to, in plants, 69–72, 73; rumen micro-organisms and, 162, 163–4, 166; as source of nitrogen for plants, 73–4; specific dynamic action of, 156, 157, 159, 161; in synthesis of amino acids, 75–8, 134
Amoeba proteus, nuclear protein of, 143
amylase, of *Endomycopsis*, 347
anticodon sites, on tRNAs, 144, 310

antinutrients, in legumes and oilseeds, 292
aquaculture systems, 263; efficiency of protein production in, 264–5, 269; practical aspects of, 265–6
aquatic ecosystems, 263; aquaculture, 264–6; different types of 'wild', 266–8; general model for, 274–7; trophodynamic methods applied to, 268–74
arginine, 7; apparent deficiency of, caused by dietary excess of lysine, 154; in cereals, 84, 85, 86, 93
Arthrobacter globiformis, percentage of substrate recovered in, 343
aspartic acid: synthesis of other amino acids from, in *E. coli*, 317
Aspergillus niger: carbohydrate utilisation by mutants of, 347; NPU value for, 353; percentage of substrate recovered in, 344
assimilation efficiency, 221; of carnivores, 226; of invertebrates, 221; of mammals, (large) 223, (small) 222
ATP: from catabolism of amino acids, 151, 152, 158, 159; hydrolysis of, to AMP and pyrophosphate, in RNA synthesis, 142; produced by rumen micro-organisms, 164–5; requirement for, in protein synthesis, 311

Bacillus megaterium: growth rate of, 307; homogenising to break cell walls of, 312; as protein supplement for rats, 308
Bacillus stearothermophilus, 325
Bacillus subtilis, cell hydrolysate from, 312
bacteria: amino acid composition of protein of, 162, 308, 352, 368; growth of, on hydrocarbons, 308, 323, 324, 325, 328–9; nitrogen fixation by, 308; nucleic acid content of, 307, 309, 310, 330, 351; nutritive value of, 332; percentage of carbohydrate converted to protein in, 344, 355; protein content of, 306, 344, 350–1; protein production by, 306–8, 356, 357, 365; protein synthesis in, 145–6, 149
bacteriophage, in large-scale production of bacteria, 367
Bacteroides ruminicola, nitrogen metabolism of, 163
barley: amino acid composition of protein of, 85, 86, 93; breeding work on, 96; Hiproly variety of, 89, 96; protein fractions of, 89; world production of, 83, 84

[377]

barley beef production, 248, 253
beans, 104; protein concentrate from, 109, 298
biological value, 6, 174, 177; of bacterial and yeast proteins, 332; of *Candida* cells and separated protein, 352; of protein in diets of developing countries, 363
biomass (animal), per unit area: of invertebrates, 222, 223, 224; of large mammals, 222, 223-4; of small mammals, 222-3
blood: protein in, 238, 239; utilisation of, 293-4
blue-green algae, nitrogen fixation by, 308
Boletus edulis, NPU value for, 353
Botrytis cinerea, attacks hydrocarbons, 323
brain, effect of undernutrition on, 363-4
Brazil, estimate of future protein requirements in, 31
breeding, improvement of crops by, 55, 90-6
Brevibacterium flavum: mutant of, producing threonine, 316
brussels sprouts, protein content of, 104, 105
buckwheat, edible seeds of, 104

calves: efficiency of protein production by, 244-6; feeding of, 243; protein content of carcasses of, 241, 289
Candida tropicalis, effect of substrate on amino acid composition of protein of, 353
Candida utilis: amino acid composition of protein of, 353; commercial production of, 340, 350, 351-2; growth of, in conjunction with *Endomycopsis*, 347; NPU value for, 353; nucleic acid content of, 346, 351; nutritive value of cells and separated protein of, 352, 355; protein content of, 346; protein production by, 356, 357
carbohydrate conversion to protein, by micro-organisms, 339-44, 365; efficiency of process, 355-8; rate of process, 344-6; types of carbohydrate used, 346-50
carboxydismutase, in chloroplasts, 135
carnivores, in ecosystem, 226, 227, 229
cassava leaves, as food in tropics, 105
castration of male animals: and protein content of carcasses, 241; and protein production, 212, 246-7, 250
cattle: efficiency of conversion of dietary protein to meat protein in, 178, 179, 208, 227, 241, 242, 289; efficiency of populations of, producing meat, 195, 196, 198; factors affecting population structure of, 184, 187; percentage of carcasses of, available as meat, 186; protein content of carcasses of, 241, 289; *see also* barley beef production, calves, dairy beef production, milk, ruminants

cellulases, of micro-organisms, 349, 350
Cellulomonas, growth rate of, 350
cellulose, as substrate for micro-organisms, 347-50
centrifuging, harvesting of micro-organisms by, 304, 313
cereals: amino acid composition of proteins of, 34, 84-6, 315; breeding work on, 89-97; competition between humans and animals for, 52, 53, 60; in diets, 286, 287; difficulties in adoption of improved varieties of, 58-9, 364; losses of harvested, 29; potential protein production by, 83-7; protein concentrates from, 107; protein content of, 84, 103; protein fractions of, 88-9; utilisation of, 291, 296; yield and protein content of, 87-8, 134; yield and protein yield of, 88; *see also individual cereals*
cheese, 290
Chenopodium, edible seeds of, 104
children, protein needs of, 9, 363
China, fish culture in, 265
chitin of microbial walls, not digested by animals, 312
Chlamydomonas mutants, without cell walls, 312, 367
Chlorella: digestibility of, 306; incorporation of ammonia into, 75, 76; potential production of, 305; protein content of, 305; source of nitrogen and rate of photosynthesis by, 74, 75; unpalatable, 313
chlorophyll, protein content of leaves proportional to content of, 105
chloroplasts, protein of, 135
cloud: allowance for, in calculation of maximum potential production of plant material, 219-20, 228, 229
cocoa, methods for increased production of, 281
coconut, protein concentrate from, 108-9
codon, of mRNA, 310
costs of production: of unconventional proteins, 114-15; variation in, 57-8; *see also* economics
cottonseed, 296; oil from, 63; protein of, 107-8
cow, percentage of protein in diet required for lactation, 119; *see also* milk
crustaceans, of ocean, 269
Cyprinus carpio, genetic selection of, 264-5
cystine, in protein of latex of upas tree, 113; *see also entry for* methionine plus cystine
cytokinins, and rate of protein synthesis, 79, 135
cytoplasm, proteins of, 135

dairy beef production, 246-50, 254, 256-7

INDEX

developing countries: economic planning in, 48, 51–3; livestock in, 60–1; policy for protein foods in, 64–5

digestibility: of algal protein, 306; of bacterial and yeast proteins, 332; of *Candida* cells and separated protein, 352; of *Fusarium*, 353

DNA: content of, in micro-organisms, 310; in protein synthesis, 78, 79, 309; 'reiteration' of desired segments of, 136; in RNA synthesis, 142, 143

dry matter: maximum production of, may achieve maximum return of nitrogen, 131, 134; percentage conversion of light energy into, 133

duck: nitrogen output of, in eggs, 16

economics: of fisheries, 276; of food processing, 299–300; of lysine supplementation of microbial protein, 331; of protein production, 45–65, (from hydrocarbons), 326–7; or urea feeding, 252

ecosystems, 217–18, 230–3; aquatic, 263–77; global land, 218–19; non-woodland, 219–27; woodland, 219, 228–30, 231

efficiency, biological, 115, 173–5; definitions of, 13–14, 24, 217, 237; factors affecting, 175–7; measures of, 178; possibilities of changing, 178–81

eggs: in diets, 286, 287; efficiency of production of, 15–17, 22–4, (by birds) 16–21, (by invertebrate pests) 21–2, 23–4; price of protein in, 55; production of, in developing countries, 52; protein of, 113; toxin from, 355; utilisation of, 290, 294; *see also* fowls, domestic

Endomycopsis fibuliger: growth of, in conjunction with *Candida*, 347

endoplasmic reticulum, ribosomes and, 146, 148

energetics: ecological, 282; of maintenance and synthesis of micro-organisms, 342–4; of production of milk and meat protein by cattle, 242; of protein synthesis, by animals, 148–51, 237–8, 311–12, by micro-organisms from hydrocarbons, 328, and by rumen micro-organisms, 164–6, 213; of protein turnover, 193

enzymes: leaf proteins as, 134, 135; in specific dynamic action, 160, 161; synthesis of, 79; in synthesis of amino acids in plants, 77–8

Escherichia coli: amino acid composition of protein of, 353; NPU value for, 353; nucleic acid content of, 310; percentage of substrate recovered in, 343; protein production by, 311, 356; as protein supplement for rats and chicks, 308;

RNA polymerase of, 143; synthesis of amino acids in, 315–16, 317

ethanol, as substrate for micro-organisms, 365

farmers: decisions to be made by, 47; politics and, 48–50

fat: in meat, 238, 239, 255; in milk, 238

feeding-stuffs for animals, 46; cereals as, 52, 53, 60; fish meal as, 243, 250–1; hydrocarbon-grown micro-organisms as, 333–4, 336, 365; leaf products as, 110, 113; possible 'stretching' of protein in, 53–4; soya beans as, 54, 368

fermentation, improvement of food quality by, 106–7, 108

fertiliser: fish meal as, 290, 363; oilseeds as, 62

fibre, in leaves, 105, 110

fish: competition between humans and animals for, 46, 50, 61, 363; in diets, 286, 287; efficiency of conversion of plant matter to, 231, 232; factors affecting populations of, 185; ratio of mass produced to water mass, in culture, 265; utilisation of, 290

fish meal, 290, 294; amino acid composition of protein of, 334; for feeding animals, 243, 250–1; as fertiliser, 290, 363; protein content of, 334

fisheries, 266–8; ecological model for, 274–7; trophodynamic methods applied to, 268–74

food: acceptability of new kinds of, 297–8, 316–17, 335, 369–70; efficiency of assimilation of, 221, 222, 223, 226; future world pattern of demand for and production of, 50–4; local surpluses of, 49–50; losses of, from pests, etc., 29, 54, 63–4, 364; preferences for different kinds of, 40–1; processing of, 285, 299–300; of rural and urban populations, 46–7; social functions of, in USA and Trobriand Islands, 39–40; unconventional sources of, 101–15, 281

forage: amount of, available to grazers on grassland and browsers in woodland, 229; drying of, by pressing, 112; leaf fibre as, 110

formaldehyde: treatment of protein with, to prevent breakdown by rumen micro-organisms, 162

formic acid, as additive to fresh grass for cattle, 256

fossil fuels, rate of accumulation of, 221

fowls, domestic: efficiency of conversion of dietary protein by, to egg protein, 178, 179, 204–5, 255, 289, and to meat protein, 205, 255, 289; efficiency of nitrogen

379

fowls (cont.)
    production by, in eggs and meat, 16–21; efficiency of populations of, producing eggs and meat, 188, 189, 191–2, 193, 195, 196, 197; factors affecting population structure of, 185, 187; *see also* eggs
fungi: adaptation of, to different carbohydrate substrates, 347; amino acid composition of protein of, 368; edible species of, 109, 354; growth of, on hydrocarbons, 323; nucleic acid content of, 351; percentages of carbohydrate substrate converted to protein in, 355, and recovered in, 343–4; potential usefulness of, as human food, 294, 355, 365, 369; protein content of, 344, 351; protein production by, 297, 356–8; woodrooting species of, 348; *see also* yeasts, *and individual species of fungi*
Fusarium semitectum: amino acid composition of protein of, 352, 353; efficiency of protein production by, 356, 357; nutritive value of, 353; protein and nucleic acid contents of, 351

gelatin, tryptophan-deficient protein, 6
genetic selection: for efficiency of growth and production, in fish, 264–5
genetic variation in efficiency: of protein production, 176; of protein utilisation, 177, 178–81
germination of seeds, food quality sometimes improved by, 106
gibberellic acid, and enzyme synthesis, 79, 135
globulin, in different cereals, 88, 89
glucose: amount of, required for synthesis of alanine and tryptophan, 314, and of peptide bond, 312; high-energy phosphate bonds from oxidation of, 150; lysine production from, 315; as substrate for micro-organisms, 353
glutamic acid: fermentation synthesis of, 315; incorporation of ammonia into, in plants, 69, 72, 75, 77
glutamine, from deamination of amino acids, 157, 158
glutelin, in different cereals, 88, 89, 91, 96
glycine: incorporation of ammonia into, in plants, 75, 77
glycogen of liver, involved in specific dynamic action, 157, 159
glyoxylate cycle, in oxidation of hydrocarbons, 307
goat, milk production by, 242, 244, 256
goose, egg and meat production by, 16, 17, 18–19, 20–1
grains, *see* cereals

grasses: cutting regime and protein production by, 126–7; dried, in ration of cattle, 249, 253; organic nitrogen production by, 127–9; potential dry matter yield of, 133; protein content of, 119–20; relations between nitrogen supplied, nitrogen uptake, and dry matter production by, 120–6
Great Lakes, USA: fisheries of, 273, 274–5
Green Revolution, 55
groundnuts, 63, 296; aflatoxin in, 288; protein per unit area from, 104
growth hormone, and protein synthesis, 155
growth rate, of micro-organisms, 316

hexadecane, as substrate for *Pseudomonas*, 353
hides, protein in, 239
histidine, 7; in cereals, 84, 85, 86, 93
histones, of nucleus, 142, 143
hormones, and protein synthesis, 155
hydrocarbons: compared with carbohydrates as substrates for protein production by micro-organisms, 358; estimates of supply of, 324–5; growth of micro-organisms on, 307, 308–9, 323–4, 353, 365; industrial processes for protein production from, 325–9, 335; nutritive value of micro-organisms grown on, 329–32; uses for protein from, 333–5, 336, 365
Hydrogenomonas: toxic effects of, on human consumers, 313

India: calorie and protein intake per head in, 363; estimates of cereal production in, 29; fish culture in, 265; future protein requirements of, 31
Indian Ocean, fisheries of, 247, 270–2
insects, nuclear protein of, 143
international aid schemes, 50
International Biological Programme, 277
International Rice Research Institute, 91
invertebrate herbivores: estimate of production by community of, 221–2, 223; possible harvest of protein from, 232, 281
Ipomoea aquatica, leaf production by, 105, 134
isoleucine, essential amino acid, 7; apparent deficiency of, caused by dietary excess of leucine, 154; in cereals, 85, 86, 93
isotopes, radioactive: in study of nitrogen metabolism in rumen, 165

Japan, fish culture in, 265

Kainji Lake, Nigeria: fish harvest potential of, 272
kwashiorkor, 93

INDEX

lambs, artificial rearing of, 250–1, 257
laminarin of seaweed, as substrate for yeast, 347
land: biomass of wild animals and protein production per unit area of, 220–5; production of leaves per unit area of, 281; production of lysine per unit area of, in milk and wheat, 254; production of protein per unit area of, 129, 133, (beef and dairy cattle), 252–4, (beef and food yeast) 358–9
leaves: amino acid composition of protein of, 34, 89, 113; fibre content of, 105, 110; incorporation of ammonia into amino acids in, 75–6; price of protein of, 56, 115; potential usefulness of, 35, 55, 297, 300; production of, per unit area of land, 281; protein concentrates from, 109–13; protein content of, 104–6, 132, 136; utilisation of, 296; 'whey' from, 110, 112–13
legumes: nitrogen-fixing bacteria in nodules on roots of, 72, 132; protein of, 35, 103–4; utilisation of, 292, 296
leucine, essential amino acid, 7; in cereals, 85, 86, 93; effects of dietary excess of, 154–5
light (solar energy): maximum potential photosynthesis from, 133, 218–19; percentage of, used for photosynthesis, 115
lignocellulose, as substrate for micro-organisms, 348–9
liver: of meat animals, parasites in, 349; ribosomes of, in protein synthesis, 149, 152, 153; as site of specific dynamic action, 156, 157, 160
livestock: cost of protein from, 55, 60; in developing countries, 60–1, 64; *see also individual species*
locust (*Schistocerca*); production of eggs and progeny by, and food consumption of, 16, 21–2, 23, 24
Lolium perenne: protein content of, 119; relations between amount of nitrogen supplied, nitrogen uptake, and dry matter production by, 120–1, 124
lysine, essential amino acid, 7; in cereals, 84, 85, 86, 88, 91, 93; in cottonseed, combined with gossypol, 108; effects of dietary excess of, 154; production of, in milk and wheat, from same area, 254; supplementation with, 59–60, 331; in yeasts and bacteria, 330–1, 353; *see also under* maize

maize: amino acid composition of protein of, 85, 86, 93, 352; high-lysine varieties of, 52, 59, 92–3, 103, 134, 364; protein concentrate from, 297; protein fractions of, 89; world production of, 83, 84

mammalian herbivores (wild): efficiency of conversion of dietary protein to meat protein by, 227, 233, 282; efficiency of conversion of plant matter to dry weight by, 232; estimates of production by community of, 223–7
marketing, of new forms of food, 42–4, 48, 369–70
meat: efficiency of populations producing, 187–96; income and demand for, 255–6, 257; percentage of carcasses available as, 186, 237, 238–9, 240, 241; price of protein in, 55; utilisation of, 286, 287, 289–90, 294; *see also* cattle, fowls, pigs, sheep, *etc.*
metals: traces of, associated with proteins, 8
methane: content of, in natural gas, 324; as substrate for bacteria, 308, 323, 324, 328–9, 365
methanol, as substrate for micro-organisms, 365
methionine, essential amino acid, 7; metabolism of chicks on diet deficient in, 152, 153, 154; supplementation with, 56, 59–60, 331, 352
methionine plus cystine, sulphur-containing amino acids, 7; amount of, in milk and wheat from same area, 254; limiting amino acids in Nigeria, 52; in protein of algae, 306, of bacteria, 368, of cereals, 84, 85, 86, 92, of plants, 113, and of yeasts, 309, 352, 368
micro-organisms, 303–4; continuous culture of, 341–2, 344–5, 355, 365–6; decomposer, in biomass, 226, 231, 232; efficiency of protein production by, 314–18, 355–9; extrapolation to animals of results obtained with, 142; nutritive value of, 329–32, 350–4; potential usefulness of, for human food, 334–5, 365–70; processing of, for food, 303, 312–14; processing of protein by, 132; of rumen, *see* rumen micro-organisms; *see also* algae, bacteria, fungi, protozoa, yeasts
milk: amino acid composition of protein of, 34, 113; in developing countries, 52, 61; efficiency of conversion of dietary protein to protein in, 178, 179, 202–3, 208, 217, 241–4; efficiency of populations producing, 191–8 *passim*, 256–7, 289; main source of dietary protein in many countries, 255, 257; price of protein in, 55, 355; production of, (with increasing percentage of dried grass in ration) 253–4, (on NPN diet) 243, 252; protein content of, 238; utilisation of, 290, 293; variation in costs of production of, 57
mitochondria, protein of, 142, 146
molasses, as substrate for micro-organisms, 340, 341, 353, 365

molluscs, 264, 269; fish feeding on, 265; raft culture of, 266
molybdenum: in nitrate reductase, 69; in nitrogenase, 73
multipurpose food (India), 56
muscle, steroids and growth of, 155
muscular work, and protein metabolism, 8
mutation: dangers of, in production of microbial protein, 367
Mycobacterium: growth of, on hydrocarbons, 323
myosin, energy requirements for synthesis of, 148–9
Mytilus edulis: yield of, per unit area, 266

net protein utilisation (NPU), 31, 174; of proteins of micro-organisms, 332, 352, 353
Neurospora crassa, rate of growth of, 344
New Guinea, protein consumption in, 30
New Zealand, markets for dairy produce of, 30
Nigeria, limiting amino acids in, 52
nitrate: amount supplied, and nitrogen content and dry matter yield of *Lolium*, 120–1; in leafy crops, 106, 120; reduction of, to nitrite, in plants, 69–72, 73
nitrate reductase, 79; induction of, 80–1; rate-limiting enzyme in protein production, 133
nitrite: reduction of, to ammonia, in plants, 69–72, 73
nitrite reductase, 69
nitrogen: danger of excess fertilisation with, 132; effects of supply of, on protein content of grass, 123–6, on type of pasture obtained, 119, on uptake and yield of nitrogen in *Lolium*, 120–3, and on yield of cereal and protein, 90; fixation of, 72–3, 74, 308
nitrogen equilibrium, dietary protein required to maintain, 4
nitrogenase, 72–3, 74
non-protein nitrogen (NPN: ammonia, urea), protein production by ruminants from, 213, 243, 251–2
nucleic acids: cell content of, increases with growth rate, 345; content of, in bacteria and yeast, 307, 309, 310, 330, 351; in diet of humans, 312–13, 351, 367–8; process for removal of, 368
nucleolus, rRNA synthesised in, 143, 149
nucleus: DNA of, 142; synthesis of protein of, 143

oats: amino acid composition of protein of, 85, 86; breeding work on, 94; protein fractions of, 89; world production of, 83, 84

ocean: estimated fisheries potential of, 267–72; estimated net primary production in, 230–1
offal, edible: of calves, 244; as percentage of carcass weight and carcass protein, 239; utilisation of, 289
oilseeds: competition between humans and animals for, 46, 50, 54, 55; protein foods from, 35, 62–3, 364; utilisation of, 287, 292, 296; *see also individual oilseeds*
Oncorhynchus tschawytscha (Chinook salmon), efficiency of protein production by, 264
oryx, possible domestication of, 282
oxalates, in leaves, 106
2-oxo-adipic acid: in yeast substrate, increases lysine content, 354

Pakistan, estimate of future protein requirements of, 31
paraffins: C_{10} to C_{20} preferentially attacked by micro-organisms, 324
pellagra, 92
Penicillium chrysogenum, carbohydrate utilisation by mutants of, 347
pentoses, as substrates for micro-organisms, 340, 344, 346
peptide bond synthesis, 145–6, 147, 310; energy cost of, 148
peptidoglycan of bacterial wall; not digested by animals, 306, 307, 312
Peru, price of fish oil in, 61
petroleum: carbon dioxide from combustion of cheap fractions of, for algal cultures, 305; microbial desulphurisation of products of, 325; as substrate for micro-organisms, 63, 303; *see also* hydrocarbons
phenylalanine plus tyrosine, 7; in cereals, 84, 85, 86, 93
phosphate bonds, high-energy: in protein synthesis, 146, 148, 149
phosphorus, associated with protein, 8
photosynthesis: net efficiency of, 219; percentage of light used for, 115; potential product of, 133, 218–19
pigs: efficiency of conversion of dietary protein to meat protein in, 178, 179, 206–7, 255, 289; protein content of carcasses of, 289
planner, economic: and protein nutrition, 48, 51–3
plants: conversion of inorganic nitrogen into protein by, 69–80; input–output system for protein production by, 131; *see also* cereals, grasses, leaves, *etc.*
polymer technology, used in food processing, 292, 293
polysomes, 146; of liver, in amino acid imbalance, 152, 153

INDEX 383

population, human: rate of growth of, 27, 28–9, 31, 36
populations of animals, 183–6, 187; efficiency of production by, 186–7, (eggs and milk) 191–2, (meat) 187–91
potatoes: in diets, 286, 287; peeling of, 107; protein content of, 102–3; use of sunlight by, 115
prices: effect of increased supply of a foodstuff on, 53; of protein in different foods, 55, 56
prolamin, in different cereals, 88, 89, 96
protein: economics of production of, 45–65; efficiency of use of, 177, 179; estimates of human requirements for, 4–9, 30–3; future demand for, 27–9, 35–44; proposals for augmenting supply of, 33–5; relative prices of, in different foods, 55, 56; turnover of, 151, 211; *see also* biological value, digestibility, food, net protein utilisation, specific dynamic action, *etc.*
protein efficiency ratio (PER), of algal and other proteins, 306
protein foods, definition of, 102; *see also* food
protein synthesis:
 in animals: effect of amino acid imbalance on, 152–5; energetics of, 148–51, 237, 311–12; hormonal regulation of, 155; mechanism of, 141–8, 309–11;
 in micro-organisms of rumen, 160–4; energetics of, 164–6, 213, 328;
 in plants, 69–80; potential of cereals for, 83–97
protozoa, 304; as possible processors of bacterial protein, 316; of rumen, amino acid composition of protein of, 162
Pseudomonas: growth of, on hydrocarbons, 323
Pseudomonas aeruginosa, amino acid composition of protein of, 353
pyrophosphate, hydrolysed in synthesis of RNA, 142

rabbits: efficiency of conversion of dietary protein to meat protein in, 205–6; efficiency of populations of, producing meat, 187–8, 194, 195, 196, 197; factors affecting population structure of, 187
rapeseed, 296
Rhizobium, in legumes, 72, 132
Rhizopus oryzae, as source of protein, 351
Rhodotorula glutinis, grows on xylose, 347
ribosomes: energy cost of synthesis of, 149; in protein synthesis, 78, 145, 146, 149, 152, 153; *see also* polysomes
ribulose-1,5-diphosphate carboxylase, in chloroplasts, 135

rice: amino acid composition of protein of, 85, 86, 87; breeding work with, 91; protein fractions of, 88, 89; world production of, 84
RNA: content of, in micro-organisms, 310; in protein synthesis, 78, 79, 309; synthesis of, 142–3
mRNA, 143, 144, 145, 146, 310; energy cost of synthesis of, 148–9; in synthesis of protein, 310, 311
rRNA, 143, 145, 310; molecular weight of, 149
tRNAs, 143–7, 310
RNA polymerase, 142, 143
roots of plants, incorporation of ammonia into amino acids in, 76
rumen micro-organisms, protein synthesis by, 160–6
ruminants: digestion of, suited for fibrous material, not concentrates, 254; efficiency of protein production by stall-fed, 237–8, 240–5; protein production from non-protein nitrogen by, 213, 243, 251–2; protein production per unit area by, 252–4; *see also* cattle, sheep
rye: amino acid composition of protein of, 85, 86; amphidiploid of wheat and, 95; world production of, 83, 84

Saccharomyces cerevisiae: carbohydrates utilised by, 346; commercial production of, 340, 341–2, 345; percentage of substrate recovered in, 343, 345
Saccharomyces fragilis, commerical production of, 340, 351
Salmo gairdneri (rainbow trout), genetic selection of, 264–5
Salmonella typhimurium, protein and nucleic acid contents of, 351
Salvelinus namaycush (lake trout), food chain of, 273
seeds: protein concentrates from, 107–9; protein content of, 103–4; storage proteins of, more easily modified genetically than leaf proteins, 134; *see also* cereals, oilseeds
serine: incorporation of ammonia into, in plants, 75, 77
sesame seed, 296
sewage effluent, cultivation of algae on, 305
Scenedesmus (alga), 305, 306
sex, and protein utilisation by cattle, 250
sheep: efficiency of conversion of dietary protein to meat protein in, 178, 179, 207–8; efficiency of populations of, producing meat, 187, 188–91, 193, 195, 196, 197–8; factors affecting population structure of, 185, 187; milk production by, 242; percentage of carcasses of,

sheep (*cont.*)
  available as meat, 186; protein content of carcasses of, 241
sigma factor, component of RNA polymerase, 143
slug (*Agriolimax*): production of eggs and progeny by, and food consumption of, 16, 21-2, 22-4
smallholdings, in developing countries, 52, 53
solvent extraction: to remove traces of hydrocarbons from micro-organisms, 326, 365; of soya bean oil, 295
sorghum: amino acid composition of protein of, 85, 86, 87, 93; breeding work on, 91-2; 160-Cernum variety of, 89, 92, 93; protein fractions of, 89; world production of, 83, 84
soya beans: amino acid composition of protein of, 334; in bread, 287, 292, 294; competition between humans and animals for, 54, 368; dominant in oilseed trade, 62-3; for feeding young animals, 243, 251; protein concentrate from, 108; protein content of, 334; supplementation of, with methionine, 56; toxin of, 355; utilisation of, 294-6, 299, 300
species, dangers of decreasing diversity of, 232-3
specific dynamic action, of amino acids and protein, 141, 155-60, 161
Spirulina maxima (alga): amino acid composition of protein of, 306; cultivation of, 305, 306, 313
Staphylococcus, nucleic acid content of, 310
starch, as substrate for micro-organisms, 346, 347
starvation, protein synthesis in, 152
steroids, and protein synthesis, 155, 212
straw, cellulose in, 348
sugar beet, use of sunlight by, 115
sugar-cane bagasse, as substrate for micro-organisms, 348, 350
sulphite liquor of paper industry, as substrate for micro-organisms, 303, 340, 353
sunflower seed, 296; oil from, 61
sweet potatoes, nitrogen fixation in intestines of population living on, 308

temperature: and protein and nucleic acid contents of cells, 345, 346; and protein production by micro-organisms, 342, 366
threonine, essential amino acid, 7; in cereals, 85, 86, 93; effects of dietary excess of, 154; in vegetable diets, 315
toxicity: of some micro-organisms, 354; tests of micro-organisms for, 333, 366
Trichoderma viride, conversion of cellulose to glucose by enzyme preparation from, 349
Triticale, wheat/rye amphidiploid, 95, 134
Trobriand Islands, social functions of food in, 39-40
trophodynamic methods, applied to aquatic ecosystems, 268-72; difficulties with, 272-4
tryptophan, essential amino acid, 7; in cereals, 85, 86, 93; glucose requirement for synthesis of, 314; produced from anthranilic acid by yeast, 316; in yeasts and bacteria, 353
tubers and roots, protein content of, 102-3
tyrosine, effects of dietary excess of, 154; *see also entry for* phenylalanine plus tyrosine

UK: calorie and protein intake per head in, 363; sources of protein consumed in (1903-13 and 1968), 286
unemployment, in developing countries, 52
unicellular organisms, *see* micro-organisms
upas tree, protein of latex of, 113
urea: from amino acids, and specific dynamic action, 157-8; energy cost of synthesis of, 152; as nitrogen source for milk production, 243, 252; supplementation of fodder for ruminants with, 54, 112, 134, 250, 256
uric acid: metabolism of, in methionine-deficient chicks, 152, 154
uric acid oxidase, not present in humans and apes, 312, 367
USA: Food and Drug Administration in, 299; population growth in, 28-9; restriction of agricultural production in, 30; social functions of food in, 39-40
USSR, exports of edible oil from, 61

valine, essential amino acid, 7; apparent deficiency of, in dietary excess of leucine, 154, 155; in cereals, 85, 86, 93
veal, production of, 244-5, 256, 257
vegetables, utilisation of, 286, 287, 292
vitamins of B group, associated with protein, 8

waste: reduction of, by food processing, 285, 288-9
water: supply of, and yield of cereal and protein, 90
wheat: amino acid composition of protein of, 85, 86; breeding work on, 94-5, 364; components of grain of, and flours from, 291; protein concentrate from, 297; protein fractions of, 89; world production of, 83, 84

whey: recovery of protein of, 290, 293; as substrate for yeasts, 303, 340

Wolffia arrhiza, as food, 306

women, protein needs of pregnant and lactating, 363

wood: fungi growing on, 349; as substrate for micro-organisms, after acid hydrolysis, 340, 349-50

woodland (oak), as ecosystem, 229–30, 231; net production of plant material in, 219; as source of protein for browsing animals, 233

wool: populations producing, 185, 189; protein required for, 251

xanthine dehydrogenase of liver, in methionine deficiency, 152, 154

xylose, as substrate for micro-organisms 347, 353

yams, protein content of, 103

yeasts: amino acid composition of protein of, 334, 352, 353, 354; commercial production of, on hydrocarbons, 325–8; nucleic acid content of, 309, 310, 330, 351; nutritive value of protein of, 332, 333–4; potential usefulness of, for human food, 303, 309, 365; protein content of, 306, 309, 334, 344, 351; protein production by, 308–9, 344, 355–8; tryptophan production by, 316; *see also Candida, Rhodotorula, Saccharomyces*

yoghurt, 290